Praise for *AI and ML for Coders in PyTorch*

A perfect hands-on guide for developers who want to actually build with AI, not just read about it. Clear, practical, and grounded in PyTorch— this is the book I wish I had when I started.

—*Dominic Monn, CEO, MentorCruise.com*

This is a book you won't want to miss if you want to become a full-stack AI practitioner. It provides a comprehensive overview and concrete examples for building a variety of AI models and applications from scratch.

—*Dr. Pin-Yu Chen, principal research scientist, IBM Research*

A must-read book for developers diving into AI/ML. You will learn generative AI through real-world coding examples in PyTorch.

—*Margaret Maynard-Reid, ML engineer at M Couture 3D*

Laurence has masterfully bridged the gap between theory and practice—*AI and ML for Coders in PyTorch* is not just a book, it's a hands-on journey through modern machine learning, from basics to large language models, all with clarity and purpose.

—*Vishwesh Shrimali, AI engineer in the automotive industry*

Laurence has done it again, distilling complex AI concepts into an approachable, coder first masterclass. This PyTorch edition makes machine learning accessible to a broader and powerful community.

—*Laura Uzcátegui, cofounder of DynG AI, Inc.*

This book is a fantastic piece for any developer or CS student looking to step into AI. Laurence pairs well-explained PyTorch code with clear foundational theory and visuals that truly boost comprehension. Covering vision, time-series, NLP, and more, it's a practical, comprehensive guide that builds solid machine learning foundations.

—*Louis-François Bouchard, AI educator; cofounder & CTO, Towards AI*

Laurence's trademark code-before-theory approach lowers the barrier for busy coders. For the millions who loved his bestselling *AI and Machine Learning for Coders*, this PyTorch sequel is your fast-track to Generative AI. Spanning LLM fine-tuning, RAG, and Stable Diffusion, this book is an essential upgrade for today's AI wave.

—*Ammar Mohanna, AI consultant & lecturer*

Laurence has done a phenomenal job crafting a book that brings developers up to speed in the ever-evolving world of generative AI. A fantastic teacher, he breaks down complex concepts with clarity, making them accessible to learners at any level. The practical code examples throughout the book provide a solid foundation to spark creativity and empower developers to start building right away.

—*Roya Kandalan, PhD, Generative AI research scientist*

AI and ML for Coders in PyTorch

*A Coder's Guide to Generative AI
and Machine Learning*

Laurence Moroney

O'REILLY®

AI and ML for Coders in PyTorch

by Laurence Moroney

Copyright © 2025 Laurence Moroney. All rights reserved.

Printed in the United States of America.

Published by O'Reilly Media, Inc., 141 Stony Circle, Suite 195, Santa Rosa, CA 95401.

O'Reilly books may be purchased for educational, business, or sales promotional use. Online editions are also available for most titles (*http://oreilly.com*). For more information, contact our corporate/institutional sales department: 800-998-9938 or *corporate@oreilly.com*.

Acquisitions Editor: Nicole Butterfield
Development Editor: Jill Leonard
Production Editor: Aleeya Rahman
Copyeditor: Doug McNair
Proofreader: Piper Content Partners

Indexer: Sue Klefstad
Cover Designer: Susan Brown
Cover Illustrator: Monica Kamsvaag
Interior Designer: David Futato
Interior Illustrator: Kate Dullea

July 2025:   First Edition

Revision History for the First Edition
2025-06-27:   First Release
2026-01-23:   Second Release

See *http://oreilly.com/catalog/errata.csp?isbn=9781098199173* for release details.

978-1-098-19917-3

[LSI]

Table of Contents

Foreword

Dear Reader,

AI is poised to transform every industry, but almost every AI application needs to be customized for its particular use. A system for reading medical records is different from one for finding defects in a factory, which is different from a product recommendation engine. For AI to reach its full potential, engineers need tools that can help them adapt to the amazing capabilities available to the millions of concrete problems we wish to solve.

When I led the Google Brain team, we started to build a C++ framework for deep learning called DistBelief. We were excited about the potential of harnessing thousands of CPUs to train a neural network (for instance, using 16,000 CPUs to train a cat detector on unlabeled YouTube videos). How far deep learning has come since then! What was once cutting-edge can now be done on your laptop! And you'll learn how to, with PyTorch, in this book. Frameworks for creating ML models to implement AI have come a long way, too.

PyTorch has made significant progress. It is designed to be easy to learn as well as powerful enough to be used by researchers. Its rich features can help one build a wide range of AI tools, from simple models, to transfer learning from others, to fine tuning the most modern generative AI models.

Today, millions of coders have become AI developers thanks to frameworks like PyTorch.

Laurence Moroney, has been a major force in helping developers succeed with AI in TensorFlow and PyTorch. I have been privileged to work together with him in teaching several specializations with deeplearning.ai and Coursera, including an upcoming one on PyTorch.

In our early days of working together, Laurence once Slacked me:

> Andrew sang a sad old song
> fainted through miss milliner
> invitation hoops
> fainted fainted
> [...]

He had trained an LSTM on lyrics of traditional Irish songs and it generated these lines. But in this post-Transformer network age, we can create far more sophisticated content—from poetry, to code, to analysis of documents! In this book, he covers both and helps you be prepared for modern AI.

And, if AI opens the door to fun like that, how could anyone not want to get involved? You can 1) work on exciting projects that move humanity forward, 2) advance your career, and 3) get free Irish poetry. I wish you the best in your journey learning PyTorch.

With Laurence as a teacher, great adventures await you.

Keep learning!

— Andrew Ng
Founder, DeepLearning.AI

Preface

Welcome to *AI and ML for Coders in PyTorch*. My machine learning (ML) journey began many years ago with languages and frameworks like Lisp and Prolog. After that, my journey took me to Google, where I helped launch and grow TensorFlow. This experience informed my previous book, *AI and Machine Learning for Coders*.

Since that book was published, whenever I met with the community to talk about AI, one question would come up: whether or not the questioner should invest their time in PyTorch. It was a strange question at first, but the more I heard it, the more I began to investigate.

That line of thought got me to this point in my career, where PyTorch, once a rival to my work, is now something I passionately embrace. Why? Because it strikes a perfect balance between having the power and flexibility to let researchers or advanced engineers push the limits and also having the simplicity for any developer to pick it up and start their journey into ML.

The goal of this book is to prepare you, as a coder, for just that—it's accessible enough if you don't fully understand ML yet, and also exposes you to the advanced concepts that will help you go deeper. The aim: to equip you to be an ML and AI developer without needing a PhD!

I hope that you'll find this book useful and that it will empower you with the confidence to get started on this wonderful and rewarding journey.

Who Should Read This Book

If you're interested in AI and ML, and you want to get up and running quickly with building models that learn from data, this book is for you. If you're interested in getting started with common AI and ML concepts—computer vision, natural language processing, sequence modeling, and more—and want to see how neural networks can be trained to solve problems in these spaces, I think you'll enjoy this book. And if you've heard all of the hoopla around generative AI, we roll our sleeves up and explore how that works with transformer and diffuser-based models.

Most of all, if you've put off entering this valuable area of computer science because of perceived difficulty, in particular believing that you'll need to dust off your old calculus books, then fear not: this book takes a code-first approach that shows you just how easy it is to get started in the world of ML and artificial intelligence using PyTorch.

Why I Wrote This Book

I first got seriously involved with artificial intelligence in the spring of 1992. A freshly minted physics graduate living in London in the midst of a terrible recession, I had been unemployed for six months. The British government started a program to train 20 people in AI technology and put out a call for applicants. I was the first participant selected. Three months later, the program failed miserably, because while there was plenty of theoretical work that could be done with AI, there was no easy way to do it practically. One could write simple inference in a language called Prolog and perform list processing in a language called Lisp, but there was no clear path to deploying them in industry. The famous "AI winter" followed.

Then, in 2016, while I was working at Google on a product called Firebase, the company offered ML training to all engineers. I sat in a room with a number of other people and listened to lectures about calculus and gradient descent. I couldn't quite match this to a practical implementation of ML, and I was suddenly transported back to 1992. I gave feedback about this, and about how we should be educating people in ML—teaching the code first to coders. Google embraced this philosophy, as did Meta with the release of PyTorch.

In particular, both emphasized high-level APIs that made it easy for developers to get started, and I realized there was a need for a book that took advantage of this and widened access to ML so that it wasn't just for mathematicians or PhDs anymore.

I believe that more people using this technology and deploying it to end users will lead to an explosion in AI and ML that will prevent another AI winter and change the world very much for the better. I'm already seeing the impact of this, from the work done by Google on diabetic retinopathy, through Penn State University, to PlantVillage building an ML model for mobile that helps farmers diagnose cassava disease, to Médecins Sans Frontières using TensorFlow models to help diagnose antibiotic resistance, and much, much more!

With the advent of generative AI, and the emergence of transformers and diffusers as libraries in their own right, the next great wave of AI is upon us. PyTorch is at the heart of all of that—so it was time for me to bring my work up-to-date and show just how easy it is for you to dip your toes in the waters of AI and ML development.

With that in mind, welcome to this book on AI and ML for coders in PyTorch. I can't wait to see what you build.

Navigating This Book

The book is written in two main parts. Part I (Chapters 1–11) talks about how to use PyTorch to build ML models for a variety of scenarios. It takes you from first principles—building a model with a neural network containing only one neuron—through computer vision, natural language processing, and sequence modeling. Part II (Chapters 12–20) then walks you through generative AI scenarios—from understanding how transformers work in applications like ChatGPT through diffusers for image generation like Midjourney. Most chapters are standalone, so you can drop in and learn something new, or, of course, you can just read the book cover to cover.

Technology You Need to Understand

The goal of the first half of the book is to help you learn how to use PyTorch to build models with a variety of architectures. The only real prerequisite to this is understanding Python, and in particular Python notation for data and array processing. You might also want to explore NumPy, a Python library for numeric calculations. If you have no familiarity with these, they are quite easy to learn, and you can probably pick up what you need as you go along (although some of the array notation might be a bit hard to grasp).

Online Resources

A variety of online resources are used by, and supported in, this book. At the very least, I would recommend that you keep an eye on O'Reilly's website for books that complement this one and for any updates and breaking changes to technologies discussed in the book. The code for this book is available on the book's GitHub page (*https://github.com/lmoroney/PyTorch-Book-FIles*), and I will keep it up to date there as the platform evolves.

Conventions Used in This Book

The following typographical conventions are used in this book:

Italic
    Indicates new terms, URLs, email addresses, filenames, and file extensions.

Constant width

> Used for program listings, as well as within paragraphs to refer to program elements such as variable or function names, databases, data types, environment variables, statements, and keywords.

Constant width bold

> Shows commands or other text that should be typed literally by the user.

Constant width italic

> Shows text that should be replaced with user-supplied values or by values determined by context.

> This element signifies a tip or suggestion.

> This element signifies a general note.

Using Code Examples

Supplemental material (code examples, exercises, etc.) is available for download at *https://github.com/lmoroney/PyTorch-Book-FIles*.

If you have a technical question or a problem using the code examples, please send email to *support@oreilly.com*.

This book is here to help you get your job done. In general, if example code is offered with this book, you may use it in your programs and documentation. You do not need to contact us for permission unless you're reproducing a significant portion of the code. For example, writing a program that uses several chunks of code from this book does not require permission. Selling or distributing examples from O'Reilly books does require permission. Answering a question by citing this book and quoting example code does not require permission. Incorporating a significant amount of example code from this book into your product's documentation does require permission.

We appreciate, but generally do not require, attribution. An attribution usually includes the title, author, publisher, and ISBN. For example: "*AI and ML for Coders with PyTorch*, by Laurence Moroney. Copyright 2025 Laurence Moroney, 978-1-098-19917-3."

If you feel your use of code examples falls outside fair use or the permission given above, feel free to contact us at *permissions@oreilly.com*.

O'Reilly Online Learning

O'REILLY® For more than 40 years, *O'Reilly Media* has provided technology and business training, knowledge, and insight to help companies succeed.

Our unique network of experts and innovators share their knowledge and expertise through books, articles, and our online learning platform. O'Reilly's online learning platform gives you on-demand access to live training courses, in-depth learning paths, interactive coding environments, and a vast collection of text and video from O'Reilly and 200+ other publishers. For more information, visit *https://oreilly.com*.

How to Contact Us

Please address comments and questions concerning this book to the publisher:

O'Reilly Media, Inc.
141 Stony Circle, Suite 195
Santa Rosa, CA 95401
800-889-8969 (in the United States or Canada)
707-827-7019 (international or local)
707-829-0104 (fax)
support@oreilly.com
https://oreilly.com/about/contact.html

We have a web page for this book, where we list errata, examples, and any additional information. You can access this page at *https://oreil.ly/ai-ml-pytorch*.

For news and information about our books and courses, visit *https://oreilly.com*.

Find us on LinkedIn: *https://linkedin.com/company/oreilly-media*.

Watch us on YouTube: *https://youtube.com/oreillymedia*.

Acknowledgments

I'd like to thank lots of people who have helped in the creation of this book.

Andrew Ng, who, as well as writing the foreword for this book, also believed in my approach to teaching AI and ML, and with whom I created several specializations at Coursera, teaching millions of people how to succeed with ML and AI. Andrew also leads a team at deeplearning.ai who were terrific at helping me be a better machine learner, including Tommy Nelson, Nick Lewis, Miguel Magana, Muhammad Mubashar, and Dapinder Dosanjh.

The team at O'Reilly that made this book possible: Jill Leonard, Nicole Butterfield, Aleeya Rahman, Kristen Brown, Doug McNair, Kate Dullea, Sue Klefstad, Kim Sandoval, and Angela Rufino, without whose hard work I never would have gotten it done!

The amazing tech review team: Louis-François Bouchard, Pin-Yu Chen, Jialin Huang, Roya Norouzi Kandalan, Margaret Maynard-Reid, Dominic Monn, Vishwesh Ravi Shrimali, and Laura Uzcátegui.

And of course, most important of all is my family, who make the most important stuff meaningful: my wife Rebecca Moroney, my daughter Claudia Moroney, and my son Christopher Moroney. Thanks to you all for making life more amazing than I ever thought it could be.

## CHAPTER 1

Introduction to PyTorch

When it comes to creating artificial intelligence (AI), machine learning (ML) and deep learning are great places to begin. When you're getting started, however, it's easy to get overwhelmed by the options and all the new terminology. This book aims to demystify things for you as a programmer. It takes you through writing code to implement concepts of ML and deep learning, and it also takes you through building models that behave more as a human does, with scenarios like computer vision, natural language processing (NLP), and more. Thus, these models become a form of synthesized, or artificial, intelligence.

But when we refer to *machine learning*, what exactly is it? Let's take a quick look at that and consider it from a programmer's perspective before we go any further. After that, in the rest of this chapter, we'll show you how to install the tools of the trade, from PyTorch itself to environments where you can code and debug your PyTorch-based models.

What Is Machine Learning?

Before we get into the ins and outs of ML, let's consider how it evolved from traditional programming. We'll start by examining what traditional programming is, and then we'll consider cases where it's limited. After that, we'll see how ML evolved to handle those cases and thus opened up new opportunities to implement new scenarios, thereby unlocking many of the concepts of AI.

Traditional programming involves writing rules that are expressed in a programming language and that act on data and give us answers. This applies just about everywhere we can program something with code.

For example, consider a game like the popular Breakout. Code determines the movement of the ball, the score, and the various conditions for winning or losing the game. Think about the scenario where the ball bounces off a brick, like in Figure 1-1.

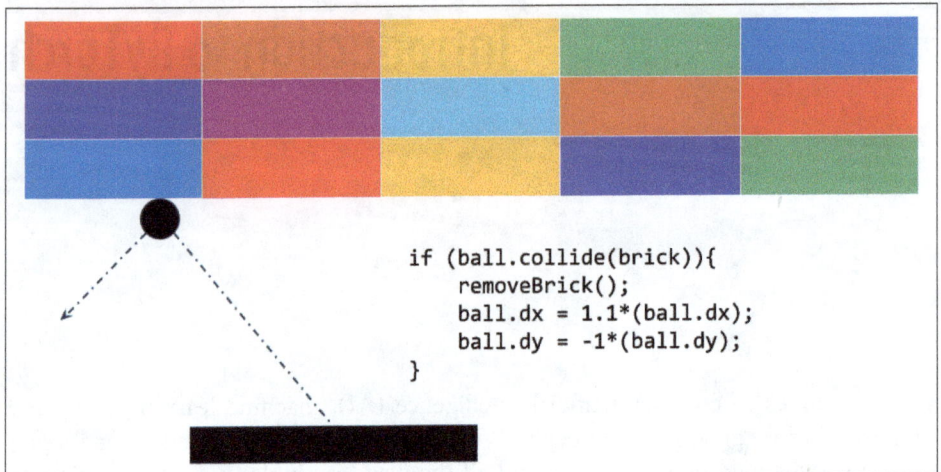

```
if (ball.collide(brick)){
    removeBrick();
    ball.dx = 1.1*(ball.dx);
    ball.dy = -1*(ball.dy);
}
```

Figure 1-1. Code in a Breakout game

Here, the motion of the ball can be determined by its dx and dy properties. When the ball hits a brick, the brick is removed, the velocity of the ball increases, and the direction of the ball's movement changes. The code acts on data about the game situation.

Alternatively, consider a financial services scenario. Say you have data about the company, such as its current stock price and earnings. By using code like that in Figure 1-2, you can calculate a valuable ratio called the *price-to-earnings ratio* (or P/E, which stands for price divided by earnings).

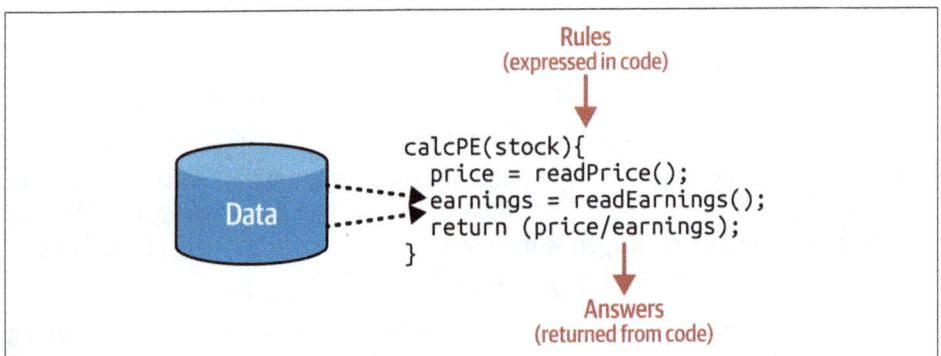

Rules
(expressed in code)

Data

```
calcPE(stock){
    price = readPrice();
    earnings = readEarnings();
    return (price/earnings);
}
```

Answers
(returned from code)

Figure 1-2. Code in a financial services scenario

Your code reads the price, reads the earnings, and returns a value that is the former divided by the latter.

If I were to try to sum up traditional programming like this in a single diagram, it might look like Figure 1-3.

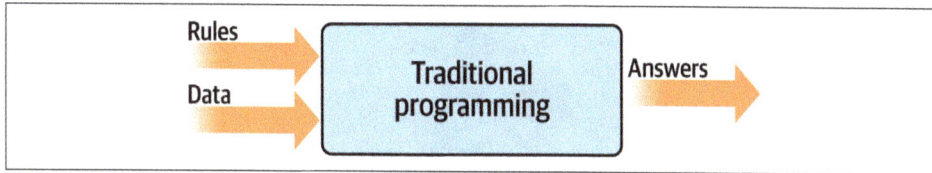

Figure 1-3. *High-level view of traditional programming*

As you can see, you have rules expressed in a programming language. These rules act on data, and the result is answers.

Limitations of Traditional Programming

The model from Figure 1-3 has been the backbone of development since its inception. But it has an inherent limitation: namely, the only scenarios that you can implement are ones for which you can derive rules. But what about other scenarios? Usually, it's unfeasible to develop them because the code is too complex. It's just not possible to write code to handle them.

Consider, for example, activity detection. Fitness monitors that can detect our activity are a recent innovation, not just because of the availability of cheap and small hardware but also because the algorithms to handle detection weren't previously feasible. Let's explore why.

Figure 1-4 shows a naive activity detection algorithm for walking. It can consider the person's speed and if that speed is less than a particular value, we can determine that they are probably walking.

```
if(speed<4){
    status=WALKING;
}
```

Figure 1-4. *Algorithm for activity detection*

Given that our data is speed, we could also extend this to detect whether they are running, as in Figure 1-5.

```
if(speed<4){
    status=WALKING;
} else {
    status=RUNNING;
}
```

Figure 1-5. Extending the algorithm for running

As you can see, going by the speed, we might say that if it is less than a particular value (say, 4 mph) the person is walking, and otherwise, they are running. It still sort of works.

Now, suppose we want to extend this to another popular fitness activity, biking. The algorithm could look like the one in Figure 1-6.

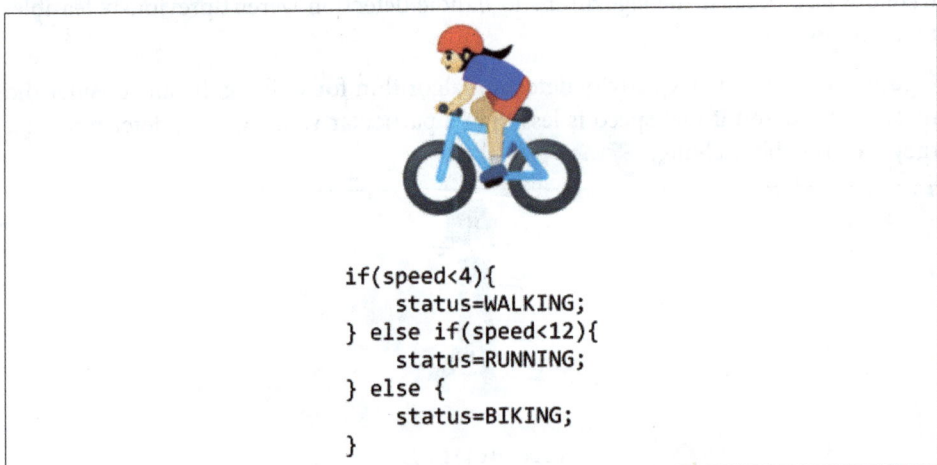

```
if(speed<4){
    status=WALKING;
} else if(speed<12){
    status=RUNNING;
} else {
    status=BIKING;
}
```

Figure 1-6. Extending the algorithm for biking

I know this algorithm is naive in that it just detects speed—some people run faster than others, and you might run downhill faster than you can cycle uphill—but on the whole, it still works. However, what happens if we want to implement another scenario, such as golfing (see Figure 1-7)?

// ???

Figure 1-7. How do we write a golfing algorithm?

Now, we're stuck. Whether a person is golfing or not, they might walk for a bit, stop, do some activity, walk for a bit more, stop, etc. So how can we use this methodology to tell whether they're playing golf?

Our ability to detect this activity using traditional rules has hit a wall. But maybe there's a better way.

Enter ML.

From Programming to Learning

Let's look back at the diagram that we used to demonstrate what traditional programming is (see Figure 1-8). Here, we have rules that act on data and give us answers. In our activity detection scenario, the data was the speed at which the person was moving—and from that, we could write rules to detect their activity, be it walking, biking, or running. However, we hit a wall when it came to golfing because we couldn't come up with rules to determine what that activity looks like.

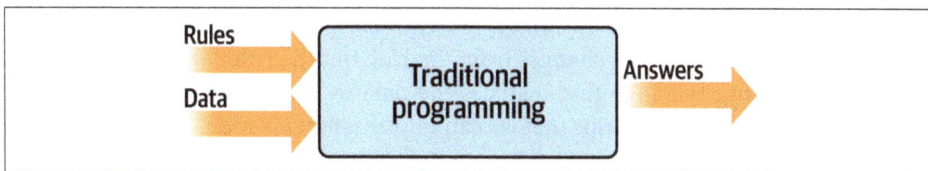

Figure 1-8. The traditional programming flow

But what would happen if we were to flip the axes around on this diagram? Instead of us coming up with the *rules*, what if we were to come up with the *answers* and, along with the data, have a way of figuring out what the rules might be?

Figure 1-9 shows what this would look like, and we can say that this high-level diagram defines *machine learning*.

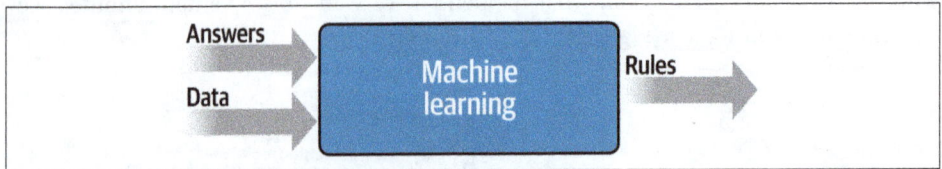

Figure 1-9. Changing the axes to get ML

So, what are the implications of this? Well, now, instead of *us* trying to figure out what the rules are, we can get lots of data about our scenario and label that data, and then the computer can figure out what the rules are that make one piece of data match a particular label and another piece of data match a different label.

How would this work for our activity detection scenario? Well, we can look at all the sensors that give us data about this person. If the person has a wearable device that detects information such as heart rate, location, and speed—and if we collect a lot of instances of this data while they're doing different activities—then we end up with a scenario of having data that says, "This is what walking looks like," "This is what running looks like," and so on (see Figure 1-10).

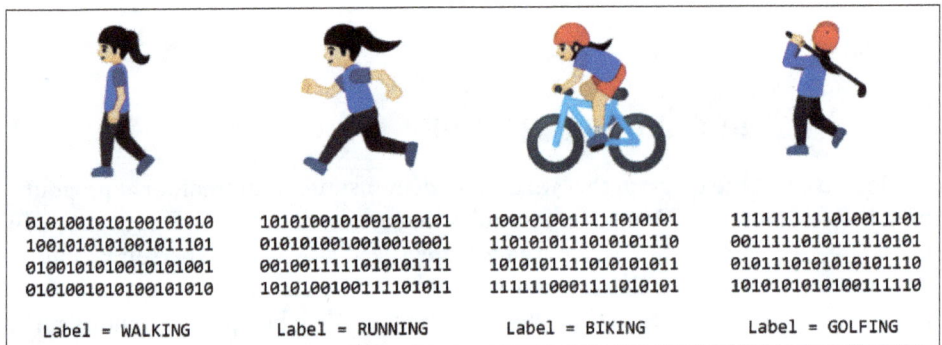

Figure 1-10. From coding to ML: gathering and labeling data

Now, our job as programmers changes from figuring out the rules, to determining the activities, to writing the code that matches the data to the labels. If we can do this, then we can expand the scenarios that we can implement with code.

ML is a technique that enables us to do this, but to get started, we'll need a framework —that's where PyTorch enters the picture. In the next section, we'll take a look at what PyTorch is and how to install it. Then, later in this chapter, you'll write your first code that learns the pattern between two values, like in the preceding scenario. It's a simple "Hello World" scenario, but it has the same foundational code pattern that's used in extremely complex ones.

The field of AI is large and abstract, encompassing everything that has to do with making computers think and act the way human beings do. One of the ways a human takes on new behaviors is through learning by example, and the discipline of ML can thus be thought of as an on-ramp to the development of AI. By way of an ML field called *computer vision*, a machine can learn to see like a human, and by way of another ML field called *natural language processing*, it can learn to read text like a human. Many more such applications of ML are possible, and we'll be covering the basics of ML in this book by using the PyTorch framework.

What Is PyTorch?

PyTorch is an ML library that is based on a previous library called *Torch*, which is an open source ML framework and scripting language that is itself based on a programming language called Lua. In 2017, development of Torch moved to PyTorch, which is a port of the framework in Python.

So, when installing PyTorch, you'll often see it referred to as "torch."

PyTorch was originally developed by Meta AI, but it was moved out to the Linux Foundation as a way of building developer confidence that it wasn't made by and for a big tech company. It's one of the two most popular ML libraries, alongside the TensorFlow/Keras ecosystem.

With the emergence of generative AI, and in particular the "open sourcing" of generative text and image models, PyTorch has exploded in popularity. It's often used for training models (which we cover in Part I of this book) as well as for inference of models (which we cover in Part II of this book).

PyTorch could also be seen as an ecosystem of libraries, each of which is tailored to specific scenarios. The important libraries and scenarios to consider are as follows:

TorchServe
> This is an easy-to-use tool that lets you deploy PyTorch models at scale. It's designed to run in multiple environments, and it's generally technology agnostic. It supports features such as multimodel serving, logging, metrics, and the easy creation of RESTful endpoints that let you do inference on models from a variety of clients.

Distributed training
> When larger models don't fit onto a single chip or machine, there are technologies and techniques that allow you to share them across multiple devices. The `torch.distributed` libraries allow you easy and native support of asynchronous execution across multiple devices.

Mobile

An important surface for inference is, of course, mobile. It's important for you to be able to deploy your AI work to Android and iOS devices, and PyTorch supports this through PyTorch Mobile.

Pretrained models

An active community of researchers and developers have created a rich ecosystem of models that you can simply use with one line of code, wrapped in the `torchvision.models` library.

Figure 1-11 provides a high-level representation of this.

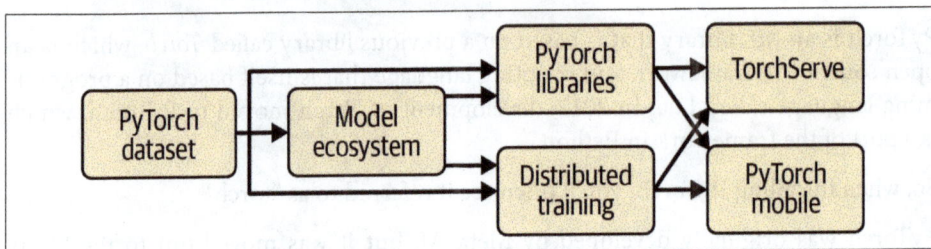

Figure 1-11. PyTorch ecosystem

The process of creating ML models is called *training*, and it's where a computer uses a set of algorithms to learn about inputs and what distinguishes them from one another. So, for example, if you want a computer to recognize cats and dogs, you can use lots of pictures of both to create a model, and the computer will use that model to try to figure out what makes a cat a cat and what makes a dog a dog. Once the model is trained, the process of having it recognize or categorize future inputs is called *inference*.

So, for training models, there are several things that you need to consider, and we will cover them in this book. Primarily, your choice will boil down to one of three things:

- Creating the model entirely from scratch yourself
- Using someone else's model because it's enough for your task
- Using parts of another person's model that have already been trained and building on top of them

The last option on the list is called *transfer learning*, and we'll cover it later in the book.

There are many ways to train a model. For the most part, you'll probably just use a single chip, whether it's a central processing unit (CPU), a graphics processing unit (GPU), or something new called a tensor processing unit (TPU). In more advanced working and research environments, you can use parallel training across multiple chips, employing a distributed training where training is intelligently spanned across multiple chips. PyTorch supports this, too, through "distributed training" libraries, as shown in Figure 1-11.

The lifeblood of any model is its data. As we discussed earlier, if you want to create a model that can recognize cats and dogs, you need to train it with lots of examples of cats and dogs. But how can you manage these examples? Over time, you'll see that this can often involve a lot more coding than the creation of the models themselves.

But luckily, the PyTorch ecosystem includes a number of built-in datasets that make this easy for you. We will also explore these throughout this book.

Beyond creating models, you'll need to be able to get them into people's hands so they can use them. To this end, PyTorch includes libraries for serving, where you can provide model inference over an HTTP connection for cloud or web users. For models to run on mobile or embedded systems, there's PyTorch Mobile, which provides tools for model inference on Android and iOS.

Next, I'll show you how to install PyTorch so that you can get started creating and using ML models with it!

Using PyTorch

In this section, we'll look at the three main ways you can install and use PyTorch. We'll start with how to install it on your developer box using the command line. Then, we'll explore using the popular PyCharm IDE to install and use PyTorch. Finally, we'll look at Google Colab and how you can use it to access your PyTorch code with a cloud-based backend in your browser.

Installing PyTorch in Python

The *Py* in PyTorch stands for Python, so it's important to have a Python environment already set up. If you don't have Python already, I strongly recommend you visit the Python website (*https://python.org*) to get up and running with it and the Learn Python website (*https://learnpython.org*) to learn the Python language syntax.

With Python, there are many ways to install frameworks, but the default one supported by the PyTorch team is pip.

So, in your Python environment, installing PyTorch is as easy as using this:

```
> pip install torch
```

Once you're up and running, you can test your PyTorch version with the following code:

```
import torch
print(torch.__version__)
```

You should then see output like that in Figure 1-12. It will print the currently running version of PyTorch—here, you can see that version 2.4.1 is installed.

```
(pytorchenv) laurencemoroney@Macmini Documents % python
Python 3.9.6 (default, Aug  9 2024, 14:24:13)
[Clang 16.0.0 (clang-1600.0.26.3)] on darwin
Type "help", "copyright", "credits" or "license" for more information.
>>> import torch
/Users/laurencemoroney/Documents/pytorchenv/lib/python3.9/site-packages/torch/_s
ubclasses/functional_tensor.py:258: UserWarning: Failed to initialize NumPy: No
module named 'numpy' (Triggered internally at /Users/runner/work/pytorch/pytorch
/pytorch/torch/csrc/utils/tensor_numpy.cpp:84.)
  cpu = _conversion_method_template(device=torch.device("cpu"))
>>> print(torch.__version__)
2.4.1
>>>
```

Figure 1-12. Running PyTorch in Python

> If you look closely at Figure 1-12, you'll see a note that shows that the torch device is "cpu." In this case, I natively installed it on my Mac, and it is configured to use the CPU. However, this is not optimal for complex models, where an accelerator like a GPU or Metal may be necessary. We will cover installation of PyTorch for accelerators later in this book.

Using PyTorch in PyCharm

I'm particularly fond of using the free community version of PyCharm (*https://oreil.ly/I2mP2*) for building models using PyTorch. PyCharm is useful for many reasons, but one of my favorites is that it makes the management of virtual environments easy. This means you can have Python environments with versions of tools such as PyTorch that are specific to your particular project. So, for example, if you want to use PyTorch 1.x in one project and PyTorch 2.x in another, you can separate them with virtual environments and not have to deal with installing/uninstalling dependencies when you switch between them. Additionally, with PyCharm, you can do step-by-step debugging of your Python code—which is a must, especially if you're just getting started!

For example, in Figure 1-13, I have a new project that's called *example1*, and I'm specifying that I'm going to create a new environment using Conda. When I create the project, I'll have a clean, new, virtual Python environment into which I can install any version of PyTorch I want.

Once you've created a project, you can open the File → Settings dialog and choose the entry for "Project: *<your project name>*" from the menu on the left. In the menu on the left, you'll see choices to change the settings for the Python Interpreter and the Project Structure. If you choose the Python Interpreter link, you'll see the interpreter

that you're using, as well as a list of packages that are installed in this virtual environment (see Figure 1-14).

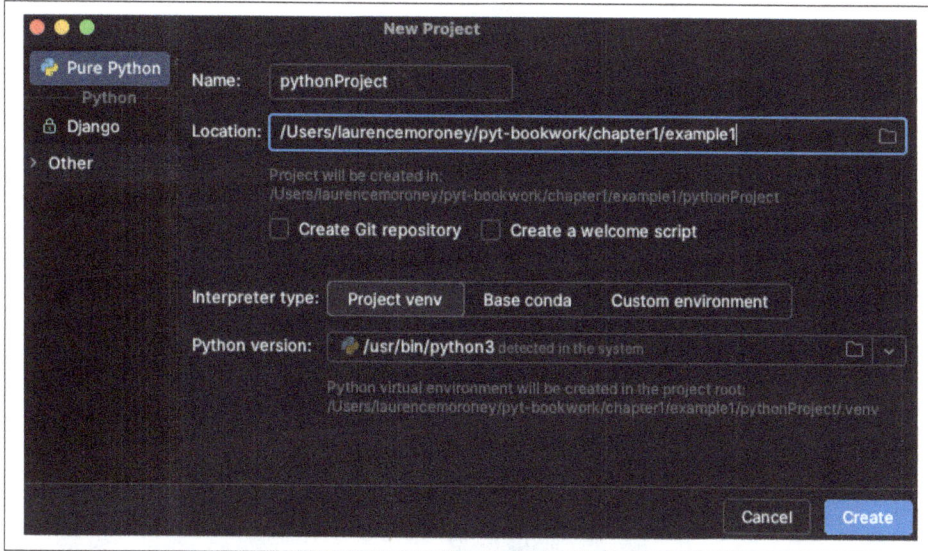

Figure 1-13. Creating a new virtual environment using PyCharm

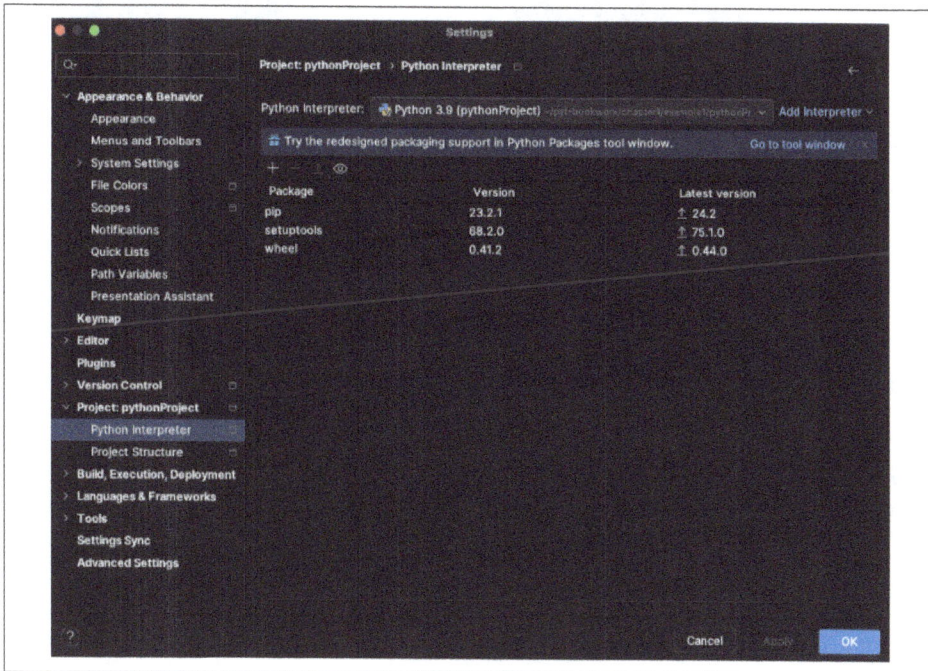

Figure 1-14. Adding packages to a virtual environment

You can then click the + button at the upper left, and a dialog will open showing the packages that are currently available. Type **torch** into the search box and you'll see all available packages with *torch* in the name (see Figure 1-15). Remember that the name of the package is *torch*, even if the technology is PyTorch.

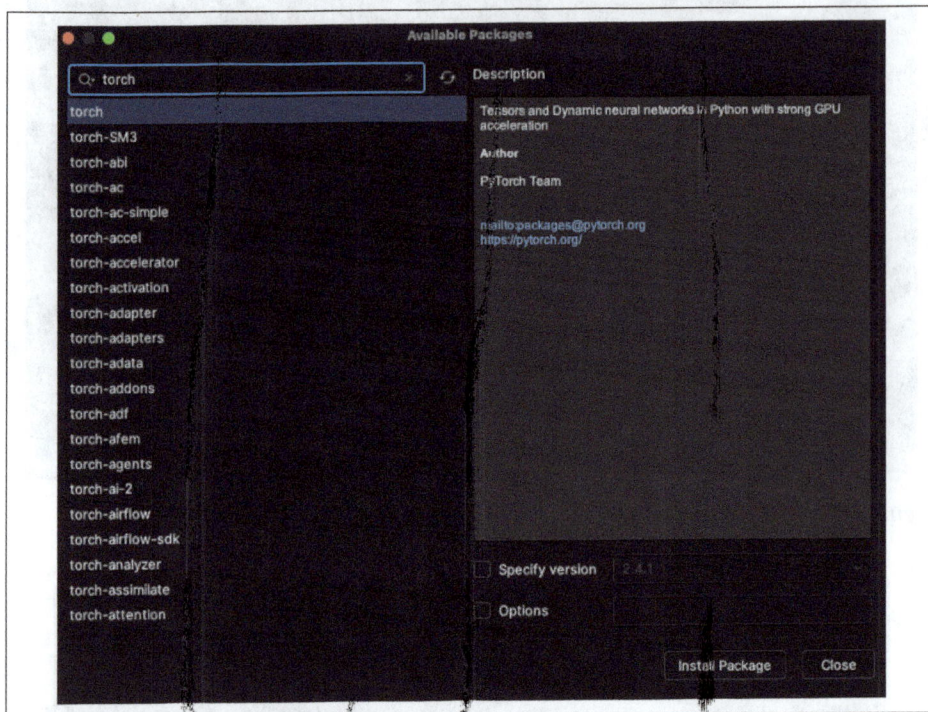

Figure 1-15. Installing torch with PyCharm

Once you've selected torch or any other package you want to install, you can click the Install Package button and PyCharm will do the rest. Then, once torch is installed, you can write and debug your PyTorch code in Python.

Using PyTorch in Google Colab

Another option, perhaps the easiest one for getting started, is to use *Google Colab* (*https://oreil.ly/c0lab*), which is a hosted Python environment that you can access via a browser. What's really neat about Colab is that it provides GPU and TPU backends so you can train models using state-of-the-art hardware at no cost.

When you visit the Colab website, you'll be given the option to open previous Colabs or start a new notebook (see Figure 1-16). If you click the + New notebook button, it will open the editor, where you can add panes of code or text (see Figure 1-17). You

---

can then execute the code by clicking the Play button (the arrow) to the left of the pane.

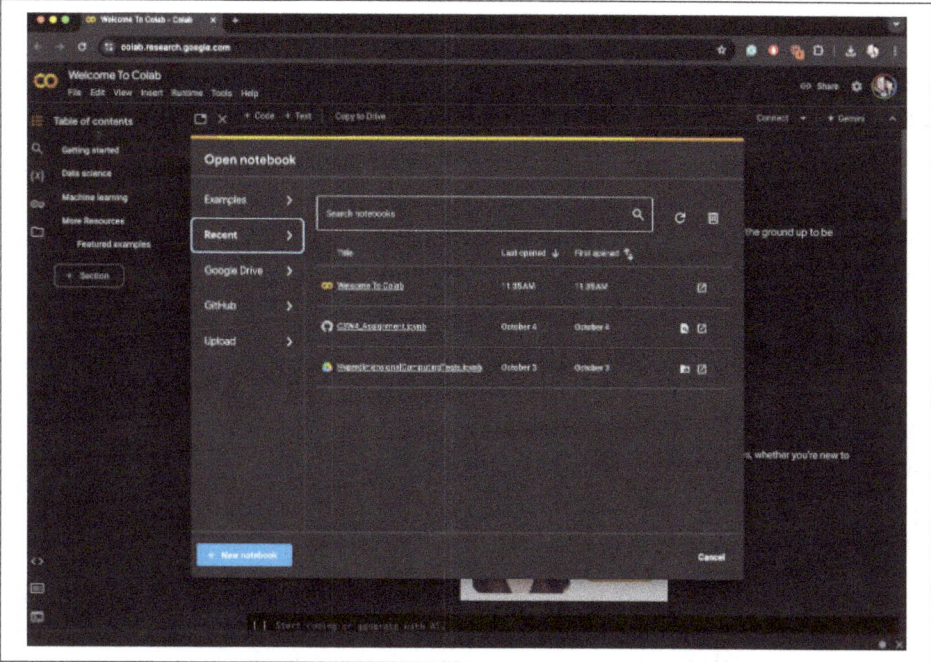

Figure 1-16. Getting started with Google Colab

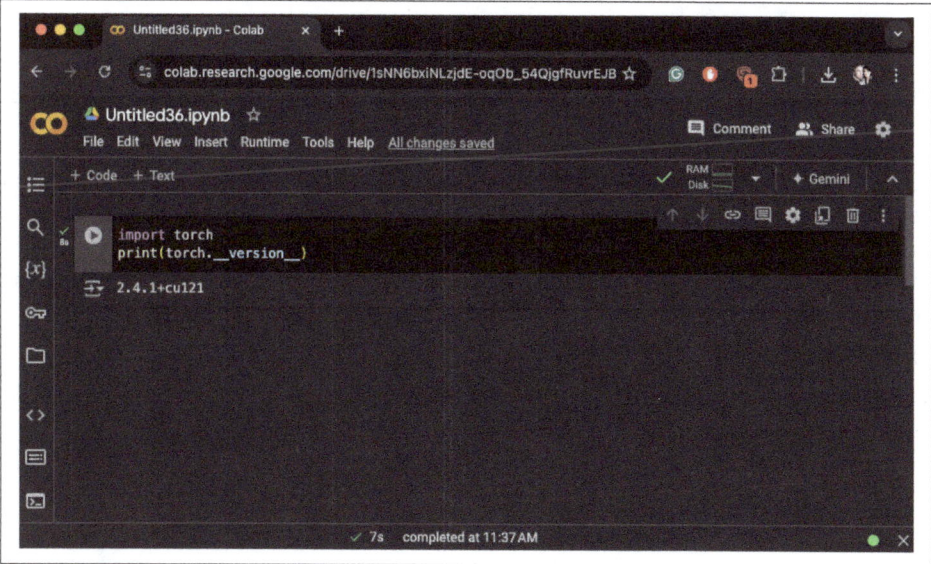

Figure 1-17. Running PyTorch code in Colab

It's always a good idea to check the PyTorch version, as shown here, to be sure you're running the correct version for the task at hand.

You can also see that in Figure 1-17 the version shown is 2.4.1+cu121, and you might want to know what the *cu121* part is! The *cu* stands for *Cuda*, which is Nvidia's library for accelerated ML on GPUs. So, the preceding message demonstrates that PyTorch 2.4.1 is installed, along with accelerators for Cuda version 12.1.

Often, Colab's built-in versions of various libraries, including PyTorch, will be a version or two behind the latest release. If that's the case, you can update it with `pip install` as shown earlier, by simply using a block of code like this, where you specify the desired version:

```
!pip install torch==<different version number>
```

Once you run this command, your current environment within Colab will use the desired version of PyTorch. However, you should be careful when doing this in Colab because the version of PyTorch you change to may not have Cuda drivers installed, meaning you could downgrade to using the CPU.

Getting Started with Machine Learning

As we saw earlier in the chapter, the ML paradigm is one in which you have data, that data is labeled, and you want to figure out the rules that match the data to the labels. The simplest possible scenario to show this in code is as follows.

Consider these two sets of numbers:

```
x = -1, 0, 1, 2, 3, 4
y = -3, -1, 1, 3, 5, 7
```

There's a relationship between the x and y values (for example, if x is –1, then y is –3; if x is 3, then y is 5; and so on). Can you see it?

After a few seconds, you probably saw that the pattern here is $y = 2x - 1$. How did you get that? Different people work it out in different ways, but I typically hear the observation that x increases by 1 in its sequence and y increases by 2; thus, $y = 2x +/-$ something. Then, they look at when $x = 0$ and see that $y = -1$, so they figure that the answer could be $y = 2x - 1$. Next, they look at the other values and see that this hypothesis "fits," and the answer is $y = 2x - 1$.

That's very similar to the ML process. Let's take a look at some code that you could write to have a neural network figure this out for you.

Here's the full code, using PyTorch. Don't worry if it doesn't make sense yet; we'll go through it line by line:

```python
import torch
import torch.nn as nn
import torch.optim as optim
import numpy as np

# Model
model = nn.Sequential(nn.Linear(1, 1))

# Loss and optimizer
criterion = nn.MSELoss()
optimizer = optim.SGD(model.parameters(), lr=0.01)

# Data
xs = torch.tensor([[-1.0], [0.0], [1.0], [2.0], [3.0], [4.0]],
                  dtype=torch.float32)
ys = torch.tensor([[-3.0], [-1.0], [1.0], [3.0], [5.0], [7.0]],
                  dtype=torch.float32)

# Train
for _ in range(500):
    optimizer.zero_grad()
    outputs = model(xs)
    loss = criterion(outputs, ys)
    loss.backward()
    optimizer.step()

# Predict
with torch.no_grad():
    print(model(torch.tensor([[10.0]], dtype=torch.float32)))
```

The first few lines are the importers that ensure the correct libraries are available, so let's jump to this line:

```python
model = nn.Sequential(nn.Linear(1, 1))
```

You've probably heard of neural networks, and you've probably seen diagrams that explain them by using layers of interconnected neurons, a little like in Figure 1-18.

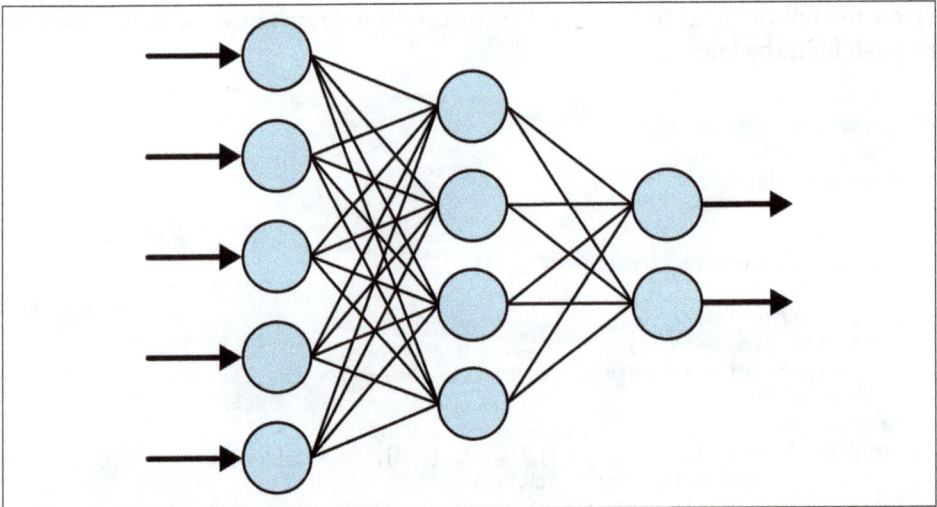

Figure 1-18. A typical neural network

When you see a neural network like this, you should consider each of the circles to be a *neuron* and each of the columns of circles to be a *layer*. So, in Figure 1-18, there are three layers: the first has five neurons, the second has four, and the third has two.

These layers are organized in a sequence through which the data flows from left to right.

Now, if we look back at our code, you'll see that we're defining a sequence of something, with what's contained in the brackets being the definition of the sequence:

```
model = nn.Sequential(nn.Linear(1, 1))
```

When using PyTorch, you define your layers by using a `Sequential`, and inside the `Sequential`, you then specify what each layer looks like. We have only one line inside our `Sequential`, so the neural network this code defines will have only one layer.

Then, you define what the layer looks like by using the `torch.nn` libraries. There are lots of different layer types, but here, we're using a `Linear` layer, in which a linear relationship (where the definition of a line is $y = wx + b$) can be defined or learned.

Our `Linear` layer has the (1,1) parameters specified, which indicates one feature "in" and one feature "out." So ultimately, we have just one layer with one neuron in our entire neural network.

In other words, the Sequential containing a Linear with the parameters (1,1) ultimately looks like Figure 1-19.

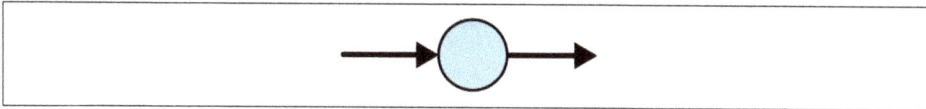

Figure 1-19. A neural network with one layer, containing one neuron

The next lines are where the fun really begins. Let's look at them again:

```
# Loss and optimizer
criterion = nn.MSELoss()
optimizer = optim.SGD(model.parameters(), lr=0.01)
```

If you've done anything with ML before, you've probably seen that it involves a lot of mathematics—and if you haven't done calculus in years, it might have seemed like a barrier to entry. Here's the part where the math comes in—it's the core of ML.

In a scenario such as this one, the computer has *no idea* what the relationship between x and y is. So, it will make a guess. Say, for example, it guesses that $y = 10x + 10$. Then, it needs to measure how good or how bad that guess is—and that's the job of the *loss function*.

The computer already knows the answers when x is –1, 0, 1, 2, 3, and 4, so the loss function can compare these to the answers for the guessed relationship. If it guessed $y = 10x + 10$, then when x is –1, y will be 0. However, the correct answer there was –3, so it's a bit off. But when x is 4, the guessed answer is 50, whereas the correct one is 7. That's really far off.

Armed with this knowledge, the computer can then make another guess. That's the job of the *optimizer*. This is where the heavy calculus is used, but with PyTorch, that can be hidden from you. You just pick the appropriate optimizer to use for different scenarios. In this case, we picked one called sgd, which stands for *stochastic gradient descent*—a complex mathematical function that, when given the values, the previous guess, and the results of calculating the errors (or loss) on that guess, can then generate another guess. Over time, its job is to minimize the loss, and by doing so bring the guessed formula closer and closer to the correct answer.

Next, we simply format our numbers into the data format that the layers expect:

```
# Data
xs = torch.tensor([[-1.0], [0.0], [1.0], [2.0], [3.0], [4.0]],
                dtype=torch.float32)

ys = torch.tensor([[-3.0], [-1.0], [1.0], [3.0], [5.0], [7.0]],
                dtype=torch.float32)
```

You'll see the word *tensor* a lot in ML; it even gives the TensorFlow framework its name. Think of a tensor as a way of storing data that's like an array that is optimized for flexibility in array size. To have PyTorch understand our data, we will load the values into tensors representing the x and y values.

The *learning* process will then begin with the training loop like this:

```
# Train
for _ in range(500):
    optimizer.zero_grad()
    outputs = model(xs)
    loss = criterion(outputs, ys)
    loss.backward()
    optimizer.step()
```

If you're new to ML, this is probably the most difficult part to understand, so let's go through it line by line.

Remember that the ML process looks like Figure 1-20.

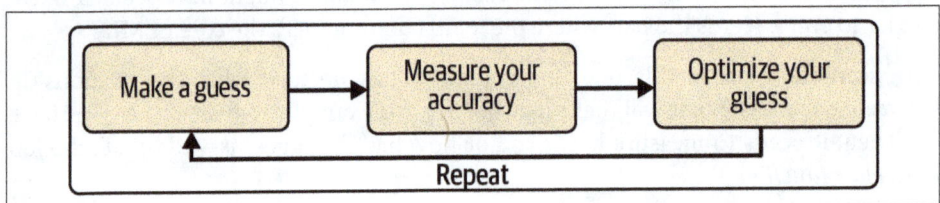

Figure 1-20. The ML process

So, the preceding code implements this as follows:

```
optimizer.zero_grad()
```

This line reads as "zero the gradients" of the optimizer. The calculus of learning involves navigating down a curve to find its minimum, and to do that, we need the gradient of the curve. The curve is calculated when we measure our accuracy, so we need to reset it at the beginning of each loop:

```
outputs = model(xs)
```

This line creates an array of the outputs that we calculate for the input x values. Even though we have given the computer the *correct* answers in our y array, we want to measure the accuracy of the guess that the computer has made for the parameters defining this line. The first time through the loop, the w and b parameters within the neuron will be randomly initialized, so our guess might be $y = 10x + 10$, for example:

```
loss = criterion(outputs, ys)
```

This line then compares the outputs (aka our guesses) with the correct answers to calculate the *loss*—which is effectively a value that tells us how good or bad the guess is:

```
loss.backward()
```

This line is the essential part of the learning process where a process called *backpropagation* happens. It's where the math from the optimizer and the loss function combine to figure out the gradients for the new set of parameters. In our case, the error from $Y = 10x + 10$ is really high and not even close to our desired values, so the calculations done in figuring out the loss will give us a *direction* or gradient in which we should go to get closer to our desired results:

```
optimizer.step()
```

This line of code finishes the job by updating the model parameters to the values based on the gradients calculated in the preceding backpropagation step.

We then repeat this process five hundred times, with the goal of finding a set of parameters for our single neuron that will give us y values that are close to our desired y values. If the set of parameters does so, it can then infer the y value for x values that the computer has never previously seen. Thus, it will have learned the relationship between the x and y values we provided.

Figure 1-21 shows a screenshot of this running in a Colab notebook. Take a look at the loss values over time.

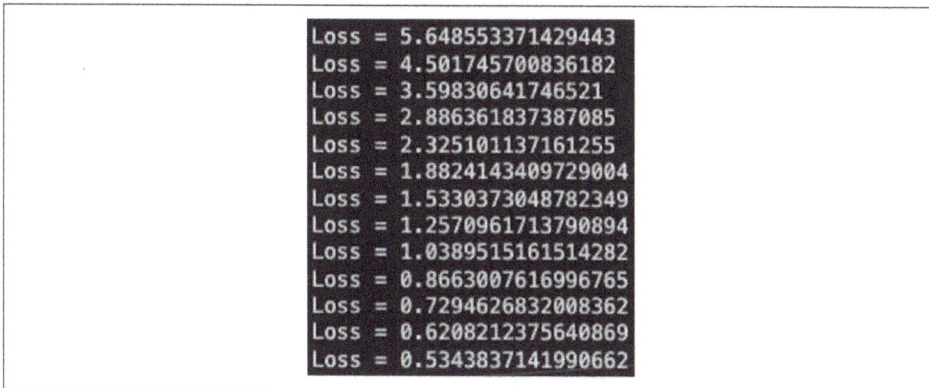

```
Loss = 5.648553371429443
Loss = 4.501745700836182
Loss = 3.59830641746521
Loss = 2.886361837387085
Loss = 2.325101137161255
Loss = 1.8824143409729004
Loss = 1.5330373048782349
Loss = 1.2570961713790894
Loss = 1.0389515161514282
Loss = 0.8663007616996765
Loss = 0.7294626832008362
Loss = 0.6208212375640869
Loss = 0.5343837141990662
```

Figure 1-21. Training the neural network

You'll see the term *epoch* often used when training models like this. Consider an epoch to be one complete pass over the data during the training process.

We can see that over the first 10 epochs, the loss went from 5.64 to 0.86. That is, after only 10 tries, the network was performing about six times better than with its initial guess.

Then take a look at what happens by the 500th epoch (see Figure 1-22).

```
Loss = 1.1965369594690856e-05
Loss = 1.1720051588781644e-05
Loss = 1.1479209206299856e-05
Loss = 1.1243268090765923e-05
Loss = 1.1012628419848625e-05
Loss = 1.0786318853206467e-05
Loss = 1.0564602234808262e-05
Loss = 1.0348104297008831e-05
Loss = 1.0135480806638952e-05
Loss = 9.927043720381334e-06
Loss = 9.723284165374935e-06
Loss = 9.52379559748806e-06
```

Figure 1-22. Training the neural network—the last few epochs

We can now see that the loss is 9.52×10^{-6}. The loss has gotten so small that the model has pretty much figured out that the relationship between the numbers is $y = 2x - 1$. This means that the *machine* has *learned* the pattern between them.

If we want our neural network to try to predict a new value, we can use code like this:

```
# Predict
with torch.no_grad():
    prediction = model(torch.tensor([[10.0]], dtype=torch.float32))
    print(prediction)
```

> The term *prediction* is typically used when dealing with ML models—but don't think of it as looking into the future! We use this term because we're dealing with a certain amount of uncertainty. Think back to the activity detection scenario we spoke about earlier. When the person was moving at a certain speed, she was *probably* walking. Similarly, when a model learns about the patterns that exist between two things, it will tell us what the answer *probably* is. In other words, it is *predicting* the answer. (Later, you'll also learn about *inference*, in which the model picks one answer among many and *infers* that it has picked the correct one.)

What do you think the answer will be when we ask the model to predict y when x is 10? You might instantly think 19, but that's not correct. The model will pick a value *very close* to 19, and there are several reasons for this. First of all, our loss wasn't 0. It was a very small amount, so we should expect any predicted answer to be off by a very small amount. Second, the neural network is trained on only a small amount of data—and in this case, it's only six pairs of (x, y) values.

The model only has a single neuron in it, and that neuron learns a *weight* and a *bias* so that $y = wx + b$. This looks exactly like the desired $y = 2x - 1$ relationship, in which we want the model to learn that $w = 2$ and $b = -1$. Given that the model was trained on only six items of data, we'd never expect the answer to be exactly these values; instead, we'd expect it to be something very close to them.

Now, run the code for yourself to see what you get. I got 18.991 when I ran it, but your answer may differ slightly because when the neural network is first initialized, there's a random element: your initial guess will be slightly different from mine and from a third person's.

Seeing What the Network Learned

This is obviously a very simple scenario in which we are matching x's to y's in a linear relationship. As mentioned in the previous section, neurons have weight and bias parameters. That makes a single neuron fine for learning a relationship like this; namely, when $y = 2x - 1$, the weight is 2 and the bias is -1.

With PyTorch, we can actually take a look at the weights and biases that are learned, with code like this:

```
# Access the first (and only) layer in the sequential model
layer = model[0]
# Get weights and bias
weights = layer.weight.data.numpy()
bias = layer.bias.data.numpy()
print("Weights:", weights)
print("Bias:", bias)
```

Once the network finishes learning, you can print out the values (or weights) that the layer learned. In my case, the output was as follows:

```
Weights: [[1.998695]]
Bias: [-0.9959542]
```

Thus, the learned relationship between x and y was $y = 1.998695\, x - 0.9959542$.

This is pretty close to what we'd expect ($y = 2x - 1$), and we could argue that it's even closer to reality because we're *assuming* that the relationship will hold for other values!

Summary

That's it for your first "Hello World" of ML. You might be thinking that this seems like massive overkill for something as simple as determining a linear relationship between two values—and you'd be right. But the cool thing about this is that the pattern of code we've created here is the same pattern that's used for far more complex scenarios. You'll see those scenarios starting in Chapter 2, where we'll explore some basic computer vision techniques—in which the machine will learn to "see" patterns in pictures and identify what's in them!

Introduction to Computer Vision

Chapter 1 introduced the basics of how machine learning works. You saw how to get started with programming using neural networks to match data to labels, and from there, you saw how to infer the rules that can be used to distinguish items.

In this chapter, we'll consider the next logical step, which is to apply these concepts to computer vision. In this process, a model learns how to recognize content in pictures so it can "see" what's in them. You'll work with a popular dataset of clothing items and build a model that can differentiate between them and thus "see" the difference between different types of clothing.

How Computer Vision Works

Computer vision is the ability of a computer to recognize items beyond just storing their pixels. For example, consider items of clothing that might look like those in Figure 2-1. They're very complex, with lots of different varieties of the same item. Take a look at the two shoes—they're very different, but they're still shoes!

Figure 2-1. Clothing examples

There are a number of different recognizable clothing items here. You understand the difference between a shirt, a coat, and a dress, and you fundamentally know what each of these items are—but how would you explain all that to somebody who has never seen clothing? How about a shoe? There are two shoes in this image, but given the major differences between them, how would you explain to someone what makes them both shoes? This is another area where the rules-based programming we spoke about in Chapter 1 can fall apart. Sometimes, it's just unfeasible to describe something with rules.

Of course, computer vision is no exception to this issue. But consider how you learned to recognize all these items—by seeing lots of different examples and gaining experience with how they're used. Can a computer learn the same way? The answer is yes, but with limitations. Throughout the rest of this chapter, we'll take a look at an example of how to teach a computer to recognize items of clothing using a well-known dataset called Fashion MNIST.

The Fashion MNIST Database

One of the foundational datasets for learning and benchmarking algorithms is the Modified National Institute of Standards and Technology (MNIST) database, which was created by Yann LeCun, Corinna Cortes, and Christopher Burges. This dataset consists of images of 70,000 handwritten digits from 0 to 9, and the images are 28 × 28 grayscale.

Fashion MNIST (*https://oreil.ly/f-mnist*) is designed to be a drop-in replacement for MNIST that has the same number of records, the same image dimensions, and the same number of classes. Rather than images of the digits 0 through 9, Fashion MNIST contains images of 10 different types of clothing.

You can see an example of the dataset contents in Figure 2-2, in which three lines are dedicated to each clothing item type.

Figure 2-2. Exploring the Fashion MNIST dataset

Fashion MNIST has a nice variety of clothing, including shirts, trousers, dresses, and lots of types of shoes! Also, as you may notice, it's monochrome, so each picture consists of a certain number of pixels with values between 0 and 255. This makes the dataset simpler to manage.

You can see a close-up of a particular image from the dataset in Figure 2-3.

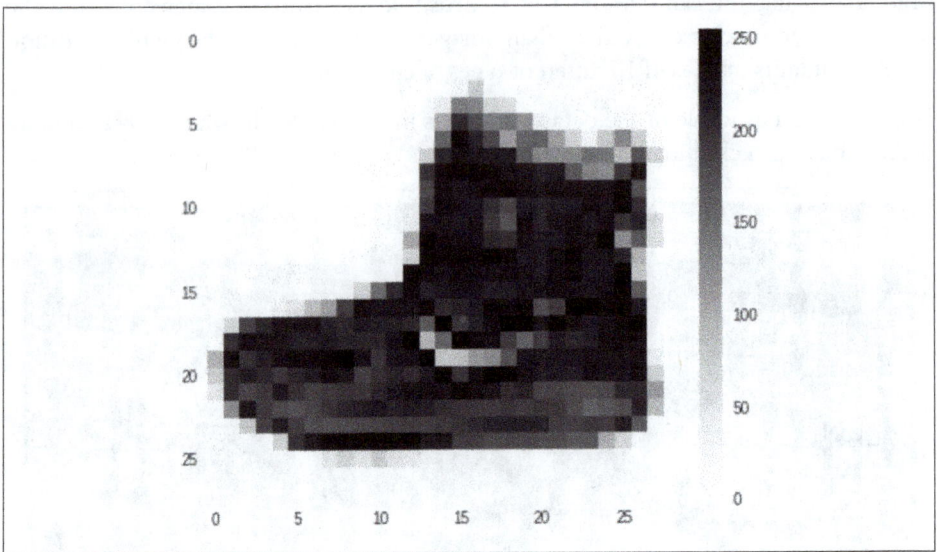

Figure 2-3. Close-up of an image in the Fashion MNIST dataset

Like any image, this one is a rectangular grid of pixels. In this case, the grid size is 28 × 28, and each pixel is a value between 0 and 255, so it is represented by a square in grayscale. To make it easier to see, I have expanded it so that it looks pixelated.

Let's now take a look at how you can use these pixel values with the functions we saw previously.

Neurons for Vision

In Chapter 1, you saw a very simple scenario in which a machine was given a set of x and y values and it learned that the relationship between them was $y = 2x - 1$. This was done using a very simple neural network with one layer and one neuron. If you were to draw that visually, it might look like Figure 2-4.

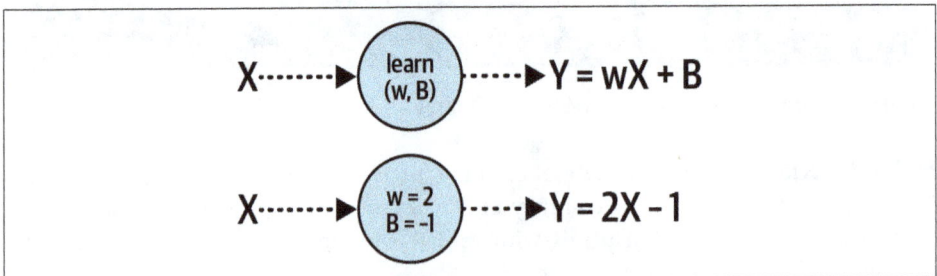

Figure 2-4. A single neuron learning a linear relationship

Each of our images is a set of 784 values (28 × 28) between 0 and 255. They can be our x. We also know that we have 10 different types of images in our dataset, so let's consider them to be our y. Now, we want to learn what the function looks like in which y is a function of x.

Given that we have 784 x values per image and our y is going to be between 0 and 9, a simple equation like $y = mx + c$ isn't going to be enough to solve the problem. That's because there's a large variety of possible values and the equation can only plot values on a line.

But what we *can* do is have several neurons working together. Each neuron will learn *parameters*, and when we have a combined function of all of these parameters working together, we can see whether we can match that pattern to our desired answer (see Figure 2-5).

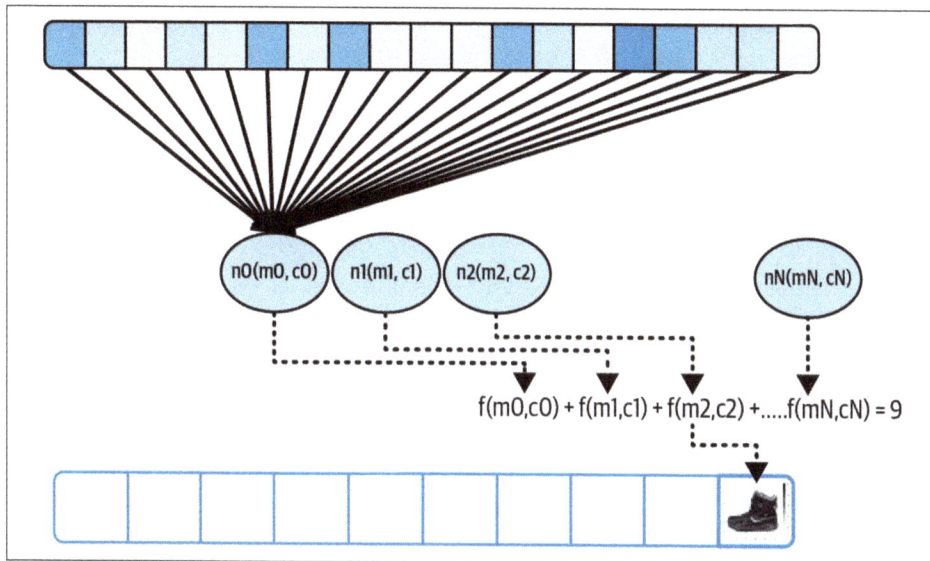

Figure 2-5. Extending our pattern for a more complex example

The gray boxes at the top of this diagram can be considered the pixels in the image, which are our X values. When we train the neural network, we load the pixels into a layer of neurons—Figure 2-5 shows them being loaded into the first neuron, but the values are loaded into just each of them. Also, consider each neuron's weight and bias (w and b) to be randomly initialized. Then, when we sum up the values of the output of each neuron, we're going to get a value. We'll do this for *every* neuron in the output layer, so neuron 0 will contain the value of the probability that the pixels will add up to label 0, neuron 1 will contain the value of the probability that the pixels will add up to label 1, etc.

Over time, we want to match that value to the desired output—which, for this image, is the number 9, which is also the label for the ankle boot that was shown in Figure 2-3. So, in other words, this neuron should have the largest value of all of the output neurons.

Given that there are 10 labels, a random initialization should get the right answer about 10% of the time. From that, the loss function and optimizer can do their job epoch by epoch to tweak the internal parameters of each neuron to improve on that 10%. And thus, over time, the computer will learn to "see" what makes a shoe a shoe or a dress a dress. You'll see this process of improvement when you run the code and your neural network effectively learns to distinguish the different items.

Designing the Neural Network

Let's take the example we just walked through and explore what it looks like in code. First, we'll look at the design of the neural network that was shown in Figure 2-5:

```python
self.linear_relu_stack = nn.Sequential(
    nn.Linear(28*28, 128),
    nn.ReLU(),
    nn.Linear(128, 10),
    nn.LogSoftmax(dim=1)
)
```

If you remember, in Chapter 1 we used a `Sequential` model, whose name suggested we could use *many* layers in Sequence. In that case, we had only one layer, but now we're using it to define multiple layers.

The first layer, a `Linear`, is a layer of neurons that learn a linear relationship between their inputs and their outputs. As before, when using a `Linear`, you give two parameters: the input shape and the output shape. Conveniently, the output shape is effectively the number of neurons you want in this layer, and we're specifying that we want 128 of them. The *input* shape is defined as (28 × 28), which is the size of the data coming into the network, and as you saw earlier, this is the dimension of a Fashion MNIST image.

The input is shown as the middle layer in Figure 2-5, and you'll often hear such layers described as *hidden layers*. The term *hidden* just means that there's no direct interface to that layer. This takes a little bit of getting used to—the middle layer is the first layer that you *define*, and in a diagram like Figure 2-5, you can see that it's in the middle of the diagram. This is because we also drew the data "coming in" to this layer. One other thing to note is that image data from datasets like Fashion MNIST is usually rectangular in shape, but a layer doesn't recognize that, so it will need to be "flattened" into a 1-D array, as shown across the top of Figure 2-5. You'll see the code for that in a moment.

With this first `Linear`, we're asking for 128 neurons to have their internal parameters randomly initialized. Often, the question I'll get asked at this point is "Why 128?" This is entirely arbitrary—there's no fixed rule for the number of neurons to use. As you design the layers, you want to pick the appropriate number of values to enable your model to actually learn. More neurons means it will run more slowly, as it has to learn more parameters. More neurons could also lead to a network that is great at recognizing the training data but not so good at recognizing data that it hasn't previously seen. (This is known as *overfitting*, and we'll discuss it later in this chapter). On the other hand, fewer neurons means that the model might not have sufficient parameters to learn.

You will need to explore this trade-off between speed of learning and accuracy of learning and do some experimentation over time to pick the right values. This process is typically called *hyperparameter tuning*. In ML, a *hyperparameter* is a value that is used to control the training, as opposed to the internal values of the neurons that get trained/learned, which are referred to as *parameters*.

When you're defining a neural network with PyTorch and using the `Sequential`, you don't just define the layers of the network and what types of neurons they may use. You can also define functions that execute on the data while it flows between the neural network layers. These are typically called *activation functions*, and an activation function is the next thing you see specified in the code as `nn.ReLU()`. An activation function is code that will execute on each neuron in the layer. PyTorch supports a number of activation functions out of the box, and a very common one in middle layers is ReLU, which stands for *rectified linear unit*. It's a simple function that returns a value only if it's greater than 0. In this case, we don't want negative values being passed to the next layer to potentially impact the summing function, so instead of writing a lot of `if-then` code, we can simply activate the layer with ReLU.

Finally, there's another `Linear` layer, which will be the *output layer*. If you look at the defined shape of (128, 10) and think of it through that "input size, output size" framework, you'll see that it has 128 "inputs" (i.e., the number of neurons in the layer above) and 10 "outputs." What are these 10? Recall that Fashion MNIST has 10 classes of clothing. Each of these neurons is effectively assigned one class, and it will end up with a probability that the input pixels match that class, so our job is to determine which one has the highest value. You might wonder how these assignments happen: where is the code that says one neuron is for a shoe and another is for a shirt? To answer that question, recall the $y = 2x - 1$ example in Chapter 1, where we had a set of input data and a set of known, correct answers that is sometimes called the *ground truth*. Fashion MNIST will work in the same way. When training the network, we provide the input images *and* their known answers as a set of what we want the output neurons to look like. Thus, the network will "learn" that when it sees a shoe, the output neurons that don't represent that shoe should have a zero value and the ones that do should have a "1" value.

We *could* also loop through the output neurons to find the highest value, but the Log Softmax activation function does that for us.

So now, when we train our neural network, we have two goals. We want to be able to feed in a 28 × 28–pixel array, and we want the neurons in the middle layer to have weights and biases (*w* and *B* values) that, when combined, will match those pixels to one of the 10 output values.

The Complete Code

Now that we've explored the architecture of the neural network, let's look at the complete code for training a model with the Fashion MNIST data.

Here's the complete code:

```python
import torch
import torch.nn as nn
import torch.optim as optim
from torchvision import datasets, transforms
from torch.utils.data import DataLoader

# Load the dataset
transform = transforms.Compose([transforms.ToTensor()])

train_dataset = datasets.FashionMNIST(root='./data', train=True,
                    download=True, transform=transform)
test_dataset = datasets.FashionMNIST(root='./data', train=False,
                    download=True, transform=transform)

train_loader = DataLoader(train_dataset, batch_size=64,
                    shuffle=True)
test_loader = DataLoader(test_dataset, batch_size=64,
                    shuffle=False)

# Define the model
class FashionMNISTModel(nn.Module):
    def __init__(self):
        super(FashionMNISTModel, self).__init__()
        self.flatten = nn.Flatten()
        self.linear_relu_stack = nn.Sequential(
            nn.Linear(28*28, 128),
            nn.ReLU(),
            nn.Linear(128, 10),
            nn.LogSoftmax(dim=1)
        )

    def forward(self, x):
        x = self.flatten(x)
        logits = self.linear_relu_stack(x)
        return logits
```

```
model = FashionMNISTModel()

# Define the loss function and optimizer
loss_function = nn.NLLLoss()
optimizer = optim.Adam(model.parameters())

# Train the model
def train(dataloader, model, loss_fn, optimizer):
    size = len(dataloader.dataset)
    model.train()
    for batch, (X, y) in enumerate(dataloader):
        # Compute prediction and loss
        pred = model(X)
        loss = loss_fn(pred, y)

        # Backpropagation
        optimizer.zero_grad()
        loss.backward()
        optimizer.step()

        if batch % 100 == 0:
            loss, current = loss.item(), batch * len(X)
            print(f"loss: {loss:>7f}
                    [{current:>5d}/{size:>5d}]")

# Training process
epochs = 5
for t in range(epochs):
    print(f"Epoch {t+1}\n-------------------------------")
    train(train_loader, model, loss_function, optimizer)
print("Done!")
```

Let's walk through this piece by piece. First, let's consider where the data comes from. In the torchvision library, there's a datasets collection, and we can load Fashion MNIST from that, addressed as follows:

```
datasets.FashionMNIST
```

So, in our first block of code, you'll see these:

```
train_dataset = datasets.FashionMNIST(root='./data', train=True,
                        download=True, transform=transform)
test_dataset = datasets.FashionMNIST(root='./data', train=False,
                        download=True, transform=transform)
```

Now, you might wonder why we're using *two* datasets. It's simple: one is for training, and one is for testing. The idea here is also simple: if you train a neural network on a set of data, it can become an expert on *that* set of data, but it may not be effective at understanding or classifying *other* data that it previously has not seen. In the case of Fashion MNIST, it might become excellent at understanding the difference between a subset of shoes and shirts, but it will do poorly when new data is presented to it. So, it's good practice to always hold back a little of your data and *not* train the neural network with it. In this case, Fashion MNIST has 70,000 items of data, but only 60,000 of them are used to train the network and the other 10,000 are used to test it. If you look at the preceding code carefully, you'll see that the difference between the two lines is the `train=` parameter. For the first one, the training set the parameter is set to True. For the other, it's set to False.

You'll also see the `transform` parameter in the datasets. It specifies a transformation to apply to the data, which was defined like this:

```
transform = transforms.Compose([transforms.ToTensor()])
```

Neural networks typically work with *normalized* values (i.e., those between 0 and 1). However, the pixels in our image are in the range of 0–255, and the values indicate their color depth, with 0 being black, 255 being white, and everything in between being shades of gray. To prepare the data for the neural network, we should map these shades to values between 0 and 1. The preceding code will automatically do that for you in PyTorch, so applying this `transform` parameter as you're loading the code will then map the pixel values from the [0, 255] integer range to a [0, 1] floating-point range and load them into an array that's suitable for the neural network (aka a Tensor).

Our job will be to fit the training images to the training labels in a manner that's similar to how we fit y to x in Chapter 1.

The math for why normalized data is better for training neural networks (*https://oreil.ly/6d_Po*) is beyond the scope of this book, but bear in mind that when you're training a neural network in PyTorch, normalization will improve performance. Often, your network will not learn and will have massive errors when dealing with nonnormalized data. You'll recall that the $y = 2x - 1$ example from Chapter 1 didn't require the data to be normalized because it was very simple, but for fun, try training it with different values of x and y where x is much larger—and you'll see it quickly fail!

Next, we define the neural network that makes up our model, as discussed earlier, but we'll flesh it out with a bit more detail—including the flattening layers and how we want the "forward" pass to work in the model.

Here's the code:

```
# Define the model
class FashionMNISTModel(nn.Module):
    def __init__(self):
        super(FashionMNISTModel, self).__init__()
        self.flatten = nn.Flatten()
        self.linear_relu_stack = nn.Sequential(
            nn.Linear(28*28, 128),
            nn.ReLU(),
            nn.Linear(128, 10),
```

```
            nn.LogSoftmax(dim=1)
        )

    def forward(self, x):
        x = self.flatten(x)
        logits = self.linear_relu_stack(x)
        return logits

model = FashionMNISTModel()
```

Some key things to note here are that the `FashionMNISTModel` class subclasses `nn.Module`, which gives you the ability to override its `forward` method. We use this method when data is passing forward through the network. Remember back in Chapter 1 when we saw the `loss.backward()` call that did backpropagation and changed the parameters of the network? You'll frequently encounter that same pattern when training models with PyTorch. You'll define functions to execute as the data moves *forward* through the network, and then you'll define others to execute as the gradients that we calculate from the loss move *backward* through the network.

So, if we look at the `init` for the class, we define two methods: `flatten`, which is set to `nn.FLatten()` (a built-in function to flatten the 2D image to 1D), and `linear_relu_stack`, which is set to the sequence of layers and operations (often abbreviated to *ops*) that define the behavior of the network.

In `forward`, we then simply define how these work. First, we flatten our data, `x`, by calling `self.flatten`, and then the results will be passed into `linear_relu_stack` to get the results. The results are called *logits*, which are log probabilities (as defined by `LogSoftmax`) that indicate the confidence the model has that each class is the correct classification.

To learn from our data, we need a loss function to calculate how good or bad our current "guess" is, and we also need an optimizer to figure out the next set of parameters for an improved guess.

Here's an example of how to define both:

```
# Define the loss function and optimizer
loss_function = nn.NLLLoss()
optimizer = optim.Adam(model.parameters())
```

First, let's look at the loss function. It's defined as `nn.NLLLoss()`, which stands for "Negative Log Likelihood Loss." Don't worry—nobody expects you to understand what that means at this point! Ultimately, as you work through learning how to do ML, you'll learn about different loss functions, and you'll experiment with which ones work well in particular scenarios. In this case, given that the output logits are log probabilities, I chose this loss function because it works particularly well for this scenario. As mentioned, over time, you'll learn a lot more about the library of loss functions, and you can choose the best ones for your scenario. But for now, just go with the flow and use this one!

For the optimizer, I've opted to use the Adam optimization algorithm. It's similar to the stochastic gradient descent that we used for the $y = 2x - 1$ model in Chapter 1, but it's generally faster and more accurate. As with the loss function, you'll learn more about optimization algorithms over time, and you'll be able to choose from the menu of optimizers that fit your scenario best. One important thing here is to note that I've passed in model.parameters() as a parameter to this. This parameter passes all the trainable parameters in the model to the optimizer so that it can adjust them to help minimize the loss calculated by the loss function.

Now, let's get down to the specifics and explore what the code we use for training the network looks like:

```python
# Train the model
def train(dataloader, model, loss_fn, optimizer):
    size = len(dataloader.dataset)
    model.train()
    for batch, (X, y) in enumerate(dataloader):
        # Compute prediction and loss
        pred = model(X)
        loss = loss_fn(pred, y)

        # Backpropagation
        optimizer.zero_grad()
        loss.backward()
        optimizer.step()

        if batch % 100 == 0:
            loss, current = loss.item(), batch * len(X)
            print(f"loss: {loss:>7f}  [{current:>5d}/{size:>5d}]")
```

While some of this will look familiar because it builds on the simple neural network from Chapter 1, there are a few new concepts here, given that we're using much more data. First, you'll see that we get the size of the dataset. We simply use this to report on progress, as shown in the very last line.

Then, we call model.train to explicitly set the model into training mode. PyTorch has optimizations that occur during training that are beyond the scope of this chapter. (To take advantage of them, you'll switch the model between training and inference modes.) Note that this is more a property of the model than a method, but the method syntax is there. Sorry if it's a little confusing!

Next up is this interesting line:

```python
for batch, (X, y) in enumerate(dataloader):
```

Let's explore this in a little more detail. We made the Fashion MNIST dataset available to our code by using a data loader. There are 60,000 records available for training, each of which is 784 pixels. That's a lot of data, and you don't necessarily need all of it in memory at once. The idea of a `batch` is to take a chunk of that data—which, by default, is 64 items—and work with it. Enumerating the data loader gives us that, so we'll train with 938 batches, 937 of 64, and the last one of 32 because you can't evenly divide 60,000 by 64!

Now, for each batch, we'll go through the same loop that we saw for the previous example. We'll get the predictions from the model, calculate the loss, backpropagate the gradients from the loss function, and optimize with new parameters.

We'll also use the term *epoch* for a training cycle with *all* of the data (i.e., every batch). We can then output the status of the training every one hundred batches so as not to overload the output console!

So, to train the network for five epochs, we can use code like this:

```
# Training process
epochs = 5
for t in range(epochs):
    print(f"Epoch {t+1}\n-------------------------------")
    train(train_loader, model, loss_function, optimizer)
print("Done!")
```

This will simply call the `train` function we specified five times—putting the network through the training loop by calculating the predictions, figuring out the loss, optimizing the parameters, and repeating five times.

Training the Neural Network

Once you've executed the code, you'll see the network train epoch by epoch. Then, after running the training, you'll see something at the end that looks like this:

```
Epoch 5
-------------------------------
loss: 0.429329 [    0/60000]
loss: 0.348756 [ 6400/60000]
loss: 0.237481 [12800/60000]
loss: 0.336960 [19200/60000]
loss: 0.435592 [25600/60000]
loss: 0.272769 [32000/60000]
loss: 0.362881 [38400/60000]
loss: 0.202799 [44800/60000]
loss: 0.354268 [51200/60000]
loss: 0.205381 [57600/60000]
Done!
```

You can see here that over time, the loss has gone down. For example, in my case, the loss value at the end of the first epoch was .345, and by the end of the fifth epoch, it was .205. This data shows us that the network is learning.

But how can we tell how *accurately* it's learning? Note that loss and accuracy, while related, don't have a direct linear relationship—for example, we can't say that if loss is 20%, then accuracy is 80%. So, we need to go a little deeper.

Recall that when we were getting the data, we got *two* datasets: one for training and one for testing. Here's a great place where we can write code to pass the test data through our network and evaluate how accurate the network is at predicting answers. We already know the correct answers, so we could do inference on all 10,000 test records, get the answers that the model predicts, and then check them against the ground truth for accuracy.

Here's the code:

```
# Function to test the model
def test(dataloader, model):
    size = len(dataloader.dataset)
    num_batches = len(dataloader)
    model.eval()  # Set the model to evaluation mode
    test_loss, correct = 0, 0
    with torch.no_grad():
        for X, y in dataloader:
            pred = model(X)
            test_loss += loss_function(pred, y).item()
            correct += (pred.argmax(1) ==
                        y).type(torch.float).sum().item()
    test_loss /= num_batches
    correct /= size
    print(f"Test Error: \n Accuracy: {(100*correct):>0.1f}%,
            Avg loss: {test_loss:>8f} \n")


# Evaluate the model
test(test_loader, model)
```

There are a few things to note in this code. First is the `model.eval()` line, which indicates that we are switching the model from training mode to inference mode. Similarly, `torch.no_grad()`will turn off gradient calculation in PyTorch to speed up inference. We're no longer *training* the model, so we don't need to do all the loss function backpropagation and optimization. We can just turn that off.

Then, as it does during training, the network just goes through every item in the data loader, gets the prediction for that item, and checks its correctness with this line:

```
correct += (pred.argmax(1) ==  y).type(torch.float).sum().item()
```

That's a bit of a mouthful, so let's break it down.

First, the `pred` value will give us the prediction from the network. The network outputs 10 values, each of which includes the probability of the class it represents being the correct one. Calling `argmax` on this will give us which one had the biggest value (i.e., the one with the probability closest to 1). The *y* value is the correct answer. For example, if we get a prediction, the neuron with the highest value is the sixth one, and *y* = 6, so we know we have a correct answer. Also, because we're dealing in batches, we want to count each time `pred.argmax(1) == y` for this batch, hence, the `sum()`.

Therefore, our accuracy value will be the sum of correct items divided by the total number of items. So, when you run this code after training the model, you should see output like this:

```
Test Error:
  Accuracy: 86.9%, Avg loss: 0.366243
```

Remarkably, after running the neural network for only five epochs, we can see that it is 86.9% accurate on data it hadn't previously seen!

At this point, you may be thinking that it's really nice to see the accuracy of the model on the test set, but you may also be asking why we've only reported loss on the training—why not also report accuracy there? It seems silly to finish training the model by only looking at minimizing loss and *then* to figure out the accuracy. And you'd be right!

Fortunately, updating the model training code to *also* report on accuracy is pretty easy to do. Here's a function called `get_accuracy()` that you can use during training:

```
# Function to calculate accuracy
def get_accuracy(pred, labels):
    _, predictions = torch.max(pred, 1)
    correct = (predictions == labels).float().sum()
    accuracy = correct / labels.shape[0]
    return accuracy
```

Then, in your training loop, you can simply call this function after the loss function call like this:

```
for batch, (X, y) in enumerate(dataloader):
    # Compute prediction and loss
    pred = model(X)
    loss = loss_fn(pred, y)
    accuracy = get_accuracy(pred, y)
```

And when you're reporting on the output of the training, you can use the accuracy metric like this:

```
if batch % 100 == 0:
    current = batch * len(X)
    avg_loss = total_loss / (batch + 1)
    avg_accuracy = total_accuracy / (batch + 1) * 100
```

```
print(f"Batch {batch}, Loss: {avg_loss:>7f},
        Accuracy: {avg_accuracy:>0.2f}%
            [{current:>5d}/{size:>5d}]")
```

Running this will give you output a bit like this:

```
Epoch 5
-------------------------------
Batch   0, Loss: 0.177518, Accuracy: 95.31% [    0/60000]
Batch 100, Loss: 0.304973, Accuracy: 88.89% [ 6400/60000]
Batch 200, Loss: 0.311628, Accuracy: 88.51% [12800/60000]
Batch 300, Loss: 0.307373, Accuracy: 88.63% [19200/60000]
Batch 400, Loss: 0.309722, Accuracy: 88.67% [25600/60000]
Batch 500, Loss: 0.310240, Accuracy: 88.60% [32000/60000]
Batch 600, Loss: 0.306988, Accuracy: 88.70% [38400/60000]
Batch 700, Loss: 0.308556, Accuracy: 88.64% [44800/60000]
Batch 800, Loss: 0.309518, Accuracy: 88.67% [51200/60000]
Batch 900, Loss: 0.311487, Accuracy: 88.59% [57600/60000]
Done!
```

Now, you're probably wondering why the accuracy for the test data (86.9%) is *lower* than the accuracy for the training data (88.59%). This is very common, and when you think about it, it makes sense: the neural network only really knows how to match the inputs it has been trained on with the outputs for those values. Our hope is that given enough data, the network will be able to generalize from the examples it has seen and thus "learn" what a shoe or a dress looks like. But there will always be examples of items that it hasn't seen that are also different enough from what it has seen to confuse it.

For example, if you grew up only ever seeing sneakers, then that's what a shoe looks like to you. So, when you first see a high-heeled shoe, you might be a little confused. From your experience, it's probably a shoe, but you don't know for sure. That's exactly what a neural network "thinks" when it "sees" inputs that are different enough from what it's been trained on.

Exploring the Model Output

Now that we've trained the model and gotten a good gauge of its accuracy by using the test set, let's explore it a little. Here's a function we can use to predict a single image:

```python
import matplotlib.pyplot as plt

def predict_single_image(image, label, model):
    # Set the model to evaluation mode
    model.eval()

    # Unsqueeze image as the model expects a batch dimension
    image = image.unsqueeze(0)
```

```
with torch.no_grad():
    prediction = model(image)
    print(prediction)
    predicted_label = prediction.argmax(1).item()

    # Display the image and predictions
    plt.imshow(image.squeeze(), cmap='gray')
    plt.title(f'Predicted: {predicted_label}, Actual: {label}')
    plt.show()

    return predicted_label

# Choose an image from the test set
image, label = test_dataset[0]  # Change index to test different images

# Predict the class for the chosen image
predicted_label = predict_single_image(image, label, model)
print(f"The model predicted {predicted_label}, and the actual label is {label}.")
```

Let's start with this code, which should look familiar to you now that you've seen the previous accuracy calculation code:

```
with torch.no_grad():
    prediction = model(image)
    print(prediction)
    predicted_label = prediction.argmax(1).item()
```

Here, we get the `image`, send it to the `model`, get back a `prediction`, and `print` it out. Then, we get the `argmax` of that to show the label. Here's an example output of the `prediction`:

```
tensor([[-12.4290, -16.0639, -14.3148, -16.2861, -13.1672,  -4.5377, -13.6284,
          -1.3124,  -8.9946,  -0.3285]])
```

These numbers may seem vague, but ultimately, our goal is simply to look for the biggest one! The `argmax` function gets the `log()` of the value, where `log(1)` is zero and the log of any value less than one is a negative value. As you look through the list, you'll notice that the value closest to 0 (–0.3285) is the very last one. This indicates that the function believes the class for this image should be class number 9. (There are 10 classes in Fashion MNIST, which are numbered 0 through 9.)

Fashion MNIST's class number 9 is "Ankle Boot," so I've also included the code to render the image in Figure 2-6.

Also, as we can see, this is an example of where the model got the prediction right. The ground truth was that it's label 9, and the prediction was for number 9. Drawing the image so that we mere humans can compare the two also gives us an ankle boot!

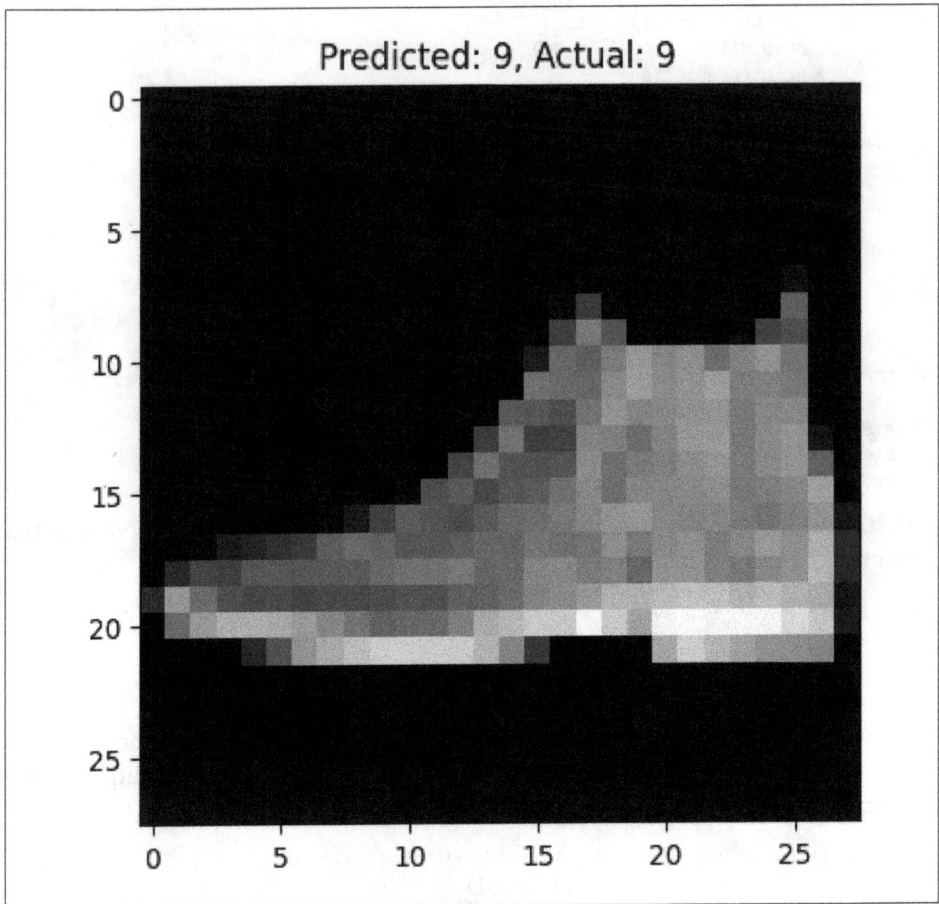

Figure 2-6. Exploring the output of the predictive model

Now, try a few different values for yourself and see if you can find anywhere the model gets it wrong.

Overfitting

In the last example, we trained for only five epochs. That is, we went through the entire training loop of having the neurons randomly initialized and checked against their labels, then that performance was measured by the loss function and updated by the optimizer five times. And the results we got were pretty good: 88.59% accuracy on the training set and 86.5% on the test set. So what happens if we train for longer?

Next, try updating it to train for 50 epochs instead of 5. In my case, I got the following accuracy figures on the training set:

```
Epoch 50
-----------------------------------
Batch 0, Loss: 0.077159, Accuracy: 96.88% [    0/60000]
Batch 100, Loss: 0.094825, Accuracy: 96.57% [ 6400/60000]
Batch 200, Loss: 0.093598, Accuracy: 96.67% [12800/60000]
Batch 300, Loss: 0.095906, Accuracy: 96.54% [19200/60000]
Batch 400, Loss: 0.096683, Accuracy: 96.48% [25600/60000]
Batch 500, Loss: 0.101872, Accuracy: 96.31% [32000/60000]
Batch 600, Loss: 0.103130, Accuracy: 96.22% [38400/60000]
Batch 700, Loss: 0.103901, Accuracy: 96.17% [44800/60000]
Batch 800, Loss: 0.104216, Accuracy: 96.15% [51200/60000]
Batch 900, Loss: 0.104010, Accuracy: 96.15% [57600/60000]
Done!
```

This is particularly exciting because we're doing much better: we're getting 96.15% accuracy!

However, for the test set, accuracy reached 89.2%:

```
Test Error:
  Accuracy: 89.2%, Avg loss: 0.433885
```

So, we got a big improvement over the training set and a smaller one over the test set. This might suggest that training our network for much longer would lead to much better results—but that's not always the case. The network is doing much better with the training data, but the model is not necessarily a better model. In fact, the divergence in the accuracy numbers shows that the model might have become overspecialized to the training data, in a process that's often called *overfitting*. As you build more neural networks, this problem is something to watch out for—and as you go through this book, you'll learn a number of techniques to avoid it!

Early Stopping

In each of the cases so far, we've hardcoded the number of epochs we're training for. While that works, we might want to train until we reach the desired accuracy instead of constantly trying different numbers of epochs and training and retraining until we get to our desired value. So, for example, if we want to train until the model is at 95% accuracy on the training set, and if we want to do it without knowing in advance how many epochs it will take. . .how can we do it?

Given that we've updated our code to check the accuracy as the model trained and to print it out, now, all we have to do is check that accuracy and end the training if it's above a certain amount—such as 95% (or 0.95 when normalized). For example, we can do this:

```
if batch % 100 == 0:
    current = batch * len(X)
    avg_loss = total_loss / (batch + 1)
    avg_accuracy = total_accuracy / (batch + 1) * 100
```

```
print(f"Batch {batch}, Loss: {avg_loss:>7f},
        Accuracy: {avg_accuracy:>0.2f}% [{current:>5d}/{size:>5d}]")

# Early stopping condition
if avg_accuracy >= 95:
    print("Reached 95% accuracy, stopping training.")
    return True  # Stop training
```

Note that if we use this code inside the `if batch % 100 == 0` block, we can break the training loop before all batches in a particular epoch have been processed. It's better to do this check at the end of the epoch, so we need to be sure to place the `if avg_accuracy >= 95` in the right place!

Now, when we're training, at the end of every epoch, the average accuracy for the epoch will be calculated—and if it hits 95%, the training will stop. Previously, I had trained the model for 50 epochs to get 96.15% accuracy, but with this early stopping, where I've defined 95% as "good enough," you can see that the model stopped training after only 37 epochs. Interestingly, accuracy was 94.99% for a couple of epochs before that, so I might have been able to stop even earlier!

This process of *early stopping* is very powerful in helping you save time as you evaluate different model architectures for solving specific problems. It helps you train your model until it's "good enough," instead of having a fixed training loop. For example, the process can look like this:

```
Epoch 36
---------------------------------
Batch 0, Loss: 0.098307, Accuracy: 96.88% [    0/60000]
Batch 100, Loss: 0.119195, Accuracy: 95.45% [ 6400/60000]
Batch 200, Loss: 0.127049, Accuracy: 95.20% [12800/60000]
Batch 300, Loss: 0.126001, Accuracy: 95.34% [19200/60000]
Batch 400, Loss: 0.127823, Accuracy: 95.25% [25600/60000]
Batch 500, Loss: 0.131262, Accuracy: 95.11% [32000/60000]
Batch 600, Loss: 0.135573, Accuracy: 94.95% [38400/60000]
Batch 700, Loss: 0.135920, Accuracy: 94.95% [44800/60000]
Batch 800, Loss: 0.135125, Accuracy: 94.99% [51200/60000]
Batch 900, Loss: 0.134854, Accuracy: 94.99% [57600/60000]
Epoch 37
---------------------------------
Batch 0, Loss: 0.104421, Accuracy: 96.88% [    0/60000]
Batch 100, Loss: 0.122693, Accuracy: 95.34% [ 6400/60000]
Batch 200, Loss: 0.124787, Accuracy: 95.26% [12800/60000]
Batch 300, Loss: 0.127841, Accuracy: 95.16% [19200/60000]
Batch 400, Loss: 0.130558, Accuracy: 95.05% [25600/60000]
Batch 500, Loss: 0.131684, Accuracy: 95.00% [32000/60000]
Batch 600, Loss: 0.132620, Accuracy: 94.95% [38400/60000]
Batch 700, Loss: 0.132498, Accuracy: 95.01% [44800/60000]
Batch 800, Loss: 0.132462, Accuracy: 95.05% [51200/60000]
Batch 900, Loss: 0.133915, Accuracy: 95.03% [57600/60000]
Reached 95% accuracy, stopping training.
```

This process can save you a lot of time you would otherwise spend manually checking on the network to see if it's learning appropriately.

Summary

In Chapter 1, you learned about how ML is based on fitting features to labels through sophisticated pattern matching with a neural network. In this chapter, you took that to the next level by going beyond a single neuron and learning how to create your first (very basic) computer vision neural network. The network was somewhat limited because of the data: all the images were 28 × 28 grayscale, with the item of clothing centered in the frame. This is a good start, but it's a very controlled scenario.

To do better at vision, you may need the computer to learn features of an image instead of learning merely the raw pixels. You can do that with a process called *convolutions*, and in the next chapter, you'll learn how to define convolutional neural networks to understand the contents of images.

## CHAPTER 3

Going Beyond the Basics: Detecting Features in Images

In Chapter 2, you learned how to get started with computer vision by creating a simple neural network that matched the input pixels of the Fashion MNIST dataset to 10 labels, each of which represented a type (or class) of clothing. And while you created a network that was pretty good at detecting clothing types, there was a clear drawback. Your neural network was trained on small monochrome images, each of which contained only a single item of clothing, and each item was centered within the image.

To take the model to the next level, you need it to be able to detect *features* in images. So, for example, instead of looking merely at the raw pixels in the image, what if we could filter the images down to constituent elements? Matching those elements, instead of raw pixels, would help the model detect the contents of images more effectively. For example, consider the Fashion MNIST dataset that we used in the last chapter. When detecting a shoe, the neural network may have been activated by lots of dark pixels clustered at the bottom of the image, which it would see as the sole of the shoe. But if the shoe were not centered and filling the frame, this logic wouldn't hold.

One method of detecting features comes from photography and image processing methodologies that you may already be familiar with. If you've ever used a tool like Photoshop or GIMP to sharpen an image, you've used a mathematical filter that works on the pixels of the image. Another word for what these filters do is *convolution*, and by using such filters in a neural network, you'll create a *convolutional neural network* (CNN).

In this chapter, you'll start by learning about how to use convolutions to detect features in an image. Then, you'll dig deeper into classifying images based on the features within. We'll also explore augmentation of images to get more features and transfer learning to take preexisting features that were learned by others, and then we'll look briefly into optimizing your models by using dropouts.

Convolutions

A *convolution* is simply a filter of weights that are used to multiply a pixel by its neighbors to get a new value for the pixel. For example, consider the ankle boot image from Fashion MNIST and the pixel values for it (see Figure 3-1).

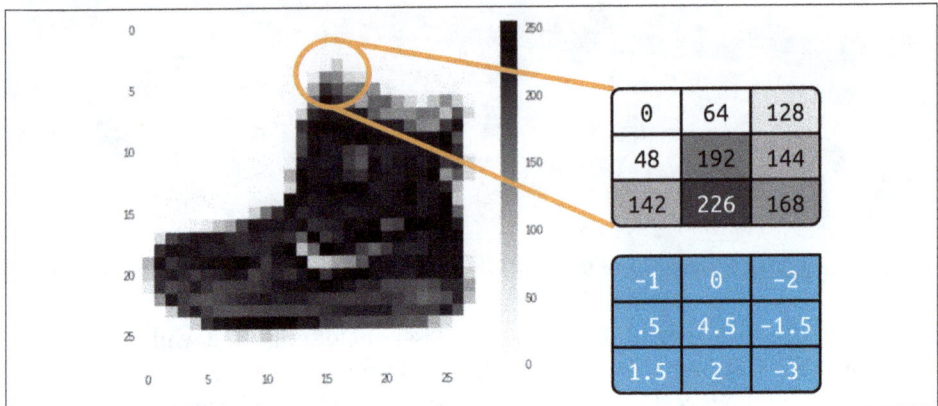

Figure 3-1. Ankle boot with convolution

If we look at the pixel in the middle of the selection, we can see that it has the value 192. (Recall that Fashion MNIST uses monochrome images with pixel values from 0 to 255.) The pixel above and to the left has the value 0, the one immediately above has the value 64, etc.

If we then define a filter in the same 3 × 3 grid, as shown below the original values, we can transform that pixel by calculating a new value for it. We do this by multiplying the current value of each pixel in the grid by the value in the same position in the filter grid and then summing up the total amount. This total will be the new value for the current pixel, and we then repeat this calculation for all pixels in the image.

So, in this case, while the current value of the pixel in the center of the selection is 192, we calculate the new value after applying the filter as follows:

```
new_val = (-1 * 0) + (0 * 64) + (-2 * 128) +
    (.5 * 48) + (4.5 * 192) + (-1.5 * 144) +
    (1.5 * 142) + (2 * 226) + (-3 * 168)
```

The result equals 577, which will be the new value for the pixel. Repeating this process for every pixel in the image will give us a filtered image.

Now, let's consider the impact of applying a filter on a more complicated image: specifically, the ascent image (*https://oreil.ly/wP8TE*) that's built into SciPy for easy testing. This is a 512 × 512 grayscale image that shows two people climbing a staircase.

Using a filter with negative values on the left, positive values on the right, and zeros in the middle will end up removing most of the information from the image except for vertical lines (see Figure 3-2).

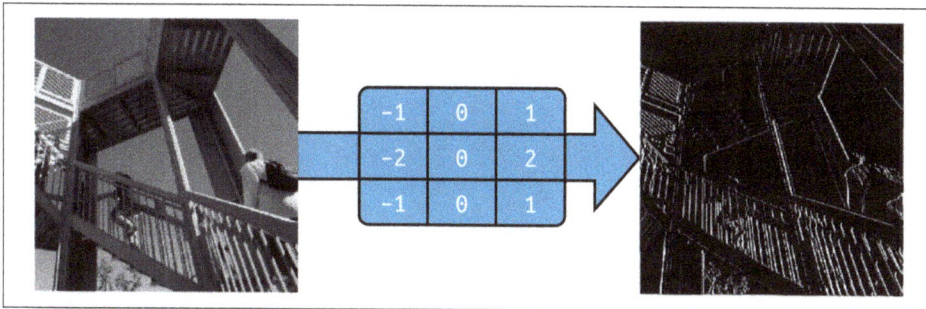

Figure 3-2. Using a filter to derive vertical lines

Similarly, a small change to the filter can emphasize the horizontal lines (see Figure 3-3).

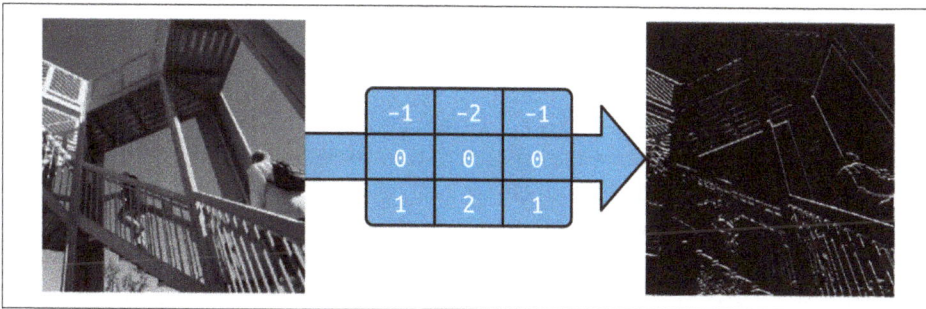

Figure 3-3. Using a filter to derive horizontal lines

These examples also show that the amount of information in the image is reduced. Therefore, we can potentially *learn* a set of filters that reduce the image to features, and those features can be matched to labels as before. Previously, we learned parameters that were used in neurons to match inputs to outputs, and similarly, we can learn the best filters to match inputs to outputs over time.

When we combine convolution with pooling, we can reduce the amount of information in the image while maintaining the features. We'll explore that next.

Pooling

Pooling is the process of eliminating pixels in your image while maintaining the semantics of the content within the image. It's best explained visually. Figure 3-4 depicts the concept of *max pooling*.

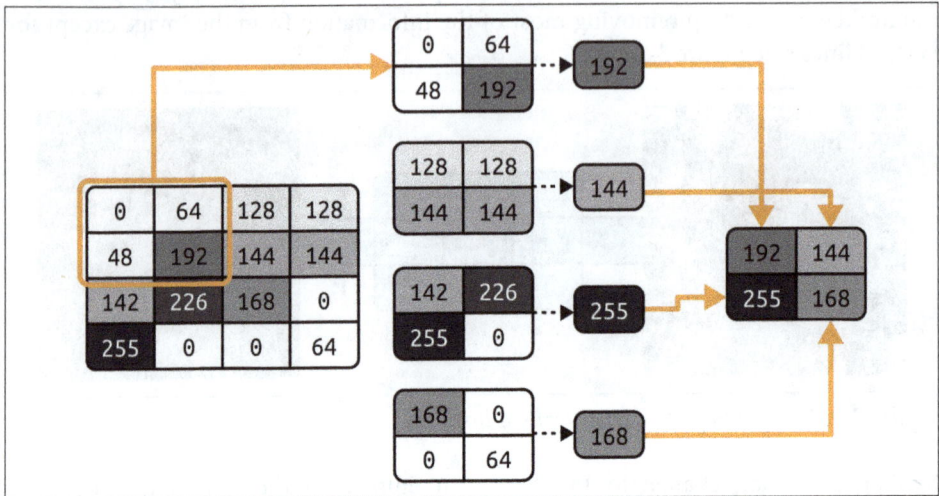

Figure 3-4. An example of max pooling

In this case, consider the box on the left to be the pixels in a monochrome image. We group them into 2 × 2 arrays, so in this case, the 16 pixels are grouped into four 2 × 2 arrays. These arrays are called *pools*.

Then, we select the *maximum* value in each of the groups and reassemble them into a new image. Thus, the pixels on the left are reduced by 75% (from 16 to 4), with the maximum value from each pool making up the new image. Figure 3-5 shows the version of ascent from Figure 3-2, with the vertical lines enhanced, after max pooling has been applied.

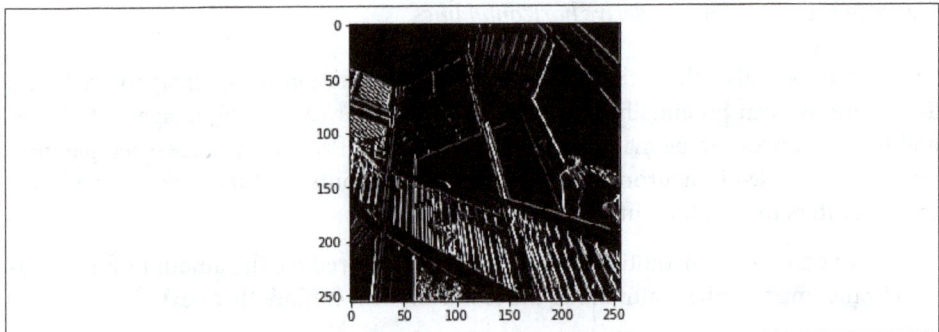

Figure 3-5. Ascent after applying vertical filter and max pooling

Note how the filtered features have not just been maintained but have been further enhanced. Also, the image size has changed from 512×512 to 256×256—making it a quarter of the original size.

> There are other approaches to pooling. These include *min pooling*, which takes the smallest pixel value from the pool, and *average pooling*, which takes the overall average value from the pool.

Implementing Convolutional Neural Networks

In Chapter 2 you created a neural network that recognized fashion images. For convenience, here's the code to define the model:

```
# Define the model
class FashionMNISTModel(nn.Module):
    def __init__(self):
        super(FashionMNISTModel, self).__init__()
        self.flatten = nn.Flatten()
        self.linear_relu_stack = nn.Sequential(
            nn.Linear(28*28, 128),
            nn.ReLU(),
            nn.Linear(128, 10),
            nn.LogSoftmax(dim=1)
        )

    def forward(self, x):
        x = self.flatten(x)
        logits = self.linear_relu_stack(x)
        return logits

model = FashionMNISTModel()

# Define the loss function and optimizer
loss_function = nn.NLLLoss()
optimizer = optim.Adam(model.parameters())

# Train the model
def train(dataloader, model, loss_fn, optimizer):
    size = len(dataloader.dataset)
    model.train()
    for batch, (X, y) in enumerate(dataloader):
        # Compute prediction and loss
        pred = model(X)
        loss = loss_fn(pred, y)

        # Backpropagation
        optimizer.zero_grad()
        loss.backward()
```

```
        optimizer.step()

        if batch % 100 == 0:
            loss, current = loss.item(), batch * len(X)
            print(f"loss: {loss:>7f}
                    [{current:>5d}/{size:>5d}]")

# Training process
epochs = 5
for t in range(epochs):
    print(f"Epoch {t+1}\n-------------------------------")
    train(train_loader, model, loss_function, optimizer)
print("Done!")
```

To convert this to a CNN, you simply use convolutional layers in our model definition on top of the current linear ones. You'll also add pooling layers.

To implement a convolutional layer, you'll use the nn.Conv2D type. It accepts as parameters the number of convolutions to use in the layer, the size of the convolutions, the activation function, etc.

For example, here's a convolutional layer that uses this type:

```
nn.Conv2d(1, 64, kernel_size=3, padding=1)
```

In this case, we want the layer to learn 64 convolutions. It will randomly initialize them, and over time, it will learn the filter values that work best to match the input values to their labels. The kernel_size = 3 indicates the size of the filter. Earlier, we showed you 3 × 3 filters, and that's what we're specifying here. The 3 × 3 filter is the most common size of filter. You can change it as you see fit, but you'll typically see an odd number of axes like 5 × 5 or 7 × 7 because of how filters remove pixels from the borders of the image, as you'll see later.

Here's how to use a pooling layer in the neural network. You'll typically do this immediately after the convolutional layer:

```
nn.MaxPool2d(kernel_size=2, stride=2)
```

In the example back in Figure 3-4, we split the image into 2 × 2 pools and picked the maximum value in each. However, we could have used the parameters that you see here to define the pool size. The kernel_size=2 parameter indicates that our pools are 2 × 2, and the stride=2 parameter means that the filter will jump over two pixels to get the next pool.

Now, let's explore the full code to define a model for Fashion MNIST with a CNN:

```
# Define the CNN model
class FashionCNN(nn.Module):
    def __init__(self):
        super(FashionCNN, self).__init__()
        self.layer1 = nn.Sequential(
```

```
        nn.Conv2d(1, 64, kernel_size=3, padding=1),
        nn.ReLU(),
        nn.MaxPool2d(kernel_size=2, stride=2))

    self.layer2 = nn.Sequential(
        nn.Conv2d(64, 64, kernel_size=3),
        nn.ReLU(),
        nn.MaxPool2d(2))  # Output: 64 x 6 x 6

    self.fc1 = nn.Linear(64 * 6 * 6, 128)
    self.fc2 = nn.Linear(128, 10)  # 10 classes

def forward(self, x):
    out = self.layer1(x)
    out = self.layer2(out)
    out = out.view(out.size(0), -1)  # Flatten the output
    out = self.fc1(out)
    out = self.fc2(out)
    return out
```

Here, we see that the class has two functions, one for initialization and one that will be called during the forward pass in each epoch during training.

The init simply defines what each of the layers in our neural network will look like. The first layer (self.layer1) will take in the one-dimensional input, have 64 convolutions, a kernel_size of 3, and padding of 1. It will then ReLU the output before max pooling it.

The next layer (self.layer2) will take the 64 convolutions of output from the previous layer and then output 64 of its own before ReLUing them and max pooling them. Its output will now be 64 × 6 × 6 because the MaxPool halves the size of the image.

The data is then fed to the next layer (self.fc1, where fc stands for *fully connected*), with the input being the shape of the output of the previous layer. The output is 128, which is the same number of neurons we used in the previous example in Chapter 2 for the neural network. This type of neural network, with multiple layers, is often referred to as a *deep* neural network or DNN.

Finally, these 128 are fed into the final layer (self.fc2) with 10 outputs—that represent the 10 classes.

In the DNN, we ran the input through a Flatten layer prior to feeding it into the first Linear layer. We've lost that in the input layer here—instead, we've just specified the 1-D input shape. Note that prior to the first Linear layer, after convolutions and pooling, the data will be flattened.

Then, we stack these layers in the forward function. We can see that we get the data x and pass it through layer1 to get out, which is passed to layer2 to get a new out. At this point, we have the convolutions that we've learned, but we need to flatten them before loading them into the Linear layers fc1 and fc2. The out = out.view(out.size(0), -1) achieves this.

If we train this network on the same data for the same 50 epochs as we used when training the network shown in Chapter 2, we will see that it works nicely. We can get to 91% accuracy on the test set quite easily:

```
Train Epoch: 44 -- Loss: 0.091689
Train Epoch: 45 -- Loss: 0.066864
Train Epoch: 46 -- Loss: 0.061322
Train Epoch: 47 -- Loss: 0.056557
Train Epoch: 48 -- Loss: 0.039695
Train Epoch: 49 -- Loss: 0.056213
Accuracy of the network on the 10000 test images: 91.31%
```

So, we can see that adding convolutions to the neural network definitely increases its ability to classify images. Next, let's take a look at the journey an image takes through the network so we can get a little bit more of an understanding of why this process works.

> If you are using the accompanying code from my GitHub, you'll notice that I'm using model.to(device) a lot. In PyTorch, if an accelerator is available, you can request that the model and/or its data use the accelerator with this command.

Exploring the Convolutional Network

With the torchsummary library, you can inspect your model. When you run it on the Fashion MNIST convolutional network we've been working on, you'll see something like this:

```
from torchsummary import summary
model = FashionCNN().to(device)
summary(model, input_size=(1, 28, 28))  # (Channels, Height, Width)
----------------------------------------------------------------
        Layer (type)               Output Shape         Param #
================================================================
            Conv2d-1          [-1, 64, 28, 28]             640
              ReLU-2          [-1, 64, 28, 28]               0
         MaxPool2d-3          [-1, 64, 14, 14]               0
            Conv2d-4          [-1, 64, 12, 12]          36,928
              ReLU-5          [-1, 64, 12, 12]               0
         MaxPool2d-6            [-1, 64, 6, 6]               0
            Linear-7                  [-1, 128]         295,040
            Linear-8                   [-1, 10]           1,290
```

```
==================================================================
Total params: 333,898
Trainable params: 333,898
Non-trainable params: 0
------------------------------------------------------------------
Input size (MB): 0.00
Forward/backward pass size (MB): 1.02
Params size (MB): 1.27
Estimated Total Size (MB): 2.30
```

Let's first take a look at the Output Shape column to get an understanding of what's going on here. Our first layer will have 28 × 28 images and apply 64 filters to them. But because our filter is 3 × 3, a one-pixel border around the image would typically be lost, reducing our overall information to 26 × 26 pixels. However, because we used the padding=1 parameter, the image was artificially inflated to 30 × 30, meaning that its output would be the correct 28 × 28 and no information would be lost.

If you don't pad the image, you'll end up with a result like the one in Figure 3-6. If we take each of the boxes as a pixel in the image, the first possible filter we can use starts in the second row and the second column. The same would happen on the right side and at the bottom of the diagram.

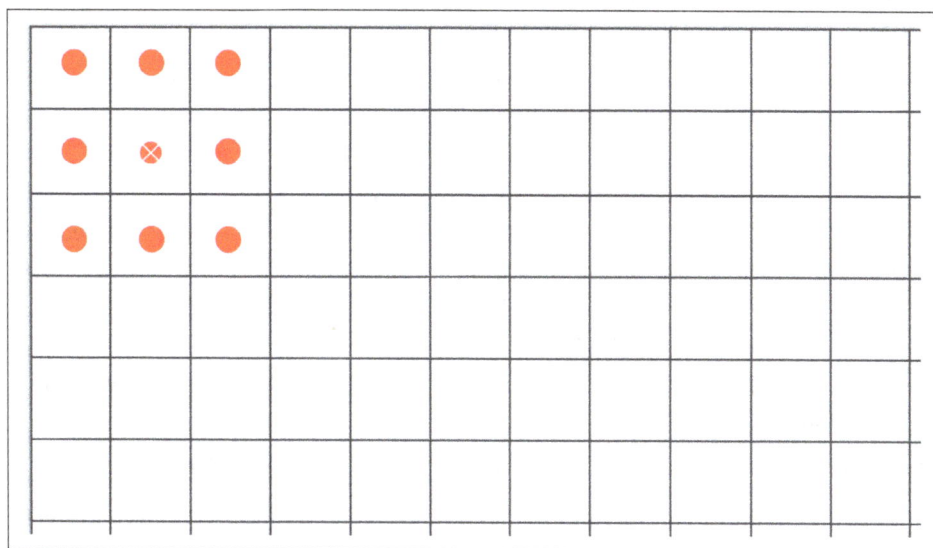

Figure 3-6. Losing pixels when running a filter

Thus, an image that is $a \times b$ pixels in shape when run through a 3 × 3 filter will become $(a - 2) \times (b - 2)$ pixels in shape. Similarly, a 5 × 5 filter would make it $(a - 4) \times (b - 4)$, and so on. As we're using a 28 × 28 image and a 3 × 3 filter, our output would now be 26 × 26. But because we padded the image up to 30 × 30 (again, to prevent loss of information), the output is now 28 × 28.

After that, the pooling layer will be 2 × 2, so the size of the image will halve on each axis, and it will then become 14 × 14. The next convolutional layer does *not* use padding, so it will reduce this further to 12 × 12, and the next pooling will output 6 x 6.

So, by the time the image has gone through two convolutional layers, the result will be many 6 × 6 images. How many? We can see that in the Param # (number of parameters) column.

Each convolution is a 3 × 3 filter, plus a bias. Remember earlier, with our fully connected Linear layers, when each layer was $y = wx + b$, where w was our parameter (aka weight) and b was our bias? This case is very similar, except that because the filter is 3 × 3, there are 9 parameters to learn. Given that we have 64 convolutions defined, we'll have 640 overall parameters. (Each convolution has 9 parameters plus a bias, for a total of 10, and there are 64 of them.)

The ReLU and MaxPooling layers don't learn anything; they just reduce the image, so there are no learned parameters there—hence, 0 are reported.

The next convolutional layer has 64 filters, but each of them is multiplied across the *previous* 64 filters, each of which has 9 parameters. We have a bias on each of the new 64 filters, so our number of parameters should be $(64 \times (64 \times 9)) + 64$, which gives us 36,928 parameters the network needs to learn.

If this is confusing, try changing the number of convolutions in the first layer to something else—for example, 10. You'll see that the number of parameters in the second layer becomes 5,824, which is $(64 \times (10 \times 9)) + 64$.

By the time we get through the second convolution, our images are 6 × 6, and we have 64 of them. If we multiply this out, we'll have 1,600 values, which we'll feed into a Linear layer of 128 neurons. Each neuron has a weight and a bias, and we'll have 128 of them, so the number of parameters the network will learn is $((6 \times 6 \times 64) \times 128) + 128$, giving us 295,040 parameters.

Then, our final Linear layer of 10 neurons will take in the output of the previous 128, so the number of parameters learned will be $(128 \times 10) + 10$, which is 1,290.

The total number of parameters will be the sum of all of these: 333,898.

Training this network requires us to learn the best set of these 333,898 parameters to match the input images to their labels. It's a slower process because there are more parameters, but as we can see from the results, it also builds a more accurate model!

Of course, with this dataset, we still have the limitation that the images are 28 × 28, monochrome, and centered. So next we'll take a look at using convolutions to explore a more complex dataset comprising color pictures of horses and humans, and we'll try to make the model determine whether an image contains one or the other. In this case, the subject won't always be centered in the image like with Fashion MNIST, so we'll have to rely on convolutions to spot distinguishing features.

Building a CNN to Distinguish Between Horses and Humans

In this section, we'll explore a more complex scenario than the Fashion MNIST classifier. We'll extend what we've learned about convolutions and CNNs to try to classify the contents of images in which the location of a feature isn't always in the same place. I've created the "Horses or Humans" dataset for this purpose.

The "Horses or Humans" Dataset

The dataset for this section (*https://oreil.ly/8VXwy*) contains over a thousand 300 × 300–pixel images. Approximately half the images are of horses, and the other half are of humans—and all are rendered in different poses. You can see some examples in Figure 3-7.

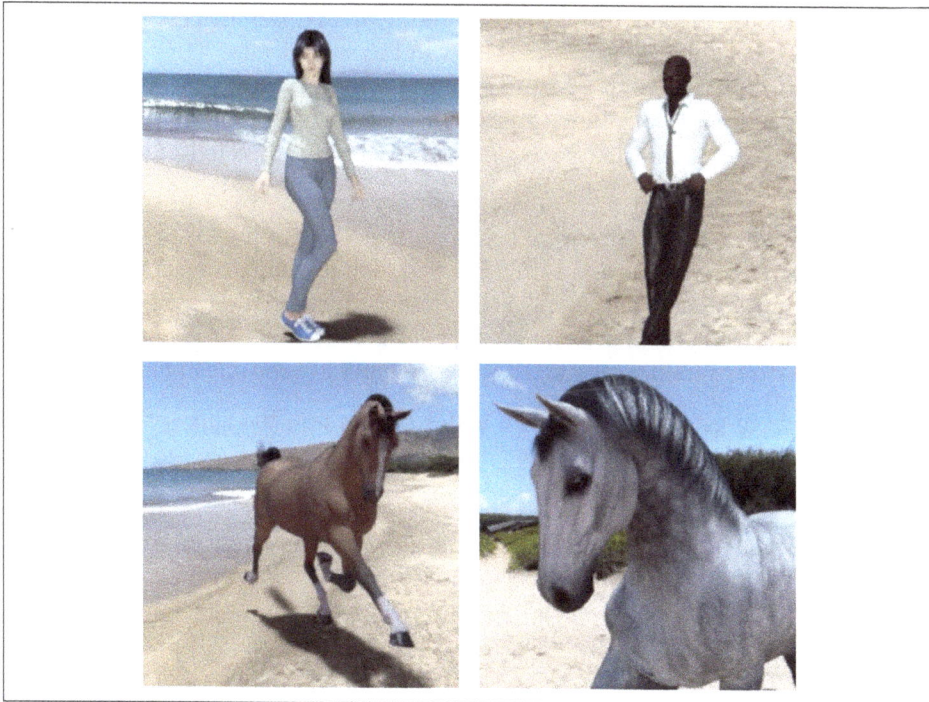

Figure 3-7. Horses and humans

As you can see, the subjects have different orientations and poses, and the image composition varies. Consider the two horses, for example—their heads are oriented differently, and one image is zoomed out (showing the complete animal), while the other is zoomed in (showing just the head and part of the body). Similarly, the humans are lit differently, have different skin tones, and are posed differently. The man has his hands on his hips, while the woman has hers outstretched. The images also contain backgrounds such as trees and beaches, so a classifier will have to determine which parts of the image are the important features that determine what makes a horse a horse and a human a human, without being affected by the background.

While the previous examples of predicting $y = 2x - 1$ or classifying small monochrome images of clothing *might* have been possible with traditional coding, it's clear that this example is far more difficult and that you are crossing the line into where ML is essential to solve a problem.

An interesting side note is that these images are all computer generated. The theory is that features spotted in a CGI image of a horse should apply to a real image, and you'll see how well this works later in this chapter.

Handling the Data

The Fashion MNIST dataset that you've been using up to this point comes with labels, and every image file has an associated file with the label details. Many image-based datasets do not have this, and "Horses or Humans" is no exception. Instead of labels, the images are sorted into subdirectories of each type, and with the DataLoader in PyTorch, you can use this structure to *automatically* assign labels to images.

First, you simply need to ensure that your directory structure has a set of named subdirectories, with each subdirectory being a label. For example, the "Horses or Humans" dataset is available as a set of ZIP files, one of which contains the training data (1,000+ images) and another of which contains the validation data (256 images). When you download and unpack them into a local directory for training and validation, you need to ensure that they are in a file structure like the one in Figure 3-8.

Here's the code to get the training data and extract it into the appropriately named subdirectories, as shown in Figure 3-8:

```
import urllib.request
import zipfile

url = "https://storage.googleapis.com/learning-datasets/
                                        horse-or-human.zip"
file_name = "horse-or-human.zip"
training_dir = 'horse-or-human/training/'
urllib.request.urlretrieve(url, file_name)

zip_ref = zipfile.ZipFile(file_name, 'r')
zip_ref.extractall(training_dir)
zip_ref.close()
```

```
url = "https://storage.googleapis.com/learning-datasets/
                    validation-horse-or-human.zip"
file_name = "validation-horse-or-human.zip"
validation_dir = 'horse-or-human/validation/'
urllib.request.urlretrieve(url, file_name)

zip_ref = zipfile.ZipFile(file_name, 'r')
zip_ref.extractall(validation_dir)
zip_ref.close()
```

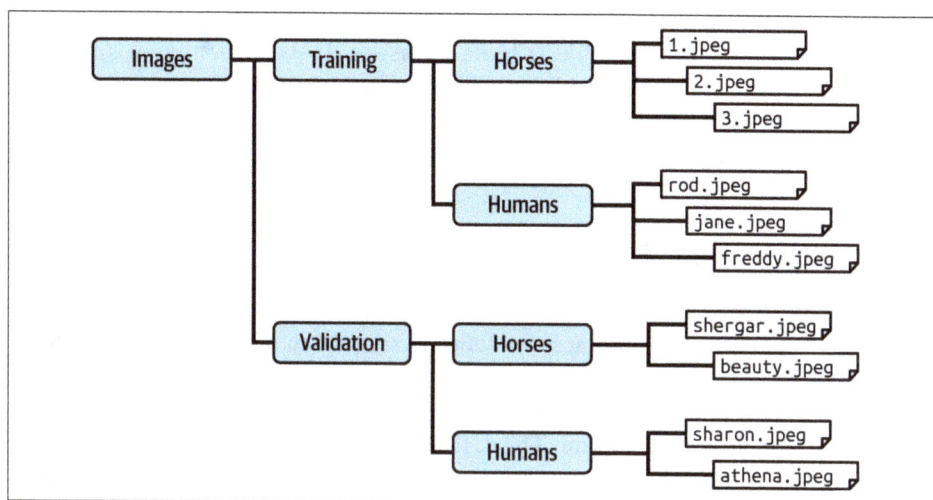

Figure 3-8. Ensuring that images are in named subdirectories

This code simply downloads the ZIP of the training data and unzips it into a directory at *horse-or-human/training*. This is the parent directory that will contain subdirectories for the image types. It does the same for the validation dataset.

Now, to use the DataLoader, we simply use the following code:

```
from torchvision import datasets, transforms
from torch.utils.data import DataLoader

# Define transformations
transform = transforms.Compose([
    transforms.Resize((150, 150)),
    transforms.ToTensor(),
    transforms.Normalize(mean=[0.5, 0.5, 0.5], std=[0.5, 0.5, 0.5])
])

# Load the datasets
train_dataset = datasets.ImageFolder(root=training_dir,
                        transform=transform)
val_dataset = datasets.ImageFolder(root=validation_dir,
                        transform=transform)
```

```
# Data loaders
train_loader = DataLoader(train_dataset, batch_size=32, shuffle=True)
val_loader = DataLoader(val_dataset, batch_size=32, shuffle=False)
```

First, we create an instance of a `transforms` object that we'll call `transform`. This will determine the rules for how we modify the images. It resizes the image to 150 × 150 and then normalizes it into a tensor. Note that the raw images are actually 300 × 300, but to make training quicker for the purposes of learning, I've resized them to 150 × 150.

Then, we specify the `dataset` objects to be `datasets.ImageFolder` types and point them to the required directory, and that will generate images for the training process by flowing them from that directory while applying the transform. The directory for training is `training_dir`, and the directory for validation is `validation_dir`, as specified earlier.

CNN Architecture for "Horses or Humans"

There are several major differences between this dataset and the Fashion MNIST one, and you have to take them into account when designing an architecture for classifying the images. First, the images are much larger—150 × 150 pixels—so more layers may be needed. Second, the images are in full color, not grayscale, so each image will have three channels instead of one. Third, there are only two image types, so we can actually classify them with only *one* output neuron. To do this, we'll drive the value of that neuron toward 0 for one of the labels and toward 1 for the other. The `sigmoid` function is ideal for this process of driving the value to one of these extremes. You can see this at the bottom of the `forward` function:

```
class HorsesHumansCNN(nn.Module):
    def __init__(self):
        super(HorsesHumansCNN, self).__init__()
        self.conv1 = nn.Conv2d(3, 16, kernel_size=3, padding=1)
        self.conv2 = nn.Conv2d(16, 32, kernel_size=3, padding=1)
        self.conv3 = nn.Conv2d(32, 64, kernel_size=3, padding=1)
        self.pool = nn.MaxPool2d(2, 2)
        self.fc1 = nn.Linear(64 * 18 * 18, 512)
        self.drop = nn.Dropout(0.25)
        self.fc2 = nn.Linear(512, 1)
    def forward(self, x):
        x = self.pool(F.relu(self.conv1(x)))
        x = self.pool(F.relu(self.conv2(x)))
        x = self.pool(F.relu(self.conv3(x)))
        x = x.view(-1, 64 * 18 * 18)
        x = F.relu(self.fc1(x))
        x = self.drop(x)
        x = self.fc2(x)
        x = torch.sigmoid(x)  # Use sigmoid to output probabilities
        return x
```

There are a number of things to note here. First of all, take a look at the very first layer. We're defining 16 filters, each of which has a kernel_size of 3, but the input shape is 3. Remember that this is because our input image is in color: there are three channels, instead of just one for the monochrome Fashion MNIST dataset we were using earlier.

At the other end, notice that there's only one neuron in the output layer. This is because we're using a binary classifier, and we can get a binary classification with just a single neuron if we activate it with a sigmoid function. The purpose of the sigmoid function is to drive one set of values toward 0 and the other toward 1, which is perfect for binary classification.

Next, notice how we stack several more convolutional layers. We do this because our image source is quite large and we want, over time, to have many smaller images, each with features highlighted. If we take a look at the results of a summary, we'll see this in action:

```
----------------------------------------------------------------
        Layer (type)             Output Shape         Param #
================================================================
          Conv2d-1           [-1, 16, 150, 150]           448
       MaxPool2d-2           [-1, 16, 75, 75]              0
          Conv2d-3           [-1, 32, 75, 75]           4,640
       MaxPool2d-4           [-1, 32, 37, 37]              0
          Conv2d-5           [-1, 64, 37, 37]          18,496
       MaxPool2d-6           [-1, 64, 18, 18]              0
         Linear-7                   [-1, 512]      10,617,344
        Dropout-8                   [-1, 512]              0
         Linear-9                     [-1, 1]             513
================================================================
Total params: 10,641,441
Trainable params: 10,641,441
Non-trainable params: 0
----------------------------------------------------------------
Input size (MB): 0.26
Forward/backward pass size (MB): 5.98
Params size (MB): 40.59
Estimated Total Size (MB): 46.83
----------------------------------------------------------------
```

Note that by the time the data has gone through all the convolutional and pooling layers, it ends up as 18 × 18 items. When a filter is applied to an input, the output data is generally referred to as a *feature map*. The theory is that these will be *activated* feature maps—where the resulting data is something we can use to classify—and that are relatively simple because they will contain just 324 pixels. We can then pass these feature maps to the fully connected neural network to match them to the appropriate labels.

This, of course, leads this network to have many more parameters than the previous network, so it will be slower to train. With this architecture, we're going to learn over 10 million parameters.

> The code in this section, as well as in many other places in this book, may require you to import Python libraries. To find the correct imports, you can check out the book's repository (*https://github.com/lmoroney/PyTorch-Book-FIles*).

To train the network, we'll have to compile it with a loss function and an optimizer. In this case, the loss function can be the BCELoss, where BCE stands for *binary cross entropy*. As the name suggests, because there are only two classes in this scenario, this is a loss function that is designed for it. For the optimizer, we can continue using the same Adam that we used earlier. Here's the code:

```
criterion = nn.BCELoss()
optimizer = optim.Adam(model.parameters(), lr=0.001)
```

We then train in the usual way:

```
def train_model(num_epochs):
    for epoch in range(num_epochs):
        model.train()
        running_loss = 0.0
        for images, labels in train_loader:
            images, labels = images.to(device), labels.to(device).float()
            optimizer.zero_grad()
            outputs = model(images).view(-1)
            loss = criterion(outputs, labels)
            loss.backward()
            optimizer.step()
            running_loss += loss.item()
```

One thing to note is that the labels are converted to floats because of the binary cross entropy, where the value of the final output node will be a float value.

Over just 15 epochs, this architecture gives us a very impressive 95%+ accuracy on the training set. Of course, this is just with the training data, and this performance isn't an indication of the network's potential performance on data that it hasn't previously seen.

Next, we'll look at using the validation set and measuring its performance to give us a good indication of how this model might perform in real life.

Using the "Horses or Humans" Validation Dataset

To add validation, you'll need a validation dataset that's separate from the training one. In some cases, you'll get a master dataset that you have to split yourself, but in the case of "Horses or Humans," there's a separate validation set that you can download. In the preceding code snippet, you've already downloaded the training and validation datasets, put them in directories, and set up data loaders for each of them. However, for training, you only used one of these datasets—the one that was set up to load the *training* data. So next, we'll switch the model into evaluation mode and explore how well it did with the *validation* data.

To download the validation set and unzip it into a different directory, you can use code that's very similar to that used for the training images.

Then, to perform the validation, you simply update your `train_model` method to perform a validation at the end of each training loop (or epoch) and report on the results. For example, you can do this:

```python
def train_model(num_epochs):
    for epoch in range(num_epochs):
        model.train()
        running_loss = 0.0
        for images, labels in train_loader:
            images, labels = images.to(device), labels.to(device).float()
            optimizer.zero_grad()
            outputs = model(images).view(-1)
            loss = criterion(outputs, labels)
            loss.backward()
            optimizer.step()
            running_loss += loss.item()

        print(f'Epoch {epoch + 1}, Loss: {running_loss /
                len(train_loader)}')

    # Evaluate on training set
        model.eval()
        with torch.no_grad():
            correct = 0
            total = 0
            for images, labels in train_loader:
                images images.to(device)
                labels = labels.to(device).float()
                outputs = model(images).view(-1)
                predicted = outputs > 0.5  # Threshold predictions
                total += labels.size(0)
                correct += (predicted == labels).sum().item()
            print(f'Training Accuracy: {100 * correct / total}%')

    # Evaluate on validation set
        model.eval()
```

```
with torch.no_grad():
    correct = 0
    total = 0
    for images, labels in val_loader:
        images = images.to(device)
        labels = labels.to(device).float()
        outputs = model(images).view(-1)
        predicted = outputs > 0.5  # Threshold predictions
        total += labels.size(0)
        correct += (predicted == labels).sum().item()
    print(f'Validation Set Accuracy: {100 * correct / total}%')

train_model(50)
```

I added code here to do *both* the training and the validation and report on accuracy. Note that this is really just for learning purposes, so you can compare. In a real-world scenario, checking the accuracy of training data is a waste of processing time!

After training for 10 epochs, you should see that your model is 99%+ accurate on the training set but only about 88% on the validation set:

```
Epoch 7, Loss: 0.0016404045829699512
Training Set Accuracy: 100.0%
Validation Set Accuracy: 88.28125%
Epoch 8, Loss: 0.0010613293736610378
Training Set Accuracy: 100.0%
Validation Set Accuracy: 89.0625%
Epoch 9, Loss: 0.0008372313717332979
Training Set Accuracy: 100.0%
Validation Set Accuracy: 86.328125%
Epoch 10, Loss: 0.0006578459407812646
Training Set Accuracy: 100.0%
Validation Set Accuracy: 87.5%
```

This is an indication that the model is overfitting, which is something we also saw in the previous chapter. It's easy to be lulled into a false sense of security by the 100% accuracy, but the other figure is more representative of how your model will behave in the real world.

Still, the performance isn't bad, considering how few images it was trained on and how diverse those images were. You're beginning to hit a wall caused by lack of data, but there are some techniques that you can use to improve your model's performance. We'll explore them later in this chapter, but before that, let's take a look at how to *use* this model.

Testing "Horses or Humans" Images

It's all very well and good to be able to build a model, but of course, you want to try it out. A major frustration of mine when I was starting my AI journey was that I could find lots of code that showed me how to build models and charts of how those models were performing, but very rarely was there code to help me kick the tires of the model myself to try it out. I'll try to help you avoid that problem in this book!

Testing the model is perhaps easiest using Colab. I've provided a "Horses or Humans" notebook on GitHub that you can open directly in Colab (*https://oreil.ly/iN9IG*).

Once you've trained the model, you'll see a section called "Running the Model." Before running it, you should find a few pictures of horses or humans online and download them to your computer. I recommend you go to Pixabay.com (*http://pixabay.com*), which is a really good site to check out for royalty-free images. It's also a good idea to get your test images together first, because the node can time out while you're searching.

Figure 3-9 shows a few pictures of horses and humans that I downloaded from Pixabay to test the model.

Figure 3-9. Test images

When they were uploaded, as you can see in Figure 3-10, the model correctly classified one image as a human and another as a horse—but despite the fact that the third image was obviously of a human, the model incorrectly classified it as a horse!

You can also upload multiple images simultaneously and have the model make predictions for all of them. You may also notice that it tends to overfit toward horses. If the human isn't fully posed (i.e., if you can't see their full body), the model can skew toward horses. That's what happened in this case. The first human model is fully posed, and the image resembles many of the poses in the dataset, so the model was able to classify her correctly. On the other hand, the second human model is facing the camera, but only her upper half is in the image. There was no training data that looked like that, so the model couldn't correctly identify her.

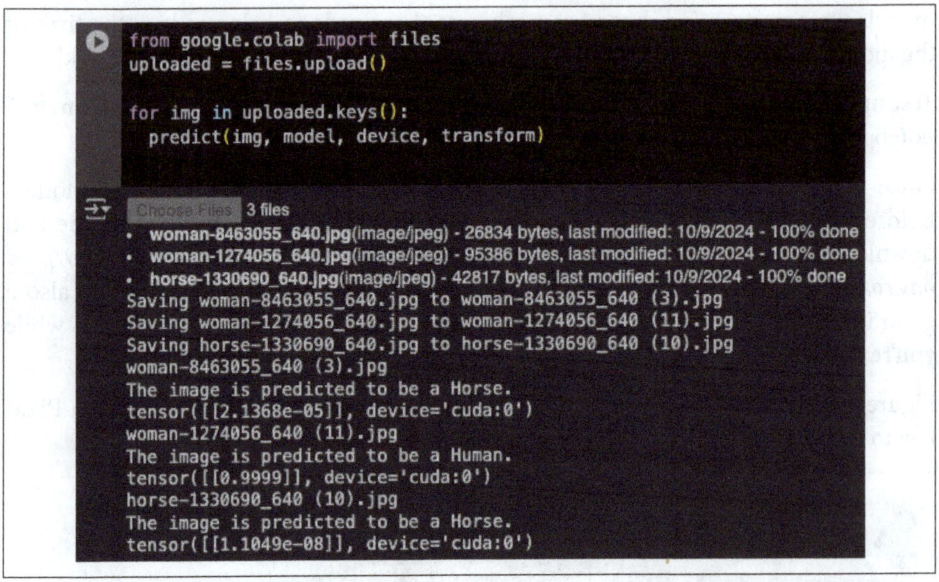

```
from google.colab import files
uploaded = files.upload()

for img in uploaded.keys():
    predict(img, model, device, transform)
```

Choose Files 3 files
- **woman-8463055_640.jpg**(image/jpeg) - 26834 bytes, last modified: 10/9/2024 - 100% done
- **woman-1274056_640.jpg**(image/jpeg) - 95386 bytes, last modified: 10/9/2024 - 100% done
- **horse-1330690_640.jpg**(image/jpeg) - 42817 bytes, last modified: 10/9/2024 - 100% done
```
Saving woman-8463055_640.jpg to woman-8463055_640 (3).jpg
Saving woman-1274056_640.jpg to woman-1274056_640 (11).jpg
Saving horse-1330690_640.jpg to horse-1330690_640 (10).jpg
woman-8463055_640 (3).jpg
The image is predicted to be a Horse.
tensor([[2.1368e-05]], device='cuda:0')
woman-1274056_640 (11).jpg
The image is predicted to be a Human.
tensor([[0.9999]], device='cuda:0')
horse-1330690_640 (10).jpg
The image is predicted to be a Horse.
tensor([[1.1049e-08]], device='cuda:0')
```

Figure 3-10. Executing the model

Let's now explore the code to see what it's doing. Perhaps the most important part is this chunk:

```
def load_image(image_path, transform):
    # Load image
    image = Image.open(image_path).convert('RGB')  # Convert to RGB
    # Apply transformations
    image = transform(image)
    # Add batch dimension, as the model expects batches
    image = image.unsqueeze(0)
    return image
```

Here, we are loading the image from the path that Colab wrote it to. Note that we specify a `transform` to apply to the image. The images being uploaded can be any shape, but if we are going to feed them into the model, they *must* be the same size that the model was trained on. So, if we use the same `transform` that we defined when performing the training, we'll know it's in the same dimensions.

At the end is this strange command: `image = image.unsqueeze(0)`.

When you look back at how the model was trained, the DataLoader objects batched the images going into it. If you think of an image as a 2D array of pixels, then the batch is an array of 2D arrays, which of course is then a 3D array.

But when we're using this code with one image at a time, there's no batch, so to make this a 3D array (which is technically a batch with one item in it), we can just unsqueeze the image along axis 0 to simulate this.

With our image in the right format, it's easy to do the classification:

```
with torch.no_grad():
    output = model(image)
```

The model then returns an array containing the classifications for the batch. Because there's only one classification in this case, it's effectively an array containing an array. You can see this back in Figure 3-10, where for the first human model, the array looks like `tensor([[2.1368e-05]], device='cuda:0')`.

So now, it's simply a matter of inspecting the value of the first element in that array. If it's greater than 0.5, we're looking at a human:

```
class_name = "Human" if prediction.item() > 0.5 else "Horse"
```

There are a few important points to consider here. First, even though the network was trained on synthetic, computer-generated imagery, it performs quite well at spotting horses and humans and differentiating them in real photographs. This is a potential boon in that you may not need thousands of photographs to train a model, and you can do it relatively cheaply with CGI.

But this dataset also demonstrates a fundamental issue you will face. Your training set cannot hope to represent *every* possible scenario your model might face in the wild, and thus, the model will always have some level of overspecialization toward the training set. We saw a clear and simple example of this earlier in this section, when the model mischaracterized the human in the center of Figure 3-9. The training set didn't include a human in that pose, and thus, the model didn't "learn" that a human could look like that. As a result, there was every chance it might see the figure as a horse, and in this case, it did.

What's the solution? The obvious one is to add more training data, with humans in that particular pose and others that weren't initially represented. That isn't always possible, though. Fortunately, there's a neat trick in PyTorch that you can use to virtually extend your dataset—it's called *image augmentation*, and we'll explore that next.

Image Augmentation

In the previous section, you built a horse-or-human classifier model that was trained on a relatively small dataset. As a result, you soon began to hit problems classifying some previously unseen images, such as the miscategorization of a woman as a horse because the training set didn't include any images of people in that pose.

One way to deal with such problems is with *image augmentation*. The idea behind this technique is that as PyTorch is loading your data, it can create additional new data by amending what it has using a number of transforms. For example, take a look at Figure 3-11. While there is nothing in the dataset that looks like the woman on the right, the image on the left is somewhat similar.

Figure 3-11. Dataset similarities

So, if you could, for example, zoom into the image on the left as you are training, as shown in Figure 3-12, you would increase the chances of the model being able to correctly classify the image on the right as a person.

Figure 3-12. Zooming in on the training set data

In a similar way, you can broaden the training set with a variety of other transformations, including the following:

- Rotation (turning the image)
- Shifting horizontally (moving the pixels horizontally with wrapping)

- Shifting vertically (moving the pixels vertically with wrapping)
- Shearing (moving the pixels either horizontally or vertically but offsetting so that the image would look like parallelogram)
- Zooming (magnifying a particular region)
- Flipping (vertically or horizontally)

Because you've been using the `datasets.ImageFolder` and a `DataLoader` to load the images, you've seen the model do a transform already—when it normalized the images like this:

```
# Define transformations
transform = transforms.Compose([
    transforms.Resize((150, 150)),
    transforms.ToTensor(),
    transforms.Normalize(mean=[0.5, 0.5, 0.5], std=[0.5, 0.5, 0.5])
])
```

Many other transforms are easily available within the torchvision.transforms library, so, for example, you could do something like this, where you can apply different sets of transforms to the training and validation datasets:

```
# Transforms for the training data
train_transforms = transforms.Compose([
    transforms.RandomHorizontalFlip(),
    transforms.RandomRotation(20),
    transforms.RandomResizedCrop(150),
    transforms.ToTensor(),
    transforms.Normalize(mean=[0.5, 0.5, 0.5], std=[0.5, 0.5, 0.5])
])

# Transforms for the validation data
val_transforms = transforms.Compose([
    transforms.Resize(150),
    transforms.CenterCrop(150),
    transforms.ToTensor(),
    transforms.Normalize(mean=[0.5, 0.5, 0.5], std=[0.5, 0.5, 0.5])
])
```

Here, in addition to rescaling the image to normalize it, you're doing the following:

- Randomly flipping horizontally
- Randomly rotating up to 20 degrees left or right
- Randomly cropping a 150 × 150 window instead of resizing

In addition, the transforms.RandomAffine library gives you the facility to do all of these things, as well as adding stuff like scaling the image (zooming in or out), shearing the image, etc. Here's an example:

```
transforms.RandomAffine(
    degrees=0,  # No rotation
    translate=(0.2, 0.2),  # Translate up to 20% vert and horizontally
    scale=(0.8, 1.2),  # Zoom in or out by 20%
    shear=20,  # Shear by up to 20 degrees
),
```

When you retrain with these parameters, one of the first things you'll notice is that training takes longer because of all the image processing. Also, your model's accuracy may not be as high as it was, because previously it was overfitting to a largely uniform set of data.

In my case, when I was training with these augmentations, my accuracy went down from 99% to 94% after 15 epochs, with validation much lower at 64%. This likely indicates overfitting in the model, but it warrants investigation by training with more epochs! One other thing to note is that random cropping might also be an issue—the CGI images generally center the subject, so random cropping will give partial subjects.

But what about the image from Figure 3-9 that the model misclassified earlier? This time, the model gets it right. Thanks to the image augmentations, the training set now has sufficient coverage for the model to understand that this particular image is a human too (see Figure 3-13). This is just a single data point, and it may not be representative of the results for real data, but it's a small step in the right direction.

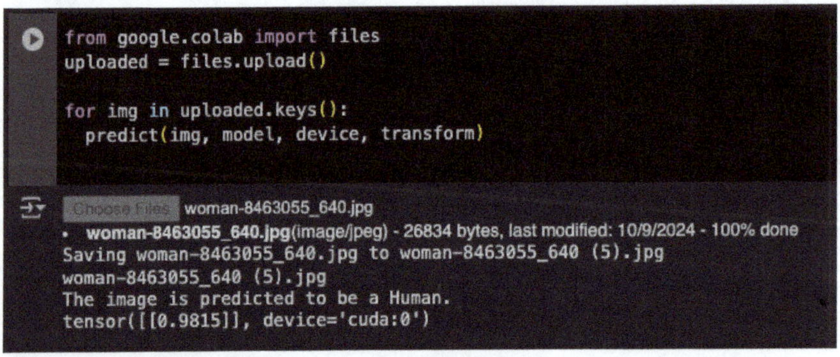

```
from google.colab import files
uploaded = files.upload()

for img in uploaded.keys():
    predict(img, model, device, transform)
```

```
Choose Files  woman-8463055_640.jpg
• woman-8463055_640.jpg(image/jpeg) - 26834 bytes, last modified: 10/9/2024 - 100% done
Saving woman-8463055_640.jpg to woman-8463055_640 (5).jpg
woman-8463055_640 (5).jpg
The image is predicted to be a Human.
tensor([[0.9815]], device='cuda:0')
```

Figure 3-13. The woman is now correctly classified

As you can see, even with a relatively small dataset like "Horses or Humans," you can start to build a pretty decent classifier. With larger datasets, you could take this further. Another way you can improve the model is by using features that the model has already learned elsewhere. Many researchers with massive resources (millions of images) and huge models that have been trained on thousands of classes have shared their models, and by using a concept called *transfer learning*, you can use the features those models learned and apply them to your data. We'll explore that next!

Transfer Learning

As we've already seen in this chapter, the use of convolutions to extract features can be a powerful tool for identifying the contents of an image. If we use this tool, we can then feed the resulting feature maps into the fully connected layers of a neural network to match them to the labels and give us a more accurate way of determining the contents of an image. Using this approach with a simple fast-to-train neural network and some image augmentation techniques, we built a model that was 80–90% accurate at distinguishing between a horse and a human when it was trained on a very small dataset.

However, we can improve our model even further by using a method called *transfer learning*. The idea behind it is simple: instead of having our model learn a set of filters from scratch for our dataset, why not have it use a set of filters that were learned on a much larger dataset, with many more features than we can "afford" to build from scratch? We can place these filters in our network and then train a model with our data using the pre-learned filters. For example, while our "Horses or Humans" dataset has only two classes, we can use an existing model that has been pretrained for one thousand classes—but at some point, we'll have to throw away some of the preexisting network and add the layers that will let us have a classifier for two classes.

Figure 3-14 shows what a CNN architecture for a classification task like ours might look like. We have a series of convolutional layers that lead to a fully connected layer, which in turn leads to an output layer.

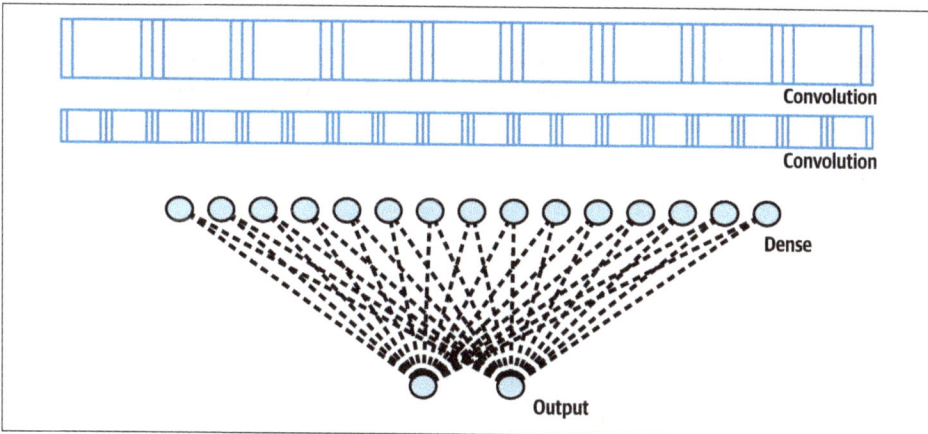

Figure 3-14. A CNN architecture

We've seen that we can build a pretty good classifier using this architecture. But what if we could use transfer learning to take the pre-learned layers from another model, freeze or lock them so that they aren't trainable, and then put them on top of our model, like in Figure 3-15?

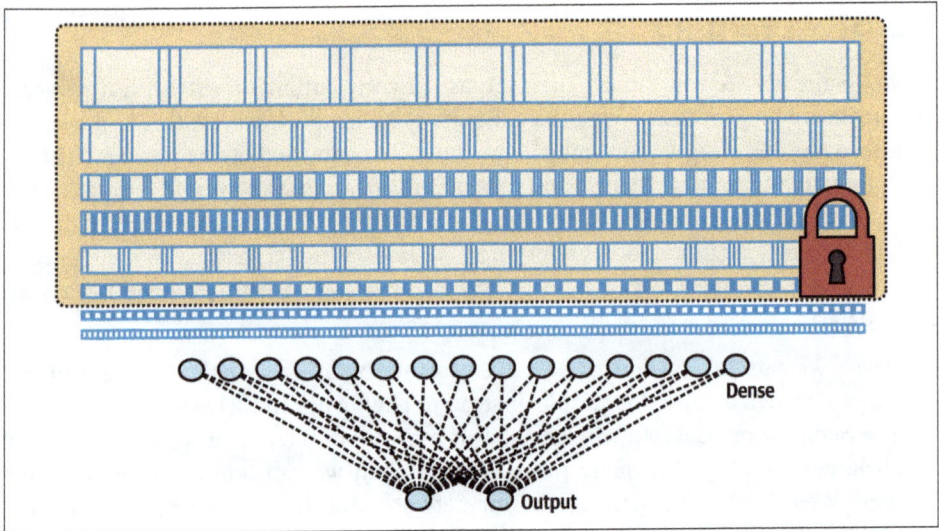

Figure 3-15. Taking and locking layers from another architecture via transfer learning

When we consider that once they've been trained, all these layers are just a set of numbers indicating the filter values, weights, and biases along with a known architecture (the number of filters per layer, the size of the filter, etc.), the idea of reusing them is pretty straightforward.

Let's look at how this would appear in code. There are several pretrained models already available from a variety of sources, so we'll use version 3 of the popular Inception model from Google, which is trained on more than a million images from a database called ImageNet. Inception has dozens of layers, and it can classify images into one thousand categories.

The torchvision.models library contains a number of models, including Inception V3, so we can easily get access to the pretrained model:

```python
import torch
import torch.nn as nn
from torchvision import models, transforms
from torch.utils.data import DataLoader
from torchvision.datasets import ImageFolder
from torch.optim import RMSprop

# Load the pretrained Inception V3 model
pre_trained_model = models.inception_v3(pretrained=True, aux_logits=True)
```

Now, we have a full Inception model that's pretrained. If you want to inspect its architecture, you can do so with this code:

```python
def print_model_summary(model):
    for name, module in model.named_modules():
```

```
    print(f"{name} : {module.__class__.__name__}")

# Example of how to use the function with your pretrained model
print_model_summary(pre_trained_model)
```

Be warned—this model is huge! Still, you should take a look through it to see the layers and their names. I like to use the one called `Mixed7_c` because its output is nice and small—it consists of 8 × 8 images—but you should feel free to experiment with others.

Next, we'll freeze the entire network from retraining and then set a variable to point to `mixed7`'s output as where we want to crop the network. We can do that with this code:

```
# Freeze all layers up to and including the 'Mixed_7c'
for name, parameter in pre_trained_model.named_parameters():
    parameter.requires_grad = False
    if 'Mixed_7c' in name:
        break
```

You'll notice that we're printing the output shape of the last layer, and you'll also see that we're getting 8 × 8 images at this point. This indicates that by the time the images have been fed through to `Mixed_7c`, the output images from the filters are 8 × 8 in size, so they're pretty easy to manage. Again, you don't have to choose that specific layer; you're welcome to experiment with others.

Now, let's see how to modify the model for transfer learning. It's pretty straightforward—if you go back to the output from the custom `print_model_summary` from a moment ago, you'll see that the *last* layer in the model is called `fc`. As you might expect, *fc* stands for *fully connected*, which is effectively a Linear layer with our densely connected neurons.

So now, it becomes as simple as replacing that layer with a new layer. We don't need to *know* the input shape for it ahead of time—we can inspect its `in_features` property to find that. So now, to create a new layer of 1,024 neurons that outputs to another layer of two neurons and replace the `fc` from Inception, all we have to do is this:

```
# Modify the existing fully connected layer
num_ftrs = pre_trained_model.fc.in_features
pre_trained_model.fc = nn.Sequential(
    nn.Linear(num_ftrs, 1024),   # New fully connected layer
    nn.ReLU(),                   # Activation layer
    nn.Linear(1024, 2)           # Final layer for binary classification
)
```

It's as simple as creating a new set of Linear layers from the last output, and feeding the results into another Linear layer. So, we then add a Linear layer of with the number of features in, and 1,024 neurons out that feeds to another layer with two neurons for our output. Also, you've probably noticed that in the previous model, we did it with one neuron and used sigmoid activation for the two classes—so you're probably wondering why we're going to two neurons in the output layer now. This was primarily a stylistic choice. Inception was designed for *n* neurons to output for *n* classes, and I wanted to keep that approach.

Training the model on this architecture over only three epochs gave us an accuracy of 99%+, with a validation accuracy of 95%+. Clearly, that's a vast improvement. Also, remember that Inception learned a massive set of features that it could use to classify the many classes it was trained on. It turns out that that feature set is also incredibly useful for learning how to classify any other images—not least, those from "Horses or Humans."

The results we got from this model are much better than those we got from our previous model, but you can continue to tweak and improve it. You can also explore how the model will work with a much larger dataset, like the famous "Dogs vs. Cats" (*https://oreil.ly/UhWMk*) from Kaggle. It's an extremely varied dataset consisting of 25,000 images of cats and dogs, often with the subjects somewhat obscured—for example, if they are held by a human.

Using the same algorithm and model design as before, you can train a "Dogs vs. Cats" classifier on Colab, using a GPU at about 3 minutes per epoch.

When I tested with very complex pictures like those in Figure 3-16, this classifier got them all correct. I chose one picture of a dog with catlike ears and one with its back turned. Both pictures of cats were nontypical.

Figure 3-16. Unusual dogs and cats that the model classified correctly

To parse the results, you can use code like this:

```python
def load_image(image_path, transform):
    # Load image
    image = Image.open(image_path).convert('RGB')  # Convert to RGB
    # Apply transformations
    image = transform(image)
    # Add batch dimension, as the model expects batches
    image = image.unsqueeze(0)
    return image


# Prediction function
def predict(image_path, model, device, transform):
    model.eval()
    image = load_image(image_path, transform)
    image = image.to(device)
    with torch.no_grad():
        output = model(image)
        print(output)
        prediction = torch.max(output, 1)
        print(prediction)
```

Note the lines where I'm printing the output of the image, calculating the prediction from that, and printing that.

When you upload some images to Colab, you can see how they predict in Figure 3-17.

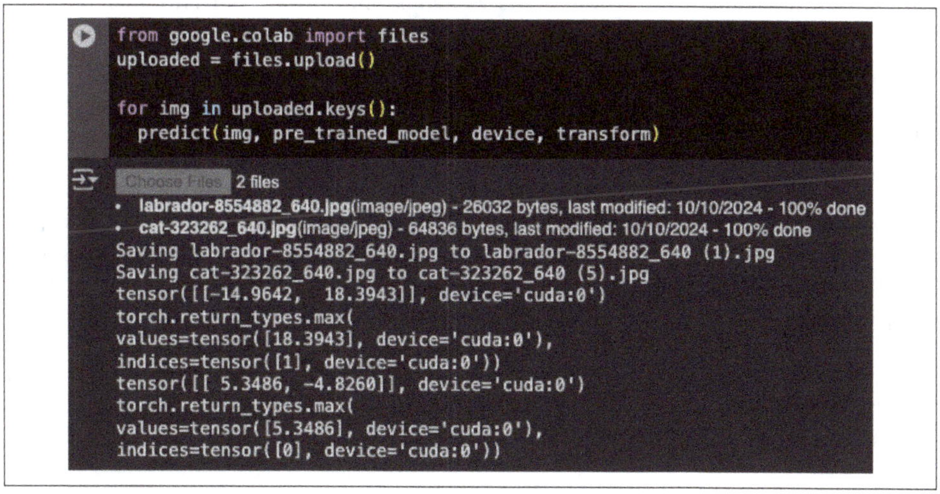

```
from google.colab import files
uploaded = files.upload()

for img in uploaded.keys():
    predict(img, pre_trained_model, device, transform)
```

```
Choose Files  2 files
  • labrador-8554882_640.jpg(image/jpeg) - 26032 bytes, last modified: 10/10/2024 - 100% done
  • cat-323262_640.jpg(image/jpeg) - 64836 bytes, last modified: 10/10/2024 - 100% done
Saving labrador-8554882_640.jpg to labrador-8554882_640 (1).jpg
Saving cat-323262_640.jpg to cat-323262_640 (5).jpg
tensor([[-14.9642,  18.3943]], device='cuda:0')
torch.return_types.max(
values=tensor([18.3943], device='cuda:0'),
indices=tensor([1], device='cuda:0'))
tensor([[ 5.3486, -4.8260]], device='cuda:0')
torch.return_types.max(
values=tensor([5.3486], device='cuda:0'),
indices=tensor([0], device='cuda:0'))
```

Figure 3-17. Classifying the cat washing its paw

The first image uploaded was "labrador," which, as its name suggests, is of a dog. The tensor returned from the model contained [−14.9642, 18.3943], meaning a very low number for the first label and a very high one for the second. Given that we used an image directory when training, the labels ended up being in alphabetical order, so it was low for cat and high for dog. Then, when we called torch.max, it gave us [1]. That indicates that neuron 1 is the one for this classification—thus, the image is a dog.

The second image had [5.3486, −4.8260], with the first neuron being higher. Thus, it detected a cat. The size of these numbers indicates the strength of the prediction. For example, it was much surer that the first image is a dog than it was that the second image is a cat.

You can find the complete code for the "Horses or Humans" and "Dogs vs. Cats" classifiers in the book's GitHub repository (*https://github.com/lmoroney/tfbook*).

Multiclass Classification

In all of the examples so far, you've been building *binary* classifiers—ones that choose between two options (horses or humans, cats or dogs). On the other hand, when you're building *multiclass classifiers*, the models are almost the same but there are a few important differences. Instead of a single neuron that is sigmoid activated or two neurons that are binary activated, your output layer will now require *n* neurons, where *n* is the number of classes you want to classify. You'll also have to change your loss function to an appropriate one for multiple categories.

A neat feature of the nn.CrossEntropyLoss loss function in PyTorch is that it can handle multiple categories, so the "Cats vs. Dogs" and "Horses or Humans" transfer learning classifiers you've built thus far in this chapter can use it without modification. But the "Horses or Humans" classifier that you built at the beginning with a *single* output neuron will not be able to because it can't handle more than two classes. This is always something to look out for, and it's a common bug when you start writing code for classification.

To go beyond two-class classification, consider, for example, the game Rock, Paper, Scissors. If you wanted to train a model to recognize the different hand gestures used in this game, you'd need to handle three categories. Fortunately, there's a simple dataset (*https://oreil.ly/VHhmS*) you can use for this.

There are two downloads: a training set of many diverse hands, with different sizes, shapes, colors, and details such as nail polish; and a testing set of equally diverse hands, none of which are in the training set. You can see some examples in Figure 3-18.

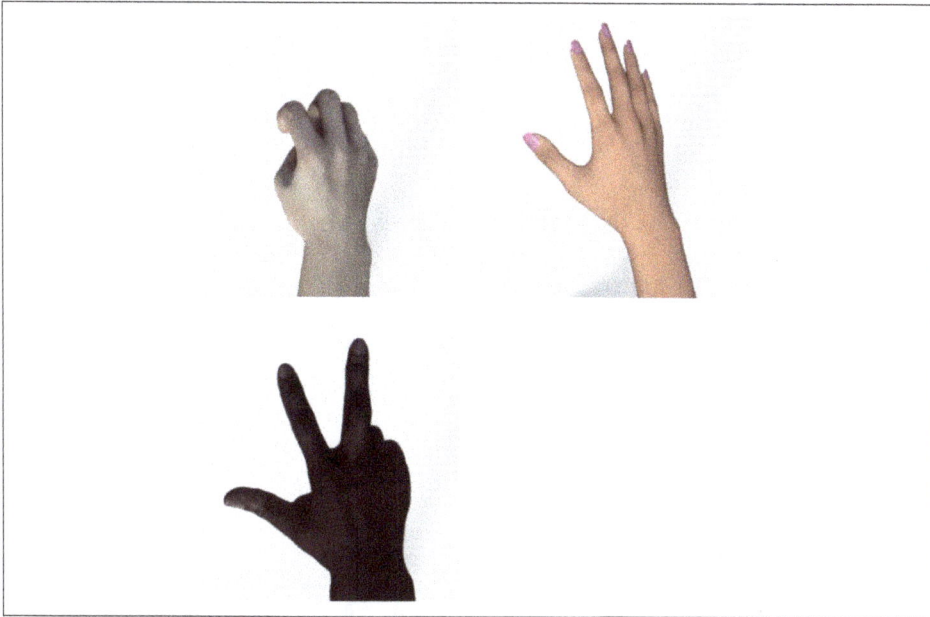

Figure 3-18. Examples of Rock, Paper, Scissors gestures

Using the dataset is simple. You can download and unzip it—the sorted subdirectories are already present in the ZIP file—and then use it to initialize an `ImageFolder`:

```
!wget --no-check-certificate \
 https://storage.googleapis.com/learning-datasets/rps.zip \
 -O /tmp/rps.zip
local_zip = '/tmp/rps.zip'
zip_ref = zipfile.ZipFile(local_zip, 'r')
zip_ref.extractall('/tmp/')
zip_ref.close()
training_dir = "/tmp/rps/"

train_dataset = ImageFolder(root=training_dir, transform=transform)
```

Be sure to use a `transform` that fits the input shape of your model. In the last few examples, we were using Inception, and it's 299 × 299.

You can use the ImageFolder for your DataLoader in the usual way:

```
train_loader = DataLoader(train_dataset, batch_size=32, shuffle=True)
```

Earlier, when we tweaked the Inception model for "Horses or Humans" or "Cats vs. Dogs," there were only *two* classes and thus *two* output neurons. Given that this data has *three* classes, we need to be sure that we change the new fully connected layer at the bottom accordingly:

```python
# Modify the existing fully connected layer
num_ftrs = pre_trained_model.fc.in_features
pre_trained_model.fc = nn.Sequential(
    nn.Linear(num_ftrs, 1024),   # New fully connected layer
    nn.ReLU(),                   # Activation layer
    nn.Linear(1024, 3)           # Final layer for classification
)
```

Now, training the model works as before: you specify the loss function and optimizer, and you call the `train_model()` function. For good repetition, this function is the same as the one used in the "Horses or Humans" and "Cats vs. Dogs" examples:

```python
# Only optimize parameters that are set to be trainable
optimizer = RMSprop(filter(lambda p: p.requires_grad,
                    pre_trained_model.parameters()), lr=0.001)

criterion = nn.CrossEntropyLoss()

# Train the model
train_model(pre_trained_model, criterion, optimizer, train_loader, num_epochs=3)
```

Your code for testing predictions will also need to change somewhat. There are now three output neurons, and they will output a high value for the predicted class and lower values for the other classes.

Note also that when you're using the `ImageFolder`, the classes are loaded in alphabetical order—so while you might expect the output neurons to be in the order of the name of the game, the order will in fact be Paper, Rock, Scissors.

Code that you can use to try out predictions in a Colab notebook will look like the following. It's very similar to what you saw earlier:

```python
def load_image(image_path, transform):
    # Load image
    image = Image.open(image_path).convert('RGB')  # Convert to RGB
    # Apply transformations
    image = transform(image)
    # Add batch dimension, as the model expects batches
    image = image.unsqueeze(0)
    return image


    # Prediction function
def predict(image_path, model, device, transform):
    model.eval()
    image = load_image(image_path, transform)
    image = image.to(device)
    with torch.no_grad():
        output = model(image)
        print(output)
        prediction = torch.max(output, 1)
        print(prediction)
```

Note that it doesn't parse the output; it just prints the classes. Figure 3-19 shows what it looks like in actual use.

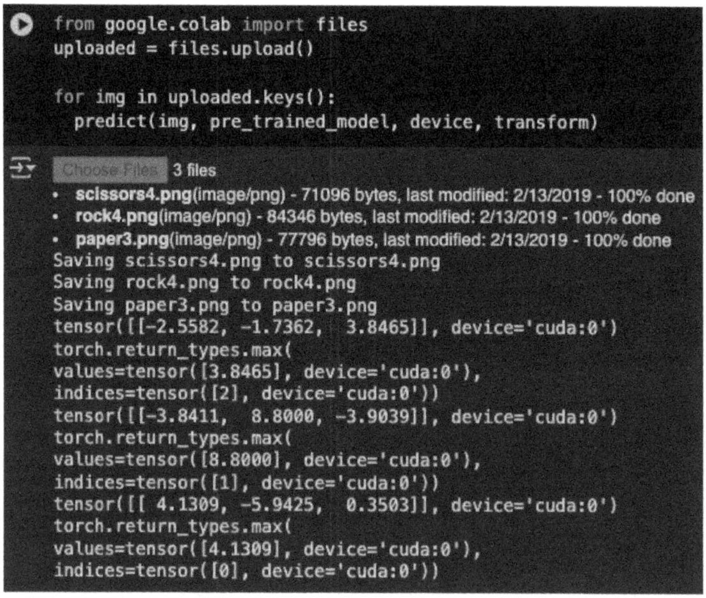

```
from google.colab import files
uploaded = files.upload()

for img in uploaded.keys():
    predict(img, pre_trained_model, device, transform)
```

```
Choose Files   3 files
  • scissors4.png(image/png) - 71096 bytes, last modified: 2/13/2019 - 100% done
  • rock4.png(image/png) - 84346 bytes, last modified: 2/13/2019 - 100% done
  • paper3.png(image/png) - 77796 bytes, last modified: 2/13/2019 - 100% done
Saving scissors4.png to scissors4.png
Saving rock4.png to rock4.png
Saving paper3.png to paper3.png
tensor([[-2.5582, -1.7362,  3.8465]], device='cuda:0')
torch.return_types.max(
values=tensor([3.8465], device='cuda:0'),
indices=tensor([2], device='cuda:0'))
tensor([[-3.8411,  8.8000, -3.9039]], device='cuda:0')
torch.return_types.max(
values=tensor([8.8000], device='cuda:0'),
indices=tensor([1], device='cuda:0'))
tensor([[ 4.1309, -5.9425,  0.3503]], device='cuda:0')
torch.return_types.max(
values=tensor([4.1309], device='cuda:0'),
indices=tensor([0], device='cuda:0'))
```

Figure 3-19. Testing the Rock, Paper, Scissors classifier

You can see from the filenames what the images were.

If you explore this a little deeper, you can see that the file named *scissors4.png* had an output of [–2.5582, –1.7362, 3.8465]. The largest number is the third one, and if you think alphabetically, you can see that the third neuron represents scissors, so it was classified correctly. Similar results were achieved for the other files.

Some images that you can use to test the dataset are available to download (*https://oreil.ly/dEUpx*). Alternatively, of course, you can try your own. Note that the training images are all done against a plain white background, though, so there may be some confusion if there is a lot of detail in the background of the photos you take.

Dropout Regularization

Earlier in this chapter, we discussed overfitting, in which a network may become too specialized in a particular type of input data and thus fare poorly on others. One technique to help overcome this is use of *dropout regularization.*

When a neural network is being trained, each individual neuron will have an effect on neurons in subsequent layers. Over time, particularly in larger networks, some neurons can become overspecialized—and that feeds downstream, potentially causing the network as a whole to become overspecialized and thus leading to overfitting. Additionally, neighboring neurons can end up with similar weights and biases, and if not monitored, this condition can lead the overall model to become overspecialized on the features activated by those neurons.

For example, consider the neural network in Figure 3-20, in which there are layers of 2, 6, 6, and 2 neurons. The neurons in the middle layers might end up with very similar weights and biases.

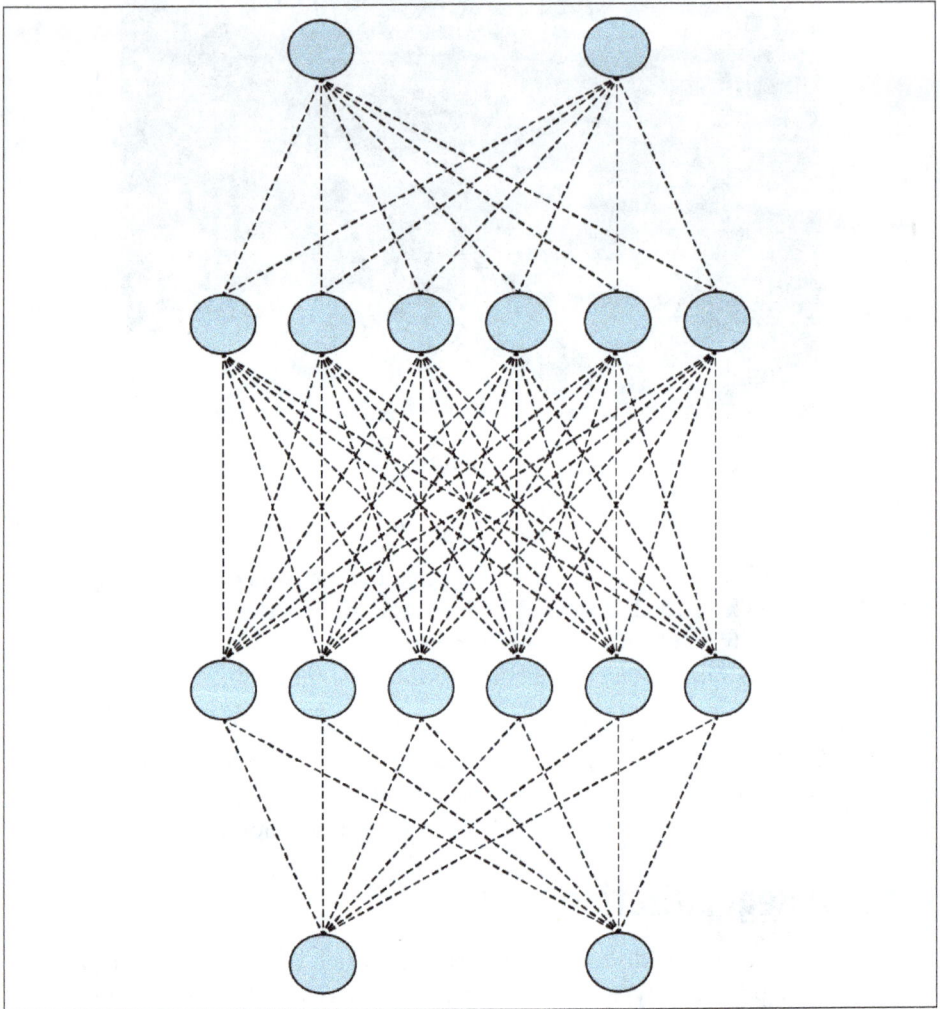

Figure 3-20. A simple neural network

While training, if you remove a random number of neurons and ignore them, then their contribution to the neurons in the next layer is temporarily blocked (see Figure 3-21). They are effectively dropped out, leading to the term *dropout regularization*.

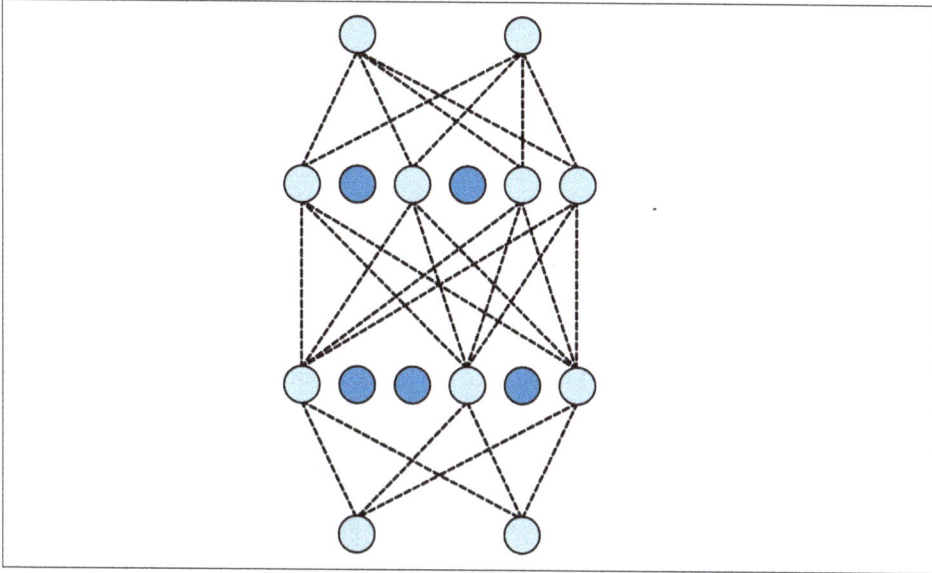

Figure 3-21. A neural network with dropouts

This reduces the chances of the neurons becoming overspecialized. The network will still learn the same number of parameters, but it should be better at generalization—that is, it should be more resilient to different inputs.

> The concept of dropouts was proposed by Nitish Srivastava et al. in their 2014 paper "Dropout: A Simple Way to Prevent Neural Networks from Overfitting (*https://oreil.ly/673CJ*)."

To implement dropouts in PyTorch, you can just use a simple layer like this:

```
nn.Dropout(0.5)
```

This will drop out, at random, the specified percentage of neurons (here, 50%) in the specified layer. Note that it may take some experimentation to find the correct percentage for your network.

For a simple example that demonstrates this, consider the new fully connected layers we added to the bottom of Inception with the transfer learning example in this chapter.

Here it is for Rock, Paper, Scissors with three output neurons:

```
num_ftrs = pre_trained_model.fc.in_features
pre_trained_model.fc = nn.Sequential(
    nn.Linear(num_ftrs, 1024),  # New fully connected layer
    nn.ReLU(),                  # Activation layer
    nn.Linear(1024, 3)          # Final layer for RPS
)
```

With dropout added, it would look like this:

```
num_ftrs = model.fc.in_features
model.fc = nn.Sequential(
    nn.Dropout(0.5),  # Adding dropout before the final FC layer
    nn.Linear(num_ftrs, 1024),  # Reduce dimensionality to 1024
    nn.ReLU(),
    nn.Dropout(0.5),  # Adding another dropout layer after ReLU activation
    nn.Linear(1024, 3)  # Final layer for RPS
)
```

The examples that we used in this chapter for transfer learning are already learning really well without the use of dropouts. However, I'd recommend that you always consider dropouts when building your models because they can greatly reduce waste in the ML process—letting your model learn just as well but much faster!

Additionally, as you design your neural networks, keep in mind that getting great results on your training set is not always a good thing because it could be a sign of overfitting. Introducing dropouts can help you remove that problem so that you can optimize your network in other areas without that false sense of security.

Summary

This chapter introduced you to a more advanced way of achieving computer vision by using convolutional neural networks. You saw how to use convolutions to apply filters that can extract features from images, and you designed your first neural networks to deal with more complex vision scenarios than those you encountered with the MNIST and Fashion MNIST datasets. You also explored techniques to improve your network's accuracy and avoid overfitting, such as the use of image augmentation and dropouts.

Before we explore further scenarios, in Chapter 4, you'll get an introduction to how to handle data in PyTorch. You'll explore the APIs that make it much easier for you to get access to data for training and testing your networks. In this chapter, you downloaded ZIP files and extracted images, but that's not always going to be possible. With PyTorch datasets, you'll be able to access lots of datasets with a standard API.

Using Data with PyTorch

In the first three chapters of this book, you trained models using a variety of data, from the Fashion MNIST dataset that was conveniently bundled via an API to the image-based "Horses or Humans" and "Dogs vs. Cats" datasets, which were available as ZIP files that you had to download and preprocess. So by now, you've probably realized that there are lots of different ways of getting the data with which to train a model.

However, many public datasets require you to learn lots of different domain-specific skills before you begin to consider your model architecture. The goal behind PyTorch domains and the tools available at the torch.utils.data.Datasets namespace is to expose datasets in a way that's easy to consume, where all the preprocessing steps of acquiring the data and getting it into PyTorch-friendly APIs are done for you.

You've already seen a little of this idea in how PyTorch handled Fashion MNIST back in Chapter 2. As a recap, all you had to do to get the data was this:

```
train_dataset = datasets.FashionMNIST(root='./data', train=True,
                        download=True, transform=transform)
```

In the case of this dataset, we also did an import from the torchvision library to get the datasets object that contained the reference to Fashion MNIST:

```
from torchvision import datasets
```

Given that it's a computer vision–oriented dataset, it makes sense that it would be in the torchvision library.

PyTorch has many other datasets of different data types that can be loaded in the same way. These include the following:

Vision
> Fashion MNIST is in the aforementioned torchvision library. It's one of the "Image Classification" built-in datasets, but there are many more for other scenarios like Image Detection, Segmentation, Optical Flow, Stereo Matching, Image Pairing, Image Captioning, Video Classification, Video Predictions, and more.

Text
> Common text datasets are available in the torchtext library. There are far too many to list here, but there are ones for Text Classification, Language Modeling, Machine Translation, Sequence Tagging, Question and Answer, and Unsupervised Learning. You can find more details on these in the PyTorch documentation (*https://oreil.ly/aFamN*). Note that this library isn't limited to the datasets; it also has many helper functions that you will use when processing text.

Audio

The torchaudio library contains many datasets that can be used in machine learning scenarios for sound or speech. Details can be found in the PyTorch documentation (*https://oreil.ly/tvDe4*).

All datasets are subclasses of `torch.utils.data.Dataset`, so it's important to take a look at this library and understand it well. That will help you not only consume existing datasets but also create your own to share with others.

Getting Started with Datasets

The `torch.utils.data.Dataset` is an abstract class that represents a dataset. To create a custom dataset, you just need to subclass it and implement these methods:

`__len__(self)`

This should return the total number of items in your dataset:

`__getitem__(self, index)`

This should return a single item from your dataset at the specified index. This item will be transformed before sending it to the model.

Here's an example:

```python
from torch.utils.data import Dataset

class CustomDataset(Dataset):
    def __init__(self, data, transforms=None):
        self.data = data
        self.transforms = transforms
```

```
def __len__(self):
    return len(self.data)

def __getitem__(self, idx):
    sample = self.data[idx]
    if self.transforms:
        sample = self.transforms(sample)
    return sample
```

That's pretty much it at a very low level. The data itself is in an array called data[ ]. Now, imagine we want to create a dataset with a linear relationship between an *x* value and a *y* value like we had in Chapter 1. How would we use it?

Let's say we start with some simple synthetic data, like this:

```
# Generate synthetic data
torch.manual_seed(0)  # For reproducibility
x = torch.arange(0, 100, dtype=torch.float32)
y = 2 * x - 1
```

Then, we could turn it into a dataset like this:

```
class CustomDataset(Dataset):
    def __init__(self, x, y):
        """
        Initialize the dataset with x and y values.
        Arguments:
        x (torch.Tensor): The input features.
        y (torch.Tensor): The output labels.
        """
        self.x = x
        self.y = y

    def __len__(self):
        """
        Return the total number of samples in the dataset.
        """
        return len(self.x)

    def __getitem__(self, idx):
        """
        Fetch the sample at index `idx` from the dataset.
        Arguments:
        idx (int): The index of the sample to retrieve.
        """
        return self.x[idx], self.y[idx]
```

Then, to use the dataset, we simply create an instance of the class, initialize it with our *x* and *y* values, pass it to a DataLoader, and enumerate that:

```
# Create an instance of CustomDataset
dataset = CustomDataset(x, y)
```

```
# Use DataLoader to handle batching and shuffling
data_loader = DataLoader(dataset, batch_size=10, shuffle=True)

# Iterate over the DataLoader
for batch_idx, (inputs, labels) in enumerate(data_loader):
    print(f"Batch {batch_idx+1}")
    print("Inputs:", inputs)
    print("Labels:", labels)
    # Break after the first batch for demonstration
    if batch_idx == 0:
        break
```

With this foundation, you can now explore the dataset classes that have been made available in the various libraries we've mentioned since the beginning of this chapter. Given that they will build on or extend this class, the APIs should look familiar.

Exploring the FashionMNIST Class

Earlier in the book, we saw the `FashionMNIST` class—which provides access to the Fashion-MNIST dataset, which is a training set of 60,000 examples of 10 classes of clothing—and an accompanying test set of 10,000 examples. Each of these examples is a 28×28 grayscale image.

In the case of this dataset, you use the *same* class whether you're using training data or test/validation data, and the data that you receive is based on the `train` parameter that you pass to it. Here's an example:

```
# Create the FashionMNIST training dataset
fashion_mnist_train = datasets.FashionMNIST(root='./data', train=True,
                                            download=True, transform=transform)
```

When you set `train=True`, the code that overrides the class init method will take the 60,000 records and return them to the caller. Other parameters are there, like specifying the root for where the data should go and even whether or not to download the data. Finally, as you'll commonly see when downloading data, there is the `transform=` parameter. As you saw in the preceding base class, this parameter will be available as an optional parameter for all datasets, and it will apply a transform when set.

Generic Dataset Classes

You may need to use some data that isn't available in the dataset classes, like `Fashion MNIST`, but you'll also want to take advantage of everything in the data ecosystem—such as the ability to transform your data, things like splitting, and all the good stuff in the `DataLoader` class you'll see later in this chapter. To that end, `torch.utils.data` provides a number of generic dataset classes you could use.

ImageFolder

In Chapter 3, we used the "Horses or Humans," "Rock, Paper, Scissors," and "Cats vs. Dogs" datasets, which are not available in a class directly but instead as a ZIP file containing the images. When we downloaded them and saved them into subdirectories for the different image types (e.g., one folder for "Horses" and another for "Humans"), the generic `ImageFolder` dataset class could act as a dataset for us.

In that case, the images were streamed (via a `DataLoader`) from the directory according to the batch size and other rules on the `DataLoader`. The labels were derived from the directory names, and the associated class indices were the labels in alphabetical order. So, "Horses" would be class 0, and "Humans" would be class 1. Please watch out for that when building and debugging because you might miss out on this ordering!

For example, we tend to say, "Rock, Paper, Scissors" in that order, and we would therefore expect them to be classes 0, 1, and 2, respectively. But in *alphabetical* order, Paper would be class 0, Rock would be class 1, and Scissors would be class 2!

One tool to use for this is to create a custom index like the one below:

```
custom_class_to_idx = {'rabbit': 0, 'dog': 1, 'cat': 2}
dataset = ImageFolder(
  root='data/animals',
  target_transform=
    lambda x: custom_class_to_idx[dataset.classes[x]]
)
dataset.class_to_idx = custom_class_to_idx
print(dataset.class_to_idx)
```

DatasetFolder

`ImageFolder` is actually a subclass of the more generic `DatasetFolder` class, one that's customized for images. The `DatasetFolder` class isn't limited to image data, and you can use it for anything. It also allows you to use directories for labels. So, for example, say you have text files that contain text of different classes with a directory structure like this:

```
root/sarcasm/document1.txt
root/sarcasm/document2.txt
root/sarcasm/document3.txt
root/factual/factdoc1.rtf
root/factual/factdoc2.doc
```

You could then use a `DatasetFolder` to stream the documents according to the correct labels. Also, because this class is document based, you could apply a transform to extract from the file!

FakeData

FakeData is a useful generic dataset that, as its name suggests, provides you with fake data. At the time of writing, it only supports creating fake image data. It's also very useful if you don't have data on hand but want to experiment with different architectures, or if you want to benchmark your system.

You can use FakeData in the same way you'd use any of the datasets you've seen so far in this book. So, for example, if you wanted to create a set of FakeData for the Mobile-Net model, which uses 224 × 224 color images, you'd do it with code like this:

```python
import torch
from torchvision.datasets import FakeData
import torchvision.transforms as transforms
from torch.utils.data import DataLoader

# Define transformations (if needed)
transform = transforms.Compose([
    transforms.ToTensor(),
    transforms.Normalize((0.5,), (0.5,))
])

# Create FakeData
fake_dataset = FakeData(size=100, image_size=(3, 224, 224),
                        num_classes=10, transform=transform)

# DataLoader
data_loader = DataLoader(fake_dataset, batch_size=10, shuffle=True)
```

This would create 100 images (containing just noise) of the desired size and span them across 10 classes. You could then use this data in a DataLoader in the same way you'd use any other dataset.

While FakeData only gives image types, you could relatively easily create your own custom data (as a CustomDataSet like we saw earlier) to provide fake data in other formats, such as numeric or sequence data.

Using Custom Splits

Up to this point, all of the data you've been using to build models has been pre-split for you into training and test sets. For example, with Fashion MNIST, you had 60,000 and 10,000 records, respectively. But what if you don't want to use those splits? What if you want to split the data yourself, according to your own needs?

Thankfully, when using the datasets namespace, you can generally do this with an easy and intuitive API.

So, for example, when you loaded the FashionMNIST class previously, you specified the train parameter to get it to give you the training data (60,000 records) or the test data (10,000 records).

To override this, you simply ignore it, and you'll get all of the data:

```
# Load the entire Fashion-MNIST dataset
dataset = datasets.FashionMNIST(root='./data',
                        download=True, transform=transform)
```

To create your own split, you can use the torch.utils.data namespace, which contains a function called random_split. So, for example, if you want to have a validation set that FashionMNIST doesn't provide, you can divide the dataset into three datasets with random_split. Here's the code that will assign 70% of the data to a training set, 15% to a testing set, and 15% to a validation set:

```
from torch.utils.data import random_split

total_count = len(dataset)
train_count = int(0.7 * total_count)
val_count = int(0.15 * total_count)

# Ensures all data is used
test_count = total_count - train_count - val_count

train_dataset, val_dataset, test_dataset =
    random_split(dataset, [train_count, val_count, test_count])
```

As you can see, this process is pretty straightforward. We get the number of records in the dataset as total_count and then calculate 70% of them (0.7 times the total count) to be the training count and 15% of them to be the validation count. When making calculations like this, you may end up with rounding errors that leave some records out—so instead of using 15% for the test count, you can just set the quotient for training to be the total minus the training and validation records. This will ensure all the data is used and none is wasted.

What's really nice about this approach is that it gives you a really simple way to get new and different slices of your dataset. As you train models, it gives you a new way to evaluate them for accuracy.

For example, one slice of the dataset may train at high accuracy while another does so at low accuracy, indicating that there are likely issues in your model architecture that make it overfit on one dataset. On the other hand, if you try multiple different splits of your data and the model training and validation results are consistent, then you'll have a signal that your architecture is sound.

I would definitely encourage you to use custom splits when training models, as it can really help you get over some gotchas!

One more thing to consider when using custom splits is that the name `ran dom_split` doesn't mean that this approach *shuffles* or randomizes your dataset. It merely splits the dataset at random points to give you different slices each time. Should you want to also shuffle the dataset, you can do it in the DataLoader, which we'll explore in the next section.

The ETL Process for Managing Data in Machine Learning

Extract, Transfer, Load (ETL) is the core pattern for training ML models, regardless of scale. We've been exploring small-scale, single-computer model building in this book, but we can use the same technology for large-scale training across multiple machines with massive datasets.

The Extract, Transfer, Load process consists of the three phases that are in the process's name:

Extract phase
    This occurs when the raw data is loaded from wherever it is stored and prepared in a way that can be transformed.

Transform phase
    This occurs when the data is manipulated in a way that makes it suitable or improved for training. For example, batching, image augmentation, mapping to features, and other such logic applied to the data can be considered part of this phase.

Load phase
    This occurs when the data is loaded into the neural network for training.

Consider the code from Chapter 3 that we used to train the "Horses or Humans" classifier. At the top of the code, you saw a chunk like this:

```python
# Define transformations
train_transform = transforms.Compose([
    transforms.Resize((150,150)),
    transforms.RandomHorizontalFlip(),
    transforms.RandomRotation(20),
    transforms.RandomAffine(
        degrees=0,  # No rotation
        translate=(0.2, 0.2),  # Translate up to 20% x and y
        scale=(0.8, 1.2),  # Zoom in or out by 20%
        shear=20,  # Shear by up to 20 degrees
    ),
    transforms.ToTensor(),
    transforms.Normalize(mean=[0.5, 0.5, 0.5], std=[0.5, 0.5, 0.5]),
])

# Load the datasets
```

```
train_dataset = datasets.ImageFolder(root=training_dir,
                            transform=train_transform)
val_dataset = datasets.ImageFolder(root=validation_dir,
                            transform=train_transform)

# Data loaders
train_loader = DataLoader(train_dataset, batch_size=32, shuffle=True)
val_loader = DataLoader(val_dataset, batch_size=32, shuffle=True)
```

This is the ETL pattern embodied in code!

The code begins by defining the transform (the "T"), but the active code doesn't begin until the lines under the Load the datasets comment. Look carefully here and you'll see that the ImageFolder is being used to *extract* the data from its location at rest on disk.

Then, as the data is extracted, the transform that we defined is applied.

Then, under the Data loaders comment, we perform the *load* of the data using the train_loader and val_loader we've defined. Strictly speaking, the actual loading doesn't take place until we execute the training loop to pull the data out of the loaders.

It's important to know that using this process can make your data pipelines less susceptible to changes in the data and the underlying schema. When you use this approach to extract data, the same underlying structure is used regardless of whether the data is small enough to fit in memory or large enough that it cannot be contained even on a simple machine. The APIs for applying the transformation are also consistent, so you can use similar ones regardless of the underlying data source. And of course, once the data is transformed, the process of loading the data is also consistent, regardless of your training backend.

However, how you load the data can have a huge impact on your training speed. Let's take a look at that next.

Optimizing the Load Phase

Let's take a closer look at the ETL process when you're training a model. We can consider the extraction and transformation of the data to be possible on any processor, including a CPU. In fact, the code you use in these phases to perform tasks like downloading data, unzipping it, and going through it record by record and processing those records is not what GPUs and TPUs are built for, so the code will likely execute on the CPU anyway. When it comes to training, however, you can get great benefits from a GPU or TPU, so it makes sense for you to use one for this phase if possible. Thus, in a situation where a GPU or TPU is available to you, you should ideally split the workload between the CPU and the GPU/TPU, with Extract and Transform taking place on the CPU and Load taking place on the GPU/TPU.

If you explore the code that you've been using in this book, you'll notice that we've used the .to(device) methodology. Whenever you're dealing with training or inference and you want the data or model to be on the accelerator, you'll see something like .to("cuda"), but for the extraction and transform, you won't see it because it would be a waste of the GPU's resources.

Suppose you're working with a large dataset. Assuming it's so large that you have to prepare the data (i.e., do the extraction and transformation) in batches, you'll end up with a situation like that shown in Figure 4-1. While the first batch is being prepared, the GPU or TPU is idle. Then, when that batch is ready, you can send it to the GPU/TPU for training, but then the CPU will be idle until the training is done and it can start preparing the second batch. There's a lot of idle time here, so we can see that there's room for optimization.

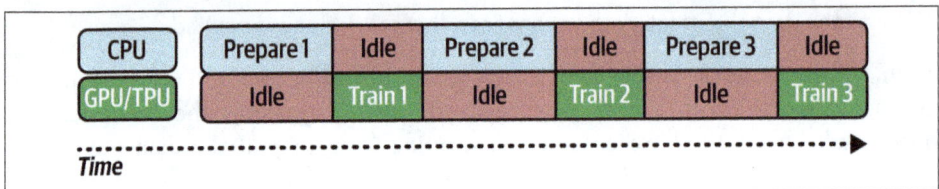

| CPU | Prepare 1 | Idle | Prepare 2 | Idle | Prepare 3 | Idle |
| GPU/TPU | Idle | Train 1 | Idle | Train 2 | Idle | Train 3 |

Time

Figure 4-1. Training on a CPU or GPU/TPU

The logical solution is to do the work in parallel, preparing and training side by side. This process is called *pipelining* and is illustrated in Figure 4-2.

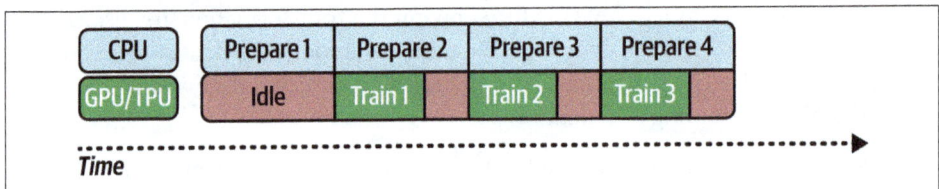

| CPU | Prepare 1 | Prepare 2 | Prepare 3 | Prepare 4 |
| GPU/TPU | Idle | Train 1 | Train 2 | Train 3 |

Time

Figure 4-2. Pipelining

In this case, while the CPU is preparing the first batch, the GPU/TPU again has nothing to work on, so it's idle. When the first batch is done, the GPU/TPU can start training—but in parallel with this, the CPU will prepare the second batch. Of course, the time it takes to train batch $n - 1$ and prepare batch n won't always be the same, and if the training time is shorter, you'll have periods of idle time on the GPU/TPU, and if the training time is longer, you'll have periods of idle time on the CPU. Choosing the correct batch size can help you optimize here—and as GPU/TPU time is likely more expensive, you'll probably want to reduce its idle time as much as possible.

This is one of the reasons why we use batching, even for the simple examples like MNIST: the pipelining model is in place so that regardless of how large your dataset is, you'll continue to use a consistent pattern for ETL on it.

Using the DataLoader Class

We've seen the `DataLoader` class many times already, but it's good to take a slightly deeper look at it now to help you get the most out of it in your ML workflows. It provides the following features that you can use.

Batching

Intuitively, you might think that the forward pass works one data item at a time. You *could* do that, but some optimizers, like stochastic gradient descent, do much better when inputs are passed in batches so that they can calculate more accurately. Batching can also speed up your training in larger scenarios, where you are using GPUs with fixed memory sizes. It's most efficient to maximize the use of that memory by having a batch of data that fits it fully. If you're using a `DataLoader`, batching is simply a matter of setting a parameter.

Shuffling

Shuffling the data is very important, particularly when you do batching. Consider the following scenario with something like Fashion MNIST.

You have 60,000 samples each for 10 classes, and they are not shuffled. You batch one thousand records at a time, so your first batch of one thousand is all for class 0, the second batch is all for class 1, and so on. In this scenario, the model may not effectively learn because each batch is biased toward a particular label—but the ability of your model to generalize will improve if the batches are shuffled, meaning your first thousand items will have varied labels, etc.

Parallel Data Loading

Often, and in particular with complex data, loading data into the model for the forward pass can be time-consuming. But the `DataLoader` class offers parallelism through Python's multiprocessing model, which can significantly speed this up.

As you saw previously, you should consider your data loading/transformation and your model learning to be two separate processes. You want to avoid scenarios where the model training has no data to work with and is sitting idle, waiting for data to be loaded, and you also want to avoid scenarios where you have tons of data piled up in memory but the model can't get to it. Parallel data loading, when well tuned, can be helpful here, and there's a skill you can learn to ensure that you're getting the most out of your training by running this at peak efficiency. You'll learn how to do this in the next section.

Custom Data Sampling

In addition to shuffling for random data sampling, you can create custom data sampling, in which you can specify how the data will be loaded. The torch.utils.data.Sampler class provides a base class that you can build a custom sampler on. This process is beyond the scope of this book, but there are many excellent examples of it online.

Parallelizing ETL to Improve Training Performance

If you're using the DataLoader class, you can easily perform parallelization by using the num_workers parameter. So, for example, say you want to train a model on the CIFAR10 dataset, and you want to use parallel training. Let's take a look at how to do this, step-by-step.

First, we'll explore the Extract and Transform steps:

```python
import torchvision.transforms as transforms
from torchvision.datasets import CIFAR10

# Define transformations
transform = transforms.Compose([
    transforms.ToTensor(),
    transforms.Normalize((0.5, 0.5, 0.5), (0.5, 0.5, 0.5))
])

# Load CIFAR10 dataset
dataset = CIFAR10(root='./data', train=True, download=True, transform=transform)
```

Then, we'll configure and create the DataLoader to the Load step:

```python
from torch.utils.data import DataLoader

# DataLoader with multiple workers
data_loader = DataLoader(dataset, batch_size=64, shuffle=True,
                         num_workers=4)
```

Note the num_workers=4 parameter, which will create four subprocesses to load the data in parallel simultaneously. Based on the hardware you have available and the number of cores, the speed of your CPU, etc., you can experiment with this number to reduce the overall bottlenecks.

What's really nice about this approach is that the ETL process is neatly encapsulated in it, so your model training loop doesn't have to change in any way, even though you're loading the data by using parallelism! Here's the code for the simple CIFAR model that uses this data:

```python
import torch

# Dummy model and optimizer setup
model = torch.nn.Sequential(
    torch.nn.Linear(3 * 32 * 32, 500),
    torch.nn.ReLU(),
    torch.nn.Linear(500, 10)
)
optimizer = torch.optim.Adam(model.parameters(), lr=0.001)
criterion = torch.nn.CrossEntropyLoss()

# Training loop
def train(model, data_loader):
    model.train()
    for batch_idx, (inputs, targets) in enumerate(data_loader):
        # Reshape inputs to match the model's expected input
        inputs = inputs.view(inputs.size(0), -1)

        # Forward pass
        outputs = model(inputs)
        loss = criterion(outputs, targets)

        # Backward pass and optimize
        optimizer.zero_grad()
        loss.backward()
        optimizer.step()

        if batch_idx % 100 == 0:
            print(f"Train Epoch: {batch_idx} Loss: {loss.item()}")

train(model, data_loader)
```

Parallelizing is another tool that's available to you when you're training your models. There's no one-size-fits-all approach, but it's a good tool to use when you experience training slowdowns. It's easy to assume that training operates slowly just because of the network architecture, but you may be surprised at how much of that time is wasted by the forward pass waiting for new data! By adding this type of parallelism, you have the potential to greatly speed up training.

Summary

This chapter covered the data ecosystem in PyTorch and introduced you to the Data set and DataLoader classes. You saw how they use a common API and a common format to help reduce the amount of code you have to write to get access to data, and you also saw how to use the ETL process, which is at the heart of the common design patterns in training models with PyTorch. In particular, we explored parallelizing the extraction, transformation, and loading of data to improve training performance.

So now that you've had a chance to look at the process, see if you can create your own dataset! Maybe do it from some photos in your albums, or some test, or just random noise like we did here.

In the next chapter, you'll take what you've learned so far and start applying it to natural language processing problems.

## CHAPTER 5

Introduction to Natural Language Processing

Natural language processing (NLP) is a technique in AI that deals with the understanding of language. It involves programming techniques to create models that can understand language, classify content, and even generate and create new compositions in language. It's also the underlying foundation to large language models (LLMs) such as GPT, Gemini, and Claude. We'll explore LLMs in later chapters, but first, we'll look at more basic NLP over the next few chapters to equip you for what's to come.

There are also lots of services that use NLP to create applications such as chatbots, but that's not in the scope of this book—instead, we'll be looking at the foundations of NLP and how to model language so that you can train neural networks to understand and classify text. In later chapters, you'll also learn how to use the predictive elements of an ML model to write some poetry. This isn't just for fun—it's also a precursor to learning how to use the transformer-based models that underpin generative AI!

We'll start this chapter by looking at how you can decompose language into numbers and how you can then use those numbers in neural networks.

Encoding Language into Numbers

Ultimately, computers deal in numbers, so to handle language, you need to convert it into numerics in a process called *encoding*.

You can encode language into numbers in many ways. The most common is to encode by letters, as is done naturally when strings are stored in your program. In memory, however, you don't store the letter *a* but an encoding of it—perhaps an ASCII or Unicode value or something else. For example, consider the word *listen*. You can encode it with ASCII into the numbers 76, 73, 83, 84, 69, and 78. This is good in that you can now use numerics to represent the word. But then consider the word *silent*, which is an anagram of *listen*. The same numbers represent that word, albeit in a different order, which might make building a model to understand the text much more difficult.

A better alternative might be to use numbers to encode entire words instead of the letters within them. In that case, *silent* could be the number *x* and *listen* could be the number *y*, and they wouldn't overlap with each other.

Using this technique, consider a sentence like "I love my dog." You could encode that with the numbers [1, 2, 3, 4]. If you then wanted to encode "I love my cat," you could do it with [1, 2, 3, 5]. By now, you've probably gotten to the point where you can tell that the sentences have a similar meaning because they're similar numerically—in other words, [1, 2, 3, 4] looks a lot like [1, 2, 3, 5].

The numbers representing words are also called *tokens*, and as a result, this process is called *tokenization*. You'll explore how to do that in code next.

Getting Started with Tokenization

The PyTorch ecosystem contains many libraries for tokenization, which takes words and turns them into tokens. A common tokenizer you might see in code samples is `torchtext`, but this has been deprecated since 2023. So, be careful when using it, especially because PyTorch versions advance but it doesn't. So, some alternatives are to use a custom tokenizer, a pretrained one from elsewhere, or (surprisingly) those from the Keras ecosystem.

Using a custom tokenizer

To give you a simple example, here's some code I used to create a custom tokenizer to turn the words of a small corpus (two sentences) into tokens:

```python
import torch

sentences = [
    'Today is a sunny day',
    'Today is a rainy day'
]

# Tokenization function
def tokenize(text):
    return text.lower().split()

# Build the vocabulary
def build_vocab(sentences):
    vocab = {}
    for sentence in sentences:
```

```
            tokens = tokenize(sentence)
        for token in tokens:
            if token not in vocab:
                vocab[token] = len(vocab) + 1
    return vocab

# Create the vocabulary index
vocab = build_vocab(sentences)

print("Vocabulary Index:", vocab)
```

> The word *corpus* is commonly used to denote a set of text items that you will use for training. It's literally the *body* of text that you'll use to train the model and create tokenizers for.

The output of this is as follows:

```
Vocabulary Index: {'today': 1, 'is': 2, 'a': 3, 'sunny': 4, 'day': 5, 'rainy': 6}
```

As you can see, the tokenizer did a really simple job of creating a list with my vocabulary, and every time it hit a unique word, it added it to the list. So the first sentence, "Today is a sunny day," yielded five tokens for the five words: "today," "is," "a," "sunny," and "day." The second sentence had *most* of these words in common, with "rainy" being the exception, so that became the sixth token.

On the other hand, you can imagine that for a very large corpus, this process would be very slow.

Using a pretrained tokenizer from Hugging Face

With that in mind, I'm going to use Hugging Face's transformers library and pre-built tokenizers from within it. In this case, because the transformers library supports many language models and these language models need tokenizers to work with their corpus of text, the tokenizer, which is trained on many millions of words, is freely available for you to use. It has bigger coverage than one you might create, and it's free and easy to use!

If you don't have this library already, you can install it with this:

```
!pip install transformers
```

Now, let's see it in action with a simple example:

```
from transformers import BertTokenizerFast

sentences = [
    'Today is a sunny day',
    'Today is a rainy day'
]
```

```python
# Initialize the tokenizer
tokenizer = BertTokenizerFast.from_pretrained('bert-base-uncased')

# Tokenize the sentences and encode them
encoded_inputs = tokenizer(sentences, padding=True, truncation=True,
                           return_tensors='pt')

# To see the tokens for each input (helpful for understanding the output)
tokens = [tokenizer.convert_ids_to_tokens(ids)
          for ids in encoded_inputs["input_ids"]]

# To get the word index similar to Keras' tokenizer
word_index = tokenizer.get_vocab()

print("Tokens:", tokens)
print("Token IDs:", encoded_inputs['input_ids'])
print("Word Index:", dict(list(word_index.items())[:10]))
# show only the first 10 for brevity
```

The output from it looks like this:

```
Tokens: [['[CLS]', 'today', 'is', 'a', 'sunny', 'day', '[SEP]'],
         ['[CLS]', 'today', 'is', 'a', 'rainy', 'day', '[SEP]']]

Token IDs: tensor([
        [ 101, 2651, 2003, 1037, 11559, 2154,  102],
        [ 101, 2651, 2003, 1037, 16373, 2154,  102]])

Word Index: {'protestant': 8330, 'initial': 3988, '##pt': 13876,
             'charters': 23010, '243': 22884, 'ref': 25416, '##dies': 18389,
             '##uchi': 15217, 'sainte': 16947, 'annette': 22521}
```

Now, let's break this down. We start by importing the `BertTokenizerFast` from the transformers library. This can be initialized with a number of pretrained tokenizers, and we choose the `'bert-base-uncased'` one. You might be wondering what on earth that is! Well, the idea here is that I wanted to take a pretrained tokenizer, and they are usually partnered with the model they were trained on. BERT (which stands for bidirectional encoder representations from transformers) is a model trained by Google on a large corpus, with a vocabulary of 30,000 words. You can find models like this in the Hugging Face model repository, and when you dig down into a model, you'll often see the transformer's code to get its tokenizer. For example, see this page that I used (*https://oreil.ly/Ok7L9*)—and while I'm not using the model, I can still get its tokenizer instead of creating a custom one.

In this case, we create a `tokenizer` object and specify the number of words that it can tokenize. This will be the maximum number of tokens to generate from the corpus of words. We also have a very small corpus here, containing only six unique words, so we'll be well under the maximum of one hundred specified.

Once I have the tokenizer, I can then just pass the text to it:

```
# Tokenize the sentences and encode them
encoded_inputs = tokenizer(sentences, padding=True, truncation=True,
                           return_tensors='pt')
```

We'll explore padding and truncation a little later in this chapter, but for now, you should note the `return_tensors='pt'` parameter. This is a nice convenience for us PyTorch developers because the return values will be `torch.Tensor` objects, which are easy for us to handle.

The BERT model uses a number of overlays on the original tokenization, such as `attention_masking`, which means it works with IDs for each word instead of raw tokens. This is beyond the scope of this chapter, but where it impacts you right now is if you don't need all that. If you just want the tokens, you have to extract the tokens in the following way, noting that your sentences were encoded as `input_Ids` within the BERT tokenizer:

```
# To see the tokens for each input (helpful for understanding the output)
tokens = [tokenizer.convert_ids_to_tokens(ids)
          for ids in encoded_inputs["input_ids"]]
```

Once you've done that, you can easily print out the following `Tokens` collection:

```
Tokens: [['[CLS]', 'today', 'is', 'a', 'sunny', 'day', '[SEP]'],
         ['[CLS]', 'today', 'is', 'a', 'rainy', 'day', '[SEP]']]
```

Now, you may be wondering what [CLS] and [SEP] are—and how the BERT model has been trained to expect sentences to begin with [CLS] (for *classifier*) and end with or be separated by [SEP] (for *separator*). These two expressions are tokenized to values 101 and 102, respectively, so when you print out the token values for your sentences, you'll see this:

```
Token IDs: tensor([
    [  101,  2651,  2003,  1037, 11559,  2154,   102],
    [  101,  2651,  2003,  1037, 16373,  2154,   102]])
```

From this, you can derive that *today* is token 2651 in BERT, *is* is token 2003, etc.

So, it really depends on you how you want to approach this. For learning with small datasets, the custom tokenizer is probably OK. But once you start getting into larger datasets, you may want to opt for a pretrained tokenizer. In that case, you may have to deal with some overhead—so for the rest of this chapter, I'm going to use custom code to tokenize and preprocess the text without the overhead of something like the BERT tokenizer.

Either way, once you have the words in your sentences tokenized, the next step is to convert your sentences into lists of numbers, with the number being the value where the word is the key. This process is called *sequencing*.

Turning Sentences into Sequences

Now that you've seen how to take words and tokenize them into numbers, the next step is to encode the sentences into sequences of numbers, which you can do as follows:

```
def text_to_sequence(text, vocab):
    return [vocab.get(token, 0) for token in tokenize(text)]
    # 0 for unknown words
```

Then, you'll be given the sequences representing the two sentences. Remember that the word index is this:

```
Vocabulary Index: {'today': 1, 'is': 2, 'a': 3, 'sunny': 4, 'day': 5, 'rainy': 6}
```

And the output will look like this:

```
[1, 2, 3, 4, 5]
[1, 2, 3, 6, 5]
```

You can then substitute the words for the numbers, and you'll see that the sentences make sense.

Now, consider what happens if you are training a neural network on a set of data. The typical pattern is that you have a set of data that's used for training but that you know won't cover 100% of your needs, but you hope it covers as much as possible. In the case of NLP, you might have many thousands of words in your training data that are used in many different contexts, but you can't have every possible word in every possible context. So when you show your neural network some new, previously unseen text that contains previously unseen words, what might happen? You guessed it—the network will get confused because it simply has no context for those words, and as a result, any prediction it gives will be negatively affected.

Using out-of-vocabulary tokens

One tool you can use to handle these situations is an *out-of-vocabulary* (OOV) *token*, which can help your neural network understand the context of the data containing previously unseen text. For example, given the previous small example corpus, suppose you want to process sentences like these:

```
test_data = [
    'Today is a snowy day',
    'Will it be rainy tomorrow?'
]
```

Remember that you're not adding this input to the corpus of existing text (which you can think of as your training data) but you're considering how a pretrained network might view this text. Say you tokenize it with the words that you've already used and your existing tokenizer, like this:

```
for test_sentence in test_data:
    test_seq = text_to_sequence(test_sentence, vocab)
    print(test_seq)
```

Then, your results will look like this:

```
[1, 2, 3, 0, 5]
[0, 0, 0, 6, 0]
```

So, the new sentences, swapping back tokens for words, would be "today is a <UNK> day" and "<UNK> <UNK> <UNK> rainy <UNK>."

Here I'm using the tag <UNK> (which stands for *unknown*) for token 0. If you check out the `text_to_sequence` code I showed previously, it uses 0 for words that aren't in its dictionary. You can, of course, use any value you like.

Understanding padding and truncation

When training neural networks, you typically need all your data to be in the same shape. Recall from earlier chapters that when you were training with images, you reformatted the images to be the same width and height. With text, you face the same issue—once you've tokenized your words and converted your sentences into sequences, they can all be different lengths. But to get them to be the same size and shape, you can use *padding*.

All the sentences we have used so far are composed of five words, so you can see that our sequences are five tokens. But what would happen if you had some sentences that were longer. Say a few had 5 words, some had 8 words, and some had 10 words. To have a neural network handle them all, they would need to be of the same length! You could convert everything to 10 words by lengthening the sentences that are shorter, convert everything to 5 words by chopping off bits of the longer ones, or follow some other strategy!

To explore padding, let's add another, much longer, sentence to the corpus:

```
sentences = [
    'Today is a sunny day',
    'Today is a rainy day',
    'Is it sunny today?',
    'I really enjoyed walking in the snow today'
]
```

When you sequence that, you'll see that your lists of numbers have different lengths. Also note that if you haven't retokenized to build the new vocabulary, the latter two sentences will be full of zeros:

```
[1, 2, 3, 4, 5]
[1, 2, 3, 6, 5]
[2, 0, 4, 0]
[0, 0, 0, 0, 0, 0, 0, 1]
```

So, don't forget to call:

```
vocab = build_vocab(sentences)
```

And then you'll have new tokens for the new words in your tokenizer, so the output will look like this:

```
[1, 2, 3, 4, 5]
[1, 2, 3, 6, 5]
[2, 7, 4, 8]
[9, 10, 11, 12, 13, 14, 15, 1]
```

Remember that when you were training neural networks in earlier chapters, your input layers of the neural network required images to have consistent sizes and shapes. It's the same with NLP, for the most part. (There's an exception for something called *ragged tensors*, but that's beyond the scope of this chapter.) So, we need a way to make our sentences the same length.

Here's a simple padding function:

```
def pad_sequences(sequences, maxlen):
    return [seq + [0] * (maxlen - len(seq)) if len(seq) < maxlen
                else seq[:maxlen] for seq in sequences]
```

This function will reshape every array in the sequence to be the same length as the maximum-length one. So, say we take our sentences and pad them after sequencing them with code like this:

```
for sentence in sentences:
    seq = text_to_sequence(sentence, vocab)
    padded_seq = pad_sequences([seq], maxlen=10)   # Example maxlen
    print(padded_seq)
```

Then, the output will look like this:

```
[[1, 2, 3, 4, 5, 0, 0, 0, 0, 0]]
[[1, 2, 3, 6, 5, 0, 0, 0, 0, 0]]
[[2, 7, 4, 8, 0, 0, 0, 0, 0, 0]]
[[9, 10, 11, 12, 13, 14, 15, 1, 0, 0]]
```

Now, each of the sequences has a length of 10 because of the maxlen parameter. It's a pretty simple implementation that you would likely want to build on if you're using this in a more serious way. For example, you might want to consider what would happen if you had a sequence that was longer than the maximum length. Right now, it would cut off everything *after* the maximum, but you might want it to exhibit different behavior! That's where *truncation* is used. As its name suggests, truncation is reducing the length to match a desired length. Typically you would truncate by cutting off the righthand side, but under some circumstances (like left-to-right languages,) you may want to truncate from the beginning.

Also note that if you're using off-the-shelf tokenizers like the BERT one we showed you earlier, much of this functionality may already be available to you, so be sure to experiment.

Removing Stopwords and Cleaning Text

In the next section you'll look at some real-world datasets, and you'll find that there's often text that you *don't* want in your dataset. You may also want to filter out so-called *stopwords*—like "the," "and," and "but"—that are too common and don't add any meaning. You may also encounter a lot of HTML tags in your text, and it would be good to have a clean way to remove them. Other things you may want to filter out include rude words, punctuation, or names. Later, we'll explore a dataset of tweets that often have somebody's user ID in them, and we'll want to filter those out.

While every task is different based on your corpus of text, there are three main things that you can do to clean up your text programmatically.

Stripping Out HTML Tags

The first thing you can do is strip out HTML tags, and fortunately, there's a library called BeautifulSoup that makes this straightforward. For example, if your sentences contain HTML tags such as `<br>`, then you can remove them by using this code:

```
from bs4 import BeautifulSoup
soup = BeautifulSoup(sentence)
sentence = soup.get_text()
```

Stripping Out Stopwords

The second thing to do is strip out stopwords, and a common way to do it is to have a stopwords list and preprocess your sentences by removing instances of stopwords. Here's an abbreviated example:

```
stopwords = ["a", "about", "above", ... "yours", "yourself", "yourselves"]
```

You can find a full stopwords list in some of the online examples for this chapter (*https://github.com/lmoroney/PyTorch-Book-FIles*).

Then, as you're iterating through your sentences, you can use code like this to remove the stopwords from your sentences:

```
words = sentence.split()
filtered_sentence = ""
for word in words:
    if word not in stopwords:
        filtered_sentence = filtered_sentence + word + " "
sentences.append(filtered_sentence)
```

Stripping Out Punctuation

The third thing you can do is strip out punctuation, and you'll want to do it because punctuation can fool a stopword remover. The one we just showed you looks for words surrounded by spaces, so it won't spot a stopword that's immediately followed by a period or a comma.

Fixing this problem is easy with the translation functions provided by the Python string library. But do be careful with this approach, as there are scenarios where it might impact NLP analysis, particularly when detecting sentiment.

The library also comes with a constant called `string.punctuation` that contains a list of common punctuation marks, so to remove them from a word, you can do the following:

```python
import string
table = str.maketrans('', '', string.punctuation)
words = sentence.split()
filtered_sentence = ""
for word in words:
    word = word.translate(table)
    if word not in stopwords:
        filtered_sentence = filtered_sentence + word + " "
sentences.append(filtered_sentence)
```

Here, before filtering for stopwords, the constant removes punctuation from each word in the sentence. So, if splitting a sentence gives you the word *it*, the word will be converted to *it* and then stripped out as a stopword. Note, however, that when doing this, you may have to update your stopwords list. It's also common for these lists to have abbreviated words and contractions like *you'll* in them, and the translator will change *you'll* to *youll*. So if you want to have those words filtered out, you'll need to update your stopwords list to include them.

Following these three steps will give you a much cleaner set of text to use. But of course, every dataset will have its idiosyncrasies that you'll need to work with.

Working with Real Data Sources

Now that you've seen the basics of getting sentences, encoding them with a word index, and sequencing the results, you can take it to the next level by taking some well-known public datasets and using the tools Python provides to get them into a format where you can easily sequence them. We'll start with a dataset in which a lot of the work has already been done for you: the IMDb dataset. After that, we'll get a bit more hands-on by processing a JSON-based dataset and a couple of comma-separated values (CSV) datasets with emotion data in them.

Getting Text Datasets

We explored some datasets in Chapter 4, so if you get stuck on any of the concepts in this section, you can get a quick review there. However, at the time of writing, accessing *text*-based datasets is a little unusual. Given that the torchtext library has been deprecated, it's not clear what will happen with its built-in datasets, so we'll get hands-on in dealing with raw data in this section.

We'll start by exploring the IMDb reviews, which is a dataset of 50,000 labeled movie reviews from the Internet Movie Database (IMDb), each of which is determined to be positive or negative in sentiment.

This code will download the raw dataset and unzip it into folders where training and test splits are already pre-made for us. These will then be stored in subdirectories, and there are further subdirectories called pos and neg in each that determine the labels of the text files they contain:

```python
import os
import urllib.request
import tarfile

def download_and_extract(url, destination):
    if not os.path.exists(destination):
        os.makedirs(destination, exist_ok=True)
    file_path = os.path.join(destination, "aclImdb_v1.tar.gz")

    if not os.path.exists(file_path):
        print("Downloading the dataset...")
        urllib.request.urlretrieve(url, file_path)
        print("Download complete.")

    if "aclImdb" not in os.listdir(destination):
        print("Extracting the dataset...")
        with tarfile.open(file_path, 'r:gz') as tar:
            tar.extractall(path=destination)
        print("Extraction complete.")

# URL for the dataset
dataset_url = "http://ai.stanford.edu/~amaas/data/sentiment/aclImdb_v1.tar.gz"
download_and_extract(dataset_url, "./data")
```

The file structure will look like the one in Figure 5-1.

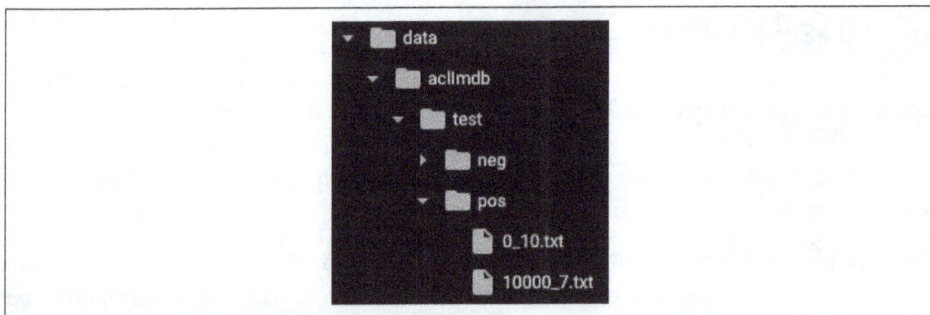

Figure 5-1. Exploring the IMDb dataset structure

In this figure you can see the *test/pos* directory and the first couple of files in it. Note that these are text files, so to create a tokenizer and vocabulary, we're going to have to read files instead of in-memory strings like in the earlier example.

Let's take a look at the code for a custom tokenizer for this:

```python
from collections import Counter
import os

# Simple tokenizer
def tokenize(text):
    return text.lower().split()

# Build vocabulary
def build_vocab(path):
    counter = Counter()
    for folder in ["pos", "neg"]:
        folder_path = os.path.join(path, folder)
        for filename in os.listdir(folder_path):
            with open(os.path.join(folder_path, filename), 'r',
                             encoding='utf-8') as file:
                counter.update(tokenize(file.read()))
    return {word: i+1 for i, word in enumerate(counter)} # Starting index from 1

vocab = build_vocab("./data/aclImdb/train/")
```

It's pretty straightforward code that just reads through each file and adds new words it discovers to the vocabulary, giving each word a new token value. Generally, you'll only want to do this for the training data, with the understanding that there will be words in the test data that aren't in the training data and that they would be tokenized with an OOV or unknown token.

The output should look like this (truncated):

```
{'a': 1, 'year': 2, 'or': 3, 'so': 4, 'ago,': 5, 'i': 6, 'was': 7…
```

This is a naive tokenizer in that the first word it sees gets the first token, the second gets the second, etc. For performance reasons, it's often better for the more frequent words in the corpus to get the earlier tokens and the less frequent ones to get the later tokens. We'll explore that in a moment.

You can then do sequencing and padding as you did earlier:

```python
def text_to_sequence(text, vocab):
    return [vocab.get(token, 0) for token in tokenize(text)]  # 0 for unknown

def pad_sequences(sequences, maxlen):
    return [seq + [0] * (maxlen - len(seq))
            if len(seq) < maxlen else seq[:maxlen] for seq in sequences]

# Example use
text = "This is an example."
seq = text_to_sequence(text, vocab)
padded_seq = pad_sequences([seq], maxlen=256)  # Example maxlen
print(seq)
```

So, for example, our sentence `This is an example` will output as `[30, 56, 144, 16040]` because those are the tokens assigned to those words. The padded sequence would have a tensor of 256 values, with these tokens as the first 4 and the next 252 being zeros!

Now, let's update the tokenizer to do the words in order of frequency. This update changes the tokenizer so that we load all of the files into memory and count the instance of each word to get a frequency table:

```python
# Build vocabulary
def build_vocab(path):
    counter = Counter()
    for folder in ["pos", "neg"]:
        folder_path = os.path.join(path, folder)
        for filename in os.listdir(folder_path):
            with open(os.path.join(folder_path, filename), 'r',
                                    encoding='utf-8') as file:
                counter.update(tokenize(file.read()))

    # Sort words by frequency in descending order
    sorted_words = sorted(counter.items(), key=lambda x: x[1], reverse=True)

    # Create vocabulary with indices starting from 1
    vocab = {word: idx + 1 for idx, (word, _) in enumerate(sorted_words)}
    vocab['<pad>'] = 0  # Add padding token with index 0
    return vocab
```

We can then output the vocabulary as this frequency table. The vocabulary is too large to show the entire index, but here are the top 20 words. Note that the tokenizer lists them in order of frequency in the dataset, so common words like *the*, *and*, and *a* are indexed:

```
{'the': 1, 'a': 2, 'and': 3, 'of': 4, 'to': 5, 'is': 6, 'in': 7, 'i': 8,
'this': 9, 'that': 10, 'it': 11, '/><br': 12, 'was': 13, 'as': 14,
'for': 15, 'with': 16, 'but': 17, 'on': 18, 'movie': 19, 'his': 20,
```

These are stopwords, as described in the previous section. Having these present can impact your training accuracy because they're the most common words and they're nondistinct (i.e., they're likely present in both positive and negative reviews), so they add noise to our training.

Also note that *br* is included in this list because it's commonly used in this corpus as the
 HTML tag.

You can also update the code to use BeautifulSoup to remove the HTML tags, and you can remove stopwords from the given list as follows. To do this, you can update the tokenizer to remove the HTML tags by using BeautifulSoup:

```
# Simple tokenizer
from bs4 import BeautifulSoup

# Note that the list of stopwords is defined in the source code.
# It's an array of words. You can define your own or just get the one from
# the book's github.

def tokenize(text):
    soup = BeautifulSoup(text, "html.parser")
    cleaned_text = soup.get_text()  # Extract text from HTML
    return [word.lower() for word in cleaned_text.split() if word.lower()
            not in stopwords]
```

Now, when you print out your word index, you'll see this:

```
{'movie': 1, 'not': 2, 'film': 3, 'one': 4, 'like': 5, 'just': 6, "it's": 7,
'even': 8, 'good': 9, 'no': 10, 'really': 11, 'can': 12, 'see': 13, '-': 14,
'get': 15, 'will': 16, 'much': 17, 'story': 18, 'also': 19, 'first': 20
```

You can see that this is much cleaner than before. There's always room to improve, however, and one thing I noted when looking at the full index was that some of the less common words toward the end were nonsensical. Often, reviewers would combine words, for example with a dash (as in *annoying-conclusion*) or a slash (as in *him/ her*), and the stripping of punctuation would incorrectly turn these combined words into a single word. Or, as you can see in the preceding code, the dash (-) character was common enough to be tokenized. You can strip that out by adding it as a stopword.

Now that you have a tokenizer for the corpus, you can encode your sentences. For example, the simple sentences we were looking at earlier in the chapter will come out like this:

```
sentences = [
    'Today is a sunny day',
    'Today is a rainy day',
    'Is it sunny today?'
]

[[1094, 6112, 246, 0, 0, 0, 0, 0]]
[[1094, 6730, 246, 0, 0, 0, 0, 0]]
[[6112, 25065, 0, 0, 0, 0, 0, 0]]
```

If you decode these, you'll see that the stopwords are dropped and you get the sentences encoded as today sunny day, today rainy day, and sunny today.

If you want to do this in code, you can create a new dict with the reversed keys and values (i.e., for a key/value pair in the word index, you can make the value the key and the key the value) and do the lookup from that. Here's the code:

```
reverse_word_index = dict(
    [(value, key) for (key, value) in vocab.items()])

decoded_review = ' '.join([reverse_word_index.get(i, '?') for i in seq])

print(decoded_review)
```

This will give the following result:

```
today sunny day
```

A common way to store labeled text data is in the comma-separated value (CSV) format. We'll discuss that next.

Getting Text from CSV Files

NLP data is also commonly available in CSV file format. Over the next couple of chapters, you'll use a CSV of data that I adapted from the open source Sentiment Analysis in Text dataset (*https://oreil.ly/7ZKEU*). The creator of this dataset sourced it from Twitter (now called X). You will use two different datasets, one where the emotions have been reduced to "positive" or "negative" for binary classification and one where the full range of emotion labels is used. Both datasets use the same structure, so I'll just show the binary version here.

While the name *CSV* seems to suggest a standard file format in which values are comma separated, there's actually a wide diversity of formats that can be considered CSV, and there's very little adherence to any particular standard. To solve this, the

Python csv library makes handling CSV files straightforward. In this case, the data is stored with two values per line. The first value is a number (0 or 1) denoting whether the sentiment is negative or positive, and the second value is a string containing the text.

The following code snippet will read the CSV and do preprocessing that's similar to what we saw in the previous section. For the full code, please check the repo for this book. The code adds spaces around the punctuation in compound words, uses BeautifulSoup to strip HTML content, and then removes all punctuation characters:

```python
import csv
sentences=[]
labels=[]
with open('/tmp/binary-emotion.csv', encoding='UTF-8') as csvfile:
    reader = csv.reader(csvfile, delimiter=",")
    for row in reader:
        labels.append(int(row[0]))
        sentence = row[1].lower()
        sentence = sentence.replace(",", " , ")
        sentence = sentence.replace(".", " . ")
        sentence = sentence.replace("-", " - ")
        sentence = sentence.replace("/", " / ")
        soup = BeautifulSoup(sentence)
        sentence = soup.get_text()
        words = sentence.split()
        filtered_sentence = ""
        for word in words:
            word = word.translate(table)
            if word not in stopwords:
                filtered_sentence = filtered_sentence + word + " "
        sentences.append(filtered_sentence)
```

This will give you a list of 35,327 sentences.

Note that this code is specific to this data. It's intended to help you understand the types of tasks you may have to take on in order to make stuff work, and it's not intended to be an exhaustive list of things that you'll have to do for every task—so your mileage may vary.

Creating training and test subsets

Now that the text corpus has been read into a list of sentences, you'll need to split it into training and test subsets for training a model. For example, if you want to use 28,000 sentences for training with the rest held back for testing, you can use code like this:

```python
training_size = 28000

training_sentences = sentences[0:training_size]
testing_sentences = sentences[training_size:]
```

```
training_labels = labels[0:training_size]
testing_labels = labels[training_size:]
```

Now that you have a training set, you can edit the tokenizer and vocabulary builder to create the word index from this corpus. As the corpus is an in-memory array of strings (`training_sentences`), the process is a lot simpler:

```python
from collections import Counter

# Assuming the tokenize function is defined elsewhere
def tokenize(text):
    # Tokenization logic, removing HTML and stopwords as discussed earlier
    soup = BeautifulSoup(text, "html.parser")
    cleaned_text = soup.get_text()
    tokens = cleaned_text.lower().split()
    filtered_tokens = [token for token in tokens if token not in stopwords]
    return filtered_tokens

def build_vocab(sentences):
    counter = Counter()
    for text in sentences:
        counter.update(tokenize(text))

    # Sort words by frequency in descending order
    sorted_words = sorted(counter.items(), key=lambda x: x[1], reverse=True)

    # Create vocabulary with indices starting from 1
    vocab = {word: idx + 1 for idx, (word, _) in enumerate(sorted_words)}
    vocab['<pad>'] = 0   # Add padding token with index 0
    return vocab


vocab = build_vocab(training_sentences)
print(vocab)
```

You can use the same helper functions to turn the text into a sequence and then padding, like this:

```python
print(testing_sentences[1])
seq = text_to_sequence(testing_sentences[1], vocab)
print(seq)
```

The results will be as follows:

```
made many new friends twitter around usa another bike across usa trip amazing
see people
[146, 259, 30, 110, 53, 198, 2161, 111, 752, 970, 2161, 407, 217, 26, 73]
```

Another common format for structured data, particularly in response to web calls, is JavaScript Object Notation (JSON). We'll explore how to read JSON data next.

Getting Text from JSON Files

JSON is an open standard file format that's used often for data interchange, particularly with web applications. It's human readable and designed to use name/value pairs, and as such, it's particularly well suited for labeled text. A quick search of Kaggle datasets for JSON yields over 2,500 results. For example, popular datasets such as the Stanford Question Answering Dataset (SQuAD) are stored in JSON.

JSON has very simple syntax in which objects are contained within braces as name/value pairs, each of which is separated by a comma. For example, a JSON object representing my name would be as follows:

```
{"firstName" : "Laurence",
 "lastName" : "Moroney"}
```

JSON also supports arrays, which are a lot like Python lists and are denoted by the square bracket syntax. Here's an example:

```
[
{"firstName" : "Laurence",
 "lastName" : "Moroney"},
{"firstName" : "Sharon",
 "lastName" : "Agathon"}
]
```

Objects can also contain arrays, so this is perfectly valid JSON:

```
[
{"firstName" : "Laurence",
 "lastName" : "Moroney",
 "emails": ["lmoroney@gmail.com", "lmoroney@galactica.net"]
},
{"firstName" : "Sharon",
 "lastName" : "Agathon",
 "emails": ["sharon@galactica.net", "boomer@cylon.org"]
}
]
```

A smaller dataset that's stored in JSON and a lot of fun to work with is the "News Headlines Dataset for Sarcasm Detection" by Rishabh Misra (*https://oreil.ly/wZ3oD*), which is available on Kaggle (*https://oreil.ly/_AScB*). This dataset collects news headlines from two sources: *The Onion* for funny or sarcastic ones and the *HuffPost* for normal headlines.

The file structure in the sarcasm dataset is very simple:

```
{"is_sarcastic": 1 or 0,
 "headline": String containing headline,
 "article_link": String Containing link}
```

The dataset consists of about 26,000 items, one per line. To make it more readable in Python, I've created a version that encloses these items in an array so the dataset can be read as a single list, which is used in the source code for this chapter.

Reading JSON files

Python's json library makes reading JSON files simple. Given that JSON uses name/value pairs, you can index the content based on the name. So, for example, for the sarcasm dataset, you can create a file handle to the JSON file, open it with the `json` library, and have an iterable go through, read each field line by line, and get the data item by using the name of the field.

Here's the code:

```python
import json
with open("/tmp/sarcasm.json", 'r') as f:
    datastore = json.load(f)
    for item in datastore:
        sentence = item['headline'].lower()
        label= item['is_sarcastic']
        link = item['article_link']
```

This makes it simple for you to create lists of sentences and labels, as you've done throughout this chapter, and then tokenize the sentences. You can also do preprocessing on the fly as you read a sentence, removing stopwords, HTML tags, punctuation, and more.

Here's the complete code to create lists of sentences, labels, and URLs while having the sentences cleaned of unwanted words and characters:

```python
with open("/tmp/sarcasm.json", 'r') as f:
    datastore = json.load(f)

sentences = []
labels = []
urls = []
for item in datastore:
    sentence = item['headline'].lower()
    sentence = sentence.replace(",", " , ")
    sentence = sentence.replace(".", " . ")
    sentence = sentence.replace("-", " - ")
    sentence = sentence.replace("/", " / ")
    soup = BeautifulSoup(sentence)
    sentence = soup.get_text()
    words = sentence.split()
    filtered_sentence = ""
    for word in words:
        word = word.translate(table)
        if word not in stopwords:
            filtered_sentence = filtered_sentence + word + " "
    sentences.append(filtered_sentence)
```

```
        labels.append(item['is_sarcastic'])
        urls.append(item['article_link'])
```

As before, you can split these into training and test sets. If you want to use 23,000 of the 26,000 items in the dataset for training, you can do the following:

```
training_size = 23000

training_sentences = sentences[0:training_size]
testing_sentences = sentences[training_size:]
training_labels = labels[0:training_size]
testing_labels = labels[training_size:]
```

Now that you have them as `in_memory` string arrays, tokenizing them and sequencing them will work exactly the same way as tokenizing and sequencing the sarcasm dataset.

Hopefully, the similar-looking code will help you see the pattern that you can follow when preparing text for neural networks to classify or generate. In the next chapter, you'll learn how to build a classifier for text using embeddings, and in Chapter 7 you'll take that a step further by exploring recurrent neural networks. Then, in Chapter 8, you'll learn how to further enhance the sequence data to create a neural network that can generate new text!

> Regular expressions (aka Regex) are terrific tools for sorting, filtering, and cleaning text. They have a syntax that's often hard to understand and difficult to learn, but I have found that generative AI tools like Gemini, Claude, and ChatGPT are really useful here.

Summary

In earlier chapters, you used images to build a classifier. Images, by definition, have well-defined dimensions—you know their width, height, and format. Text, on the other hand, can be far more difficult to work with. It is often unstructured, can contain undesirable content such as formatting instructions, doesn't always contain what you want, and often has to be filtered to remove nonsensical or irrelevant content.

In this chapter, you saw how to take text and convert it to numbers using word tokenization, and you then explored how to read and filter text in a variety of formats. With these skills in hand, you're now ready to take the next step and learn how *meaning* can be inferred from words—which is the first step in understanding natural language.

## CHAPTER 6

Making Sentiment Programmable by Using Embeddings

In Chapter 5, you saw how to take words and encode them into tokens. Then, you saw how to encode sentences full of words into sequences full of tokens, padding or truncating them as appropriate to end up with a well-shaped set of data that you can use to train a neural network. However, in none of that was there any type of modeling of the *meaning* of a word. And while it's true that there's no absolute numeric encoding that could encapsulate meaning, there are relative ones.

In this chapter, you'll learn about techniques to encapsulate meaning, and in particular, the concept of *embeddings*, in which vectors in high-dimensional space are created to represent words. The directions of these vectors can be learned over time, based on the use of the words in the corpus. Then, when you're given a sentence, you can investigate the directions of the word vectors, sum them up, and from the overall direction of the summation, establish the sentiment of the sentence as a product of its words. Also, related to this, as the model scans the sentences, the positioning of the words in the sentence can also help train an appropriate embedding.

In this chapter, we'll also explore how that works. Using the News Headlines Dataset for Sarcasm Detection dataset from Chapter 5, you'll build embeddings to help a model detect sarcasm in a sentence. You'll also work with some cool visualization tools that help you understand how words in a corpus get mapped to vectors so you can see which words determine the overall classification.

Establishing Meaning from Words

Before we get into the higher-dimensional vectors for embeddings, let's use some simple examples to try to visualize how meaning can be derived from numerics.

Consider this: using the sarcasm dataset from Chapter 5, what would happen if you encoded all of the words that make up sarcastic headlines with positive numbers and those that make up realistic headlines with negative numbers?

A Simple Example: Positives and Negatives

Take, for example, this sarcastic headline from the dataset:

```
christian bale given neutered male statuette named oscar
```

Assuming that all words in our vocabulary start with a value of 0, we could add 1 to the value for each of the words in this sentence, and we would end up with this:

```
{ "christian" : 1, "bale" : 1, "given" : 1, "neutered": 1, "male" : 1,
  "statuette": 1, "named" : 1, "oscar": 1}
```

> This isn't the same as the *tokenization* of words that you did in the last chapter. You could consider replacing each word (e.g., *christian*) with the token representing it that is encoded from the corpus, but I'll leave the words in for now to make the code easier to read.

Then, in the next step, consider an ordinary headline (not a sarcastic one), like this:

```
gareth bale scores wonder goal against germany
```

Because this is a different sentiment, we could instead subtract 1 from the current value of each word, so our value set would look like this:

```
{ "christian" : 1, "bale" : 0, "given" : 1, "neutered": 1, "male" : 1,
  "statuette": 1, "named" : 1, "oscar": 1, "gareth" : -1, "scores": -1,
  "wonder" : -1, "goal" : -1, "against" : -1, "germany" : -1}
```

Note that the sarcastic `bale` (from `christian bale`) has been offset by the nonsarcastic `bale` (from `gareth bale`), so its score ends up as 0. Repeat this process thousands of times and you'll end up with a huge list of words from your corpus that are scored based on their usage.

Now, imagine we want to establish the sentiment of this sentence:

```
neutered male named against germany, wins statuette!
```

Using our existing value set, we could look at the scores of each word and add them up. We would get a score of 2, indicating (because it's a positive number) that this is a sarcastic sentence.

For what it's worth, the word *bale* is used five times in the Sarcasm dataset, twice in a normal headline and three times in a sarcastic one. So, in a model like this, the word *bale* would be scored –1 across the whole dataset.

Going a Little Deeper: Vectors

Hopefully, the previous example has helped you understand the mental model of establishing some form of *relative* meaning for a word, through its association with other words in the same "direction." In our case, while the computer doesn't understand the meanings of individual words, it can move labeled words from a known sarcastic headline in one direction (by adding 1) and move labeled words from a known normal headline in another direction (by subtracting 1). This gives us a basic understanding of the meaning of the words, but it does lose some nuance.

But what if we increased the dimensionality of the direction to try to capture some more information? For example, suppose we were to look at characters from the Jane Austen novel *Pride and Prejudice*, considering the dimensions of gender and nobility. We could plot the former on the *x*-axis and the latter on the *y*-axis, with the length of the vector denoting each character's wealth (see Figure 6-1).

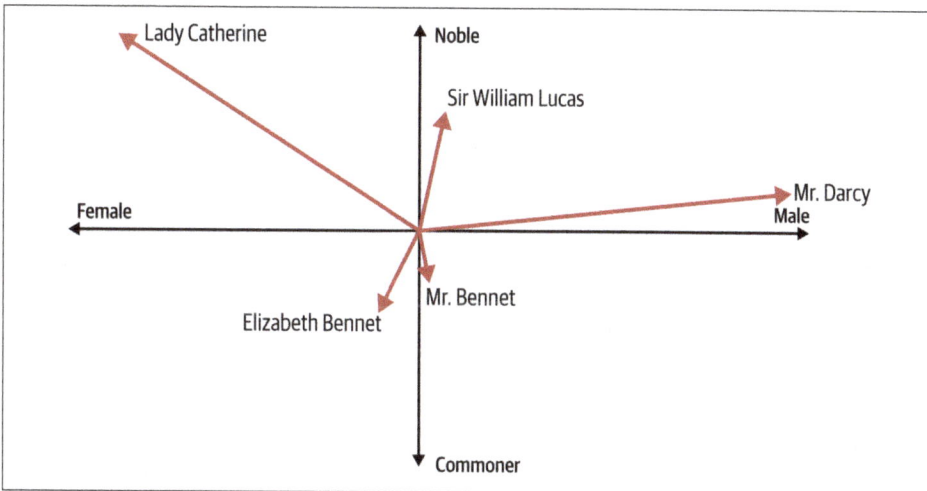

Figure 6-1. *Characters in* Pride and Prejudice *as vectors*

From an inspection of the graph, you can derive a fair amount of information about each character. Three of them are male. Mr. Darcy is extremely wealthy, but his nobility isn't clear (he's called "Mister," unlike the less wealthy but apparently more noble Sir William Lucas). The other "Mister," Mr. Bennet, is clearly not nobility and is struggling financially. Elizabeth Bennet, his daughter, is similar to him but female. Lady Catherine, the other female character in our example, is noble and incredibly wealthy. The romance between Mr. Darcy and Elizabeth causes tension—with *preju-dice* coming from the noble side of the vectors toward the less-noble.

As this example shows, by considering multiple dimensions, we can begin to see real meaning in the words (which are character names here). Again, we're not talking about concrete definitions but more about a *relative* meaning based on the axes and the relationship between the vector for one word and the other vectors.

This leads us to the concept of an *embedding*, which is simply a vector representation of a word that is learned while training a neural network. We'll explore that next.

Embeddings in PyTorch

Much like you've seen with `Linear` and `Conv2D`, PyTorch implements embeddings by using a layer. This creates a lookup table that maps from an integer to an embedding table, the contents of which are the coefficients of the vector representing the word identified by that integer. So, in the *Pride and Prejudice* example from the previous section, the x and y coordinates would give us the embeddings for a particular character from the book. Of course, in a real NLP problem, we'll use far more than two dimensions. Thus, the direction of a vector in the vector space could be seen as encoding the "meaning" of a word, and words with similar vectors (i.e., pointing in roughly the same direction) could be considered related to that word.

The embedding layer will be initialized randomly—that is, the coordinates of the vectors will be completely random to start with and will be learned during training by using backpropagation. When training is complete, the embeddings will roughly encode similarities between words, allowing us to identify words that are somewhat similar based on the direction of the vectors for those words.

This is all quite abstract, so I think the best way to understand how to use embeddings is to roll up your sleeves and give them a try. Let's start with a sarcasm detector using the Sarcasm dataset from Chapter 5.

Building a Sarcasm Detector by Using Embeddings

In Chapter 5, you loaded and did some preprocessing on a JSON dataset called the News Headlines Dataset for Sarcasm Detection (the sarcasm dataset, for short). By the time you were done, you had lists of training and testing data and labels:

```
training_size = 28000
training_sentences = sentences[0:training_size]
testing_sentences = sentences[training_size:]
training_labels = labels[0:training_size]
testing_labels = labels[training_size:]
```

For the training data, you created a `build_vocab` helper function to create a dictionary of the frequency of each word, sorted in order of the most frequent. The size of this dictionary is the `vocab_size`.

To get an embedding layer in PyTorch, you can use the `nn.Embedding` layer type, like this, by specifying the desired vocab size and the number of embedding dimensions:

```
nn.Embedding(vocab_size, embedding_dim)
```

This will initialize a vector with `embedding_dim` axes for each word. So, for example, if `embedding_dim` is 16, then every word in the vocabulary will be assigned a 16-dimensional vector.

Over time, the attributes for each token (encoded as values for the vector in each of its dimensions) will be learned through backpropagation as the network learns by matching the training data to its labels.

An important next step is feeding the output of the embedding layer into a dense layer. The easiest way to do this, similar to how you would when using a convolutional neural network, is to use pooling. In this instance, the dimensions of the embeddings are averaged out to produce a fixed-length output vector, and `Adaptive AvePool1d(1)` reduces the input along the length of the sequence to a fixed vector size of 1.

As an example, consider this model architecture:

```
self.embedding = nn.Embedding(vocab_size, embedding_dim)
self.global_pool = nn.AdaptiveAvgPool1d(1)
self.fc1 = nn.Linear(embedding_dim, 24)
self.fc2 = nn.Linear(24, 1)
self.relu = nn.ReLU()
self.sigmoid = nn.Sigmoid()
```

Here, an embedding layer is defined, and it's given the vocab size and an embedding dimension. Let's take a look at the number of trainable parameters in the network, using `torchinfo.summary`:

```
=================================================================
Layer (type:depth-idx)              Output Shape            Param #
=================================================================
TextClassificationModel             [32, 1]                 --
├─Embedding: 1-1                    [32, 100, 100]          2,429,200
├─AdaptiveAvgPool1d: 1-2            [32, 100, 1]            --
├─Linear: 1-3                       [32, 24]                2,424
├─ReLU: 1-4                         [32, 24]                --
├─Linear: 1-5                       [32, 1]                 25
├─Sigmoid: 1-6                      [32, 1]                 --
=================================================================
Total params: 2,431,649
Trainable params: 2,431,649
Non-trainable params: 0
```

```
Total mult-adds (M): 77.81
=================================================================
Input size (MB): 0.03
Forward/backward pass size (MB): 2.57
Params size (MB): 9.73
Estimated Total Size (MB): 12.32
=================================================================
```

The vocabulary size is 24,292 words, and as the embedding has 100 dimensions, the total number of trainable parameters in the embedding layer will be 2,429,200. The first linear layer has 100 values in with 24 values out, so that's a total of 2,400 weights, but each of the neurons also has a bias, so add 24 to get to 2,424.

Similarly, the last linear has 24 values in, with just a single neuron out. For a total of 24 parameters, plus one for the bias, this equals 25. The entire network has 2,431,649 parameters to learn. Note that the average pooling layer has 0 trainable parameters, as it's just averaging the parameters in the embedding layer before it to get a single 16-value vector.

If we train this model, we'll get a pretty decent training accuracy of 99%+ after 30 epochs—but our validation accuracy will be below 80% (see Figure 6-2).

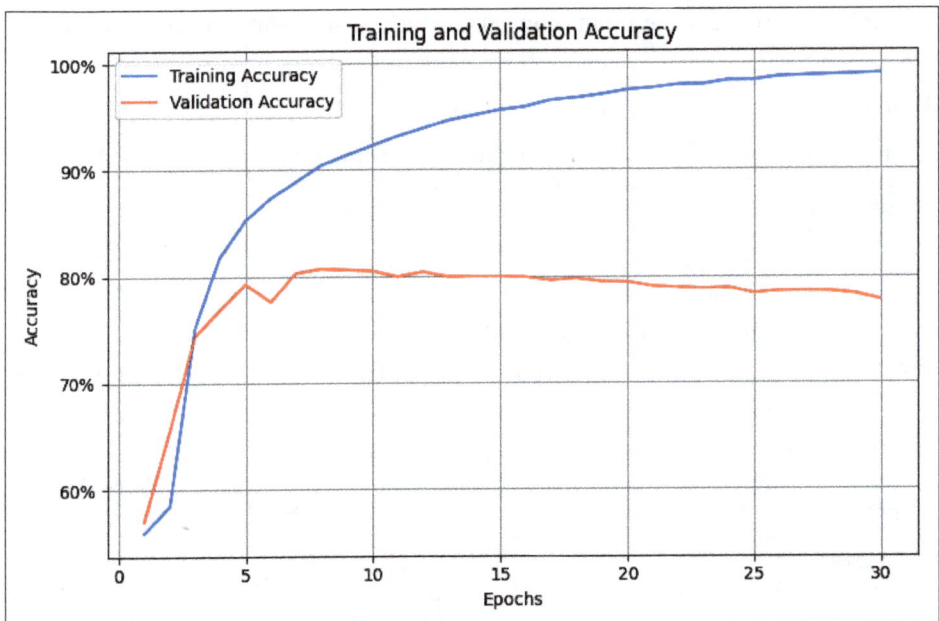

Figure 6-2. Training accuracy versus validation accuracy

That might seem to be a reasonable curve, given that the validation data likely contains many words that aren't present in the training data. However, if you examine the loss curves for training versus validation over one hundred epochs, you'll see a problem. Although you would expect to see that the training accuracy is higher than the validation accuracy, a clear indicator of overfitting is that while the validation accuracy is dropping a little over time (as shown in Figure 6-2), its loss is increasing sharply, as shown in Figure 6-3.

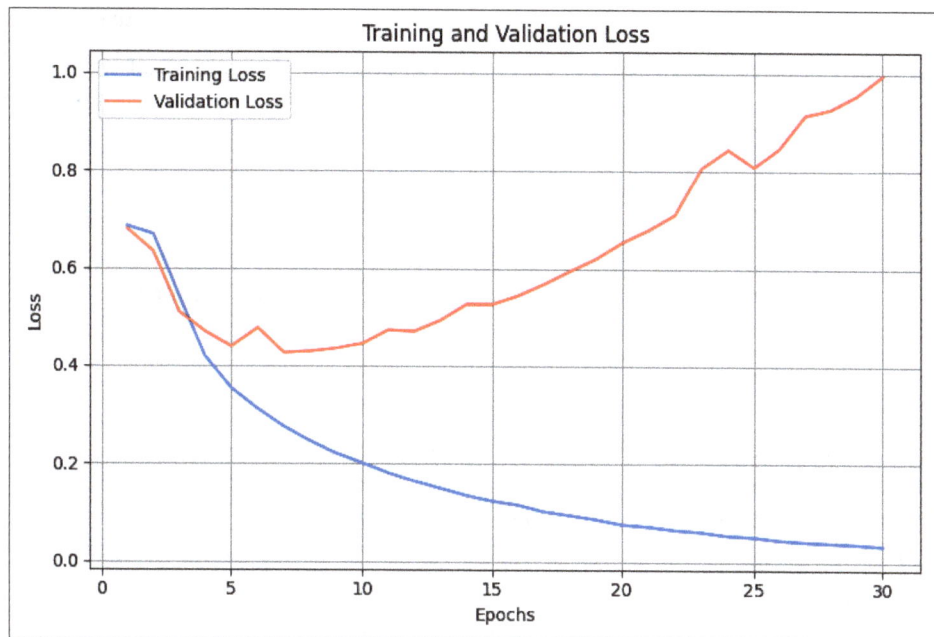

Figure 6-3. *Training loss versus validation loss*

Overfitting like this is common with NLP models, due to the somewhat unpredictable nature of language. In the next sections, we'll look at how to reduce this effect by using a number of techniques.

Reducing Overfitting in Language Models

Overfitting happens when the network becomes overspecialized to the training data, and one part of this involves the network becoming very good at matching patterns in "noisy" data in the training set that doesn't exist anywhere else. Because this particular noise isn't present in the validation set, the better the network gets at matching it, the worse the loss of the validation set will be. This can result in the escalating loss that you saw in Figure 6-3.

In this section, we'll explore several ways to generalize the model and reduce overfitting.

Adjusting the learning rate

A hyperparameter of the optimizer is the learning rate (LR). The details of this parameter are beyond the scope of this chapter, but consider it to be a value that if too high will cause the network to potentially learn too quickly and miss nuance. The flipside is also true—if you set it too low, your network may not learn effectively.

Perhaps the biggest factor that can lead to overfitting is whether the LR of your optimizer is too high. If it is, then your network learns *too quickly*. For this example, the code to define the optimizer was as follows:

```
optimizer = optim.Adam(model.parameters(), lr=0.001,
                    betas=(0.9, 0.999), amsgrad=False)
```

These are the defaults for the Adam optimizer. One thing to experiment with is the learning rate parameter (lr), and in the following code, you'll see the results of an instance when I reduced by an order of 10 to 0.0001, like this:

```
optimizer = optim.Adam(model.parameters(), lr=0.0001,
                    betas=(0.9, 0.999), amsgrad=False)
```

The betas values stay at their defaults, as does amsgrad. Also note that both beta values must be between 0 and 1, and typically, both are close to 1. Amsgrad is an alternative implementation of the Adam optimizer that was introduced in the paper "On the Convergence of Adam and Beyond" by Sashank Reddi, Satyen Kale, and Sanjiv Kumar (*https://oreil.ly/FhTDi*).

This much lower LR has a profound impact on the network. Figure 6-4 shows the accuracy of the network over one hundred epochs. The lower LR can be seen in the first 10 epochs or so, where it appears that the network isn't learning, before it "breaks out" and starts to learn quickly.

Exploring the loss (as illustrated in Figure 6-5), we can see that even while the accuracy wasn't going up for the first few epochs, the loss was going down. You could therefore be confident that the network would eventually start to learn, if you were watching it epoch by epoch.

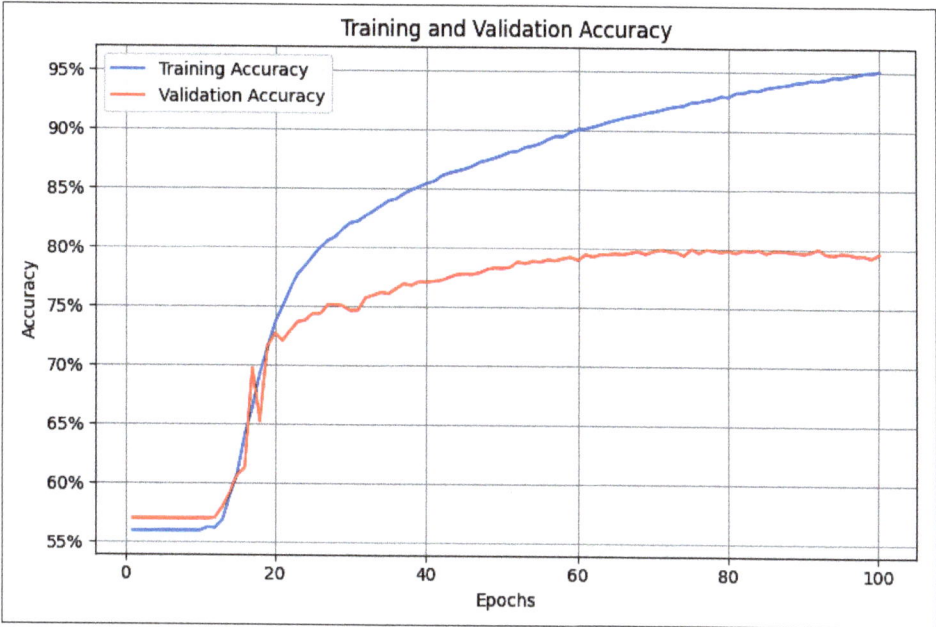

Figure 6-4. Accuracy with a lower LR

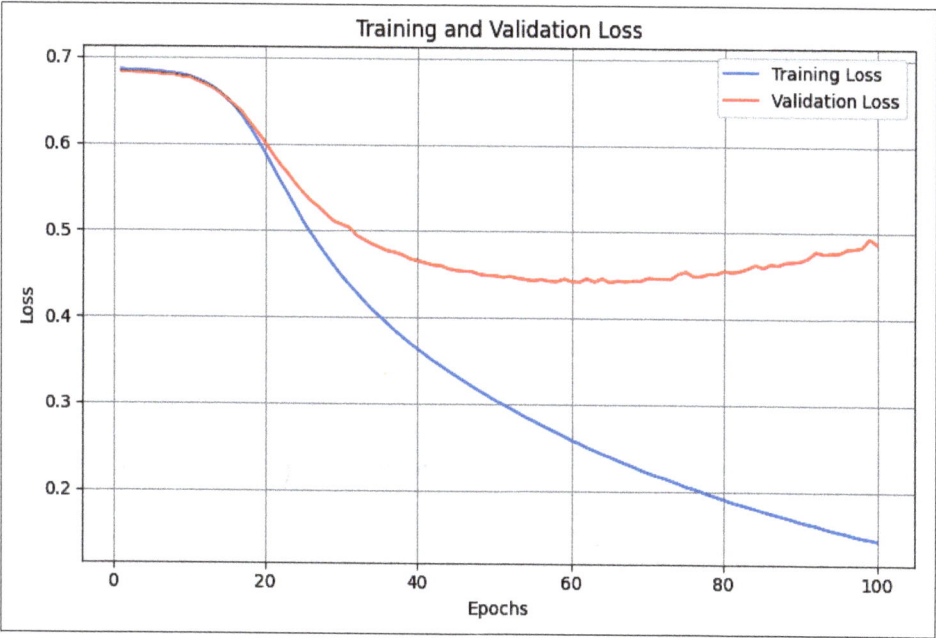

Figure 6-5. Loss with a lower LR

And while the loss does start to show the same curve of overfitting that you saw in Figure 6-3, note that it happens much later and at a much lower rate. By epoch 30, the loss is at about 0.49, whereas with the higher LR in Figure 6-3, it was more than double that amount. And while it takes the network longer to get to a good accuracy rate, it does so with less loss, so you can be more confident in the results. With these hyperparameters, the loss on the validation set started to increase at about epoch 60, at which point, the training set had 90% accuracy and the validation set had about 81% accuracy, showing that we have quite an effective network.

Of course, it's easy to just tweak the optimizer and then declare victory, but there are a number of other methods you can use to improve your model. You'll learn about those in the next few sections, and for them, I've reverted back to using the default Adam optimizer so the effects of tweaking the LR won't hide the benefits offered by these other techniques.

Exploring vocabulary size

The sarcasm dataset deals with words, so if you explore the words in the dataset and in particular their frequency, you might get a clue that helps fix the overfitting issue.

I've provided a `word_frequency` helper function that lets you explore the frequency of words in the vocabulary. It looks like this:

```python
def word_frequency(sentences, word_dict):
    frequency = {word: 0 for word in word_dict}

    for sentence in sentences:
        words = sentence.lower().split()
        for word in words:
            if word in frequency:
                frequency[word] += 1

    return frequency
```

You can run it with code like this:

```python
word_freq = word_frequency(training_sentences, word_index)
print(word_freq)
```

You'll then see results like this: a dictionary containing the frequency of each word, starting with the most frequently used one, and moving on from there. Here are the first few words:

```python
{'new': 1318, 'trump': 1117, 'man': 1075, 'not': 634, 'just': 501,
 'will': 484, 'one': 469, 'year': 440, …
```

If you want to plot this, you can iterate through each item in the list and make the *x* value the ordinal of where you are (1 for the first item, 2 for the second item, etc.). The *y* value will then be a `newlist[item]`, which you can plot with `matplotlib`. Here's the code:

```
import matplotlib.pyplot as plt
from collections import OrderedDict
newlist = (OrderedDict(sorted(word_freq.items(), key=lambda t: t[1],
                       reverse=True)))

xs=[]
ys=[]
curr_x = 1
for item in newlist:
  xs.append(curr_x)
  curr_x=curr_x+1
  ys.append(newlist[item])

print(ys)
plt.plot(xs,ys)
```

The result is shown in Figure 6-6.

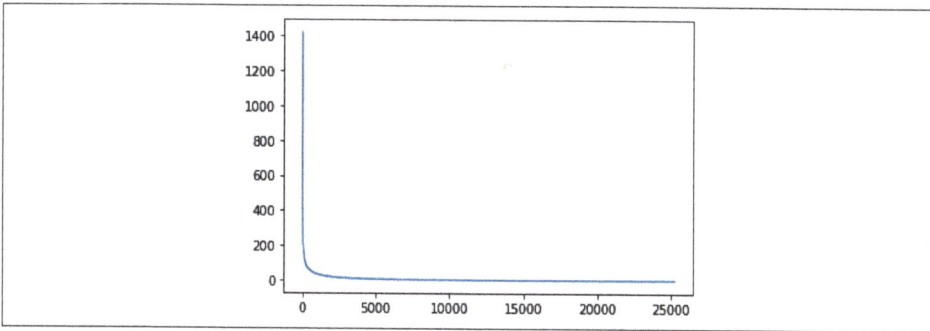

Figure 6-6. Exploring the frequency of words

This "hockey stick" curve shows us that very few words are used many times, whereas most words are used very few times. But every word is effectively weighted equally because every word has an "entry" in the embedding. Given that we have a relatively large training set in comparison with the validation set, we're ending up in a situation where there are many words present in the training set that aren't present in the validation set.

You can zoom in on the data by changing the axis of the plot just before calling plt.show. For example, to look at the volume of words from 300 to 10,000 on the *x*-axis with the scale from 0 to 100 on the *y*-axis, you can use this code:

```
plt.plot(xs,ys)
plt.axis([300,10000,0,100])
plt.show()
```

The result is in Figure 6-7.

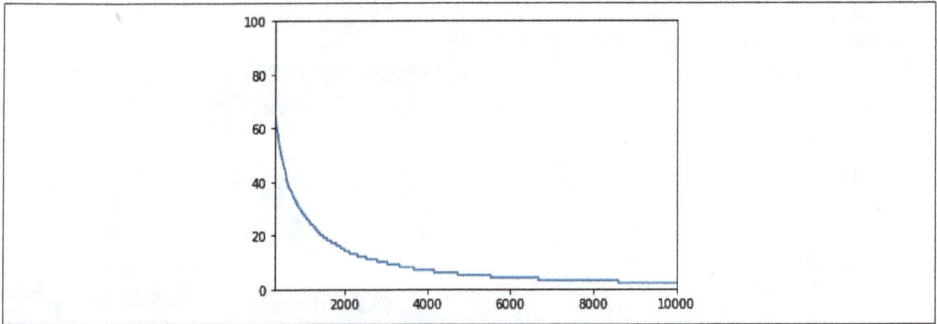

Figure 6-7. Frequency of words from 300 to 10,000

There are almost 25,000 words in the corpus, and the code is set up to only train for all of them! But if we look at the words in positions 2,000 onward, which is over 90% of our vocabulary, we'll see that they're each used fewer than 20 times in the entire corpus!

This could explain the overfitting, so the logical next step is to see if we can reduce the vocabulary we are training for. Within the `build_vocab` helper function, we can add a parameter for the maximum vocab size we're interested in, like this:

```python
def build_vocab(sentences, max_vocab_size=10000):
    counter = Counter()
    for text in sentences:
        counter.update(tokenize(text))

    # Take only the top max_vocab_size-1 most frequent words
    # (leave room for special tokens)
    most_common = counter.most_common(max_vocab_size - 2)
    # -2 for <pad> and <unk>

    # Create vocabulary with indices starting from 2
    vocab = {word: idx + 2 for idx, (word, _) in enumerate(most_common)}
    vocab['<pad>'] = 0  # Add padding token
    vocab['<unk>'] = 1  # Add unknown token
    return vocab
```

Then, when building our `word_index`, we can specify a maximum vocab size that we're interested in exploring:

```
vocab_size = 2000
word_index = build_vocab(training_sentences, max_vocab_size=vocab_size)
```

The embedding layer was already initialized with the vocab size, so the model architecture doesn't need to change. Indeed, with the reduced vocab size, the number of learned parameters drops sharply, giving us a simpler network that learns faster:

```
==================================================================
Layer (type:depth-idx)              Output Shape           Param #
==================================================================
TextClassificationModel             [32, 1]                --
├─Embedding: 1-1                    [32, 100, 100]         200,100
├─AdaptiveAvgPool1d: 1-2            [32, 100, 1]           --
├─Linear: 1-3                       [32, 24]               2,424
├─ReLU: 1-4                         [32, 24]               --
├─Linear: 1-5                       [32, 1]                25
├─Sigmoid: 1-6                      [32, 1]                --
==================================================================
Total params: 202,549
Trainable params: 202,549
Non-trainable params: 0
Total mult-adds (M): 6.48
==================================================================
Input size (MB): 0.03
Forward/backward pass size (MB): 2.57
Params size (MB): 0.81
Estimated Total Size (MB): 3.40
==================================================================
```

The model has shrunk from 2.4 million parameters to only 202,549.

After retraining and exploring the smaller model, we can see that the results have changed.

Figure 6-8 shows the accuracy metrics. Now, the training set accuracy is about 82% and the validation accuracy is about 76%. They're closer to each other and not diverging, which is a good sign that we've gotten rid of most of the overfitting.

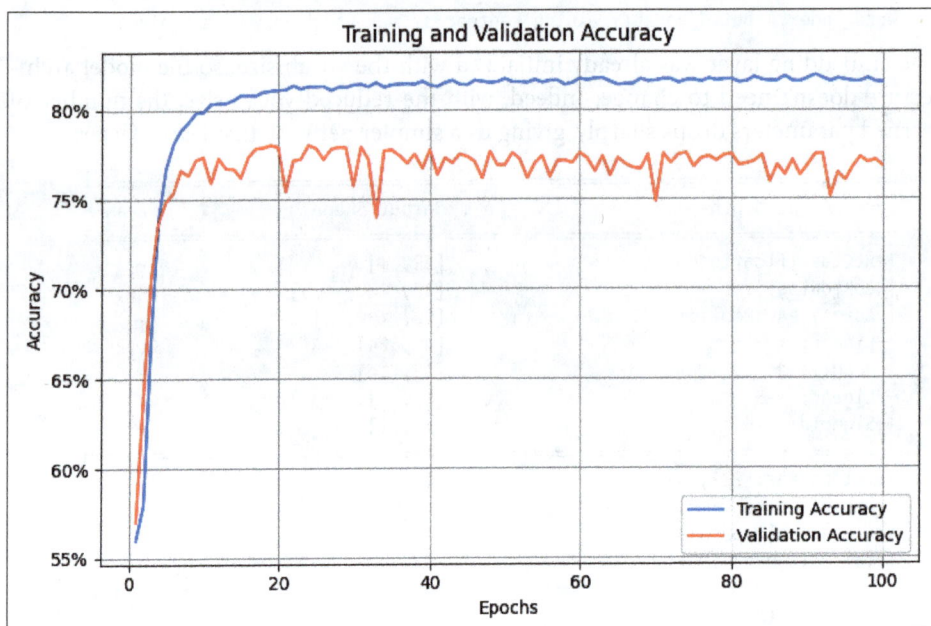

Figure 6-8. Accuracy with a two thousand–word vocabulary

This is somewhat reinforced by the loss plot in Figure 6-9. The loss on the validation set is rising but much slower than before, so reducing the size of the vocabulary to prevent the training set from overfitting on low-frequency words that were possibly only present in the training set appears to have worked.

It's worth experimenting with different vocab sizes, but remember that you can also have too small of a vocab size and overfit to that. You'll need to find a balance. In this case, my choice of taking words that appear 20 times or more was purely arbitrary.

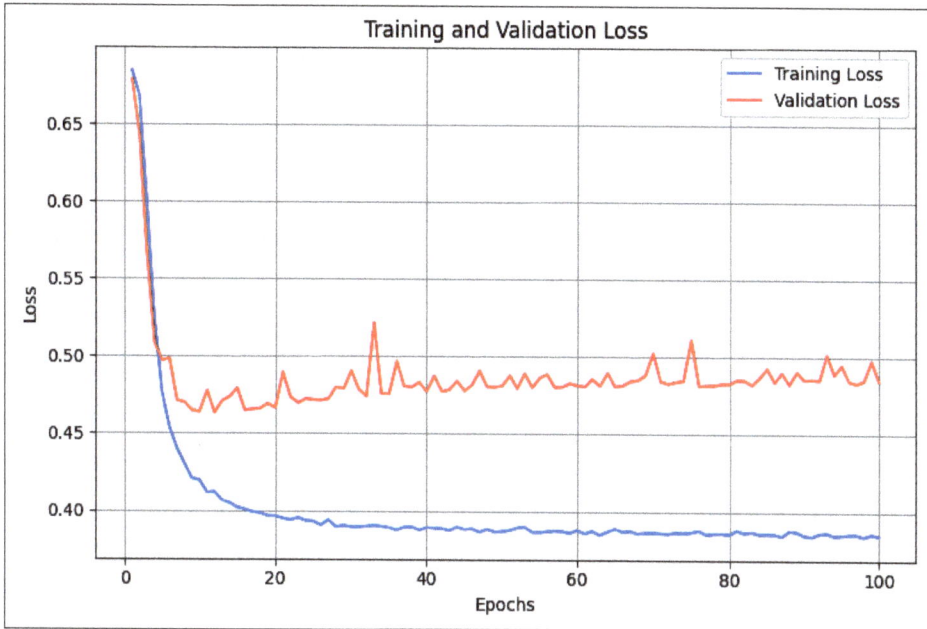

Figure 6-9. Loss with a two thousand–word vocabulary

Exploring embedding dimensions

For this example, I arbitrarily chose an embedding dimension of 16. In this instance, words are encoded as vectors in 16-dimensional space, with their directions indicating their overall meaning. But is 16 a good number? With only two thousand words in our vocabulary, it might be on the high side, leading to a high degree of sparseness of direction.

I believe that the best way to think about sparseness is to project into three dimensions. Think of it like the earth, with one thousand vectors pointing from the core to a place on the surface. The vectors are in three dimensions, *x*, *y*, and *z*. There's a lot of surface area for them to cover, but if many of them are missing *x* and *y*, meaning they're just zero, a lot of them will be pointing to (0, 0, *z*) and a whole lot of the earth's surface will be untouched! Thus, there will be a total lack of distinctiveness.

Research has shown that a best practice for embedding size is to have it be the fourth root of the vocabulary size. The fourth root of 2,000 is 6.687, so let's explore what happens if we round this up and change the embedding dimension to 7.

You can see the result of training for one hundred epochs in Figure 6-10. The training set's accuracy stabilized at about 83%, and the validation set's accuracy stabilized at about 77%. Despite some jitters, the lines are pretty flat, showing that the model has converged. This isn't much different from the results in Figure 6-8, but reducing the embedding dimensionality allows the model to train significantly faster.

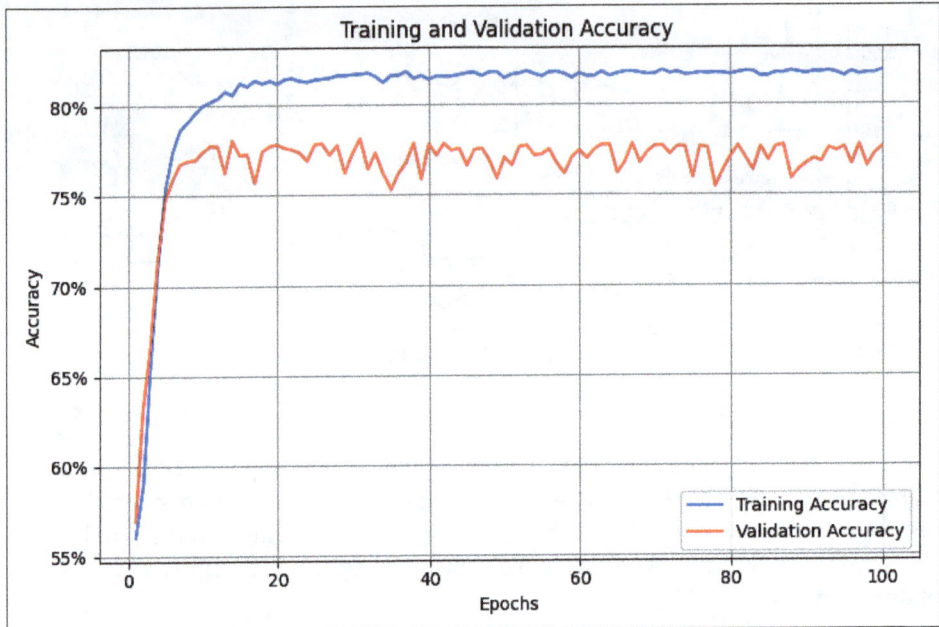

Figure 6-10. Training versus validation accuracy for seven dimensions

Figure 6-11 shows the loss in training and validation. While it initially appeared that the loss was climbing at about epoch 20, it soon flattened out. Again, a good sign!

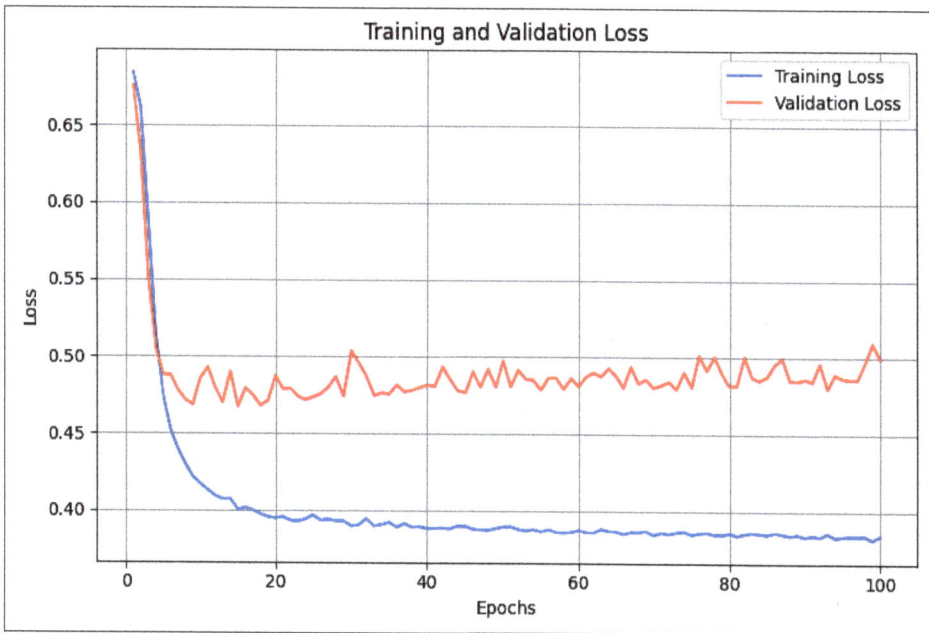

Figure 6-11. Training versus validation loss for seven dimensions

Now that the dimensionality has been reduced, we can do a bit more tweaking of the model architecture.

Exploring the model architecture

After the optimizations in the previous sections, the model architecture looks like this:

```
class TextClassificationModel(nn.Module):
    def __init__(self, vocab_size, embedding_dim, hidden_dim=24):
        super(TextClassificationModel, self).__init__()
        self.embedding = nn.Embedding(vocab_size, embedding_dim)
        self.global_pool = nn.AdaptiveAvgPool1d(1)
        self.fc1 = nn.Linear(embedding_dim, hidden_dim)
        self.fc2 = nn.Linear(hidden_dim, 1)
        self.relu = nn.ReLU()
        self.sigmoid = nn.Sigmoid()
```

One thing that comes to mind is the dimensionality—the `GlobalAveragePooling1D` layer now emits just 7 dimensions, but they're being fed into a hidden layer of 24 neurons, which is overkill. Let's explore what happens when this is reduced to 8 neurons and trained for 100 epochs.

You can see the training versus validation accuracy in Figure 6-12. When compared to Figure 6-7, where 24 neurons were used, the overall result is quite similar, but the model was somewhat faster to train.

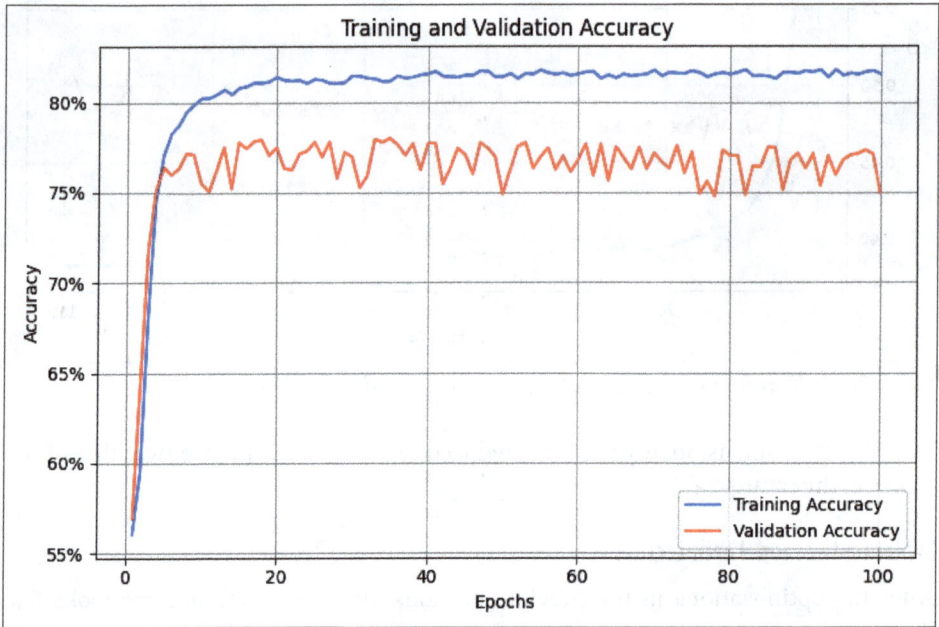

Figure 6-12. Reduced dense-architecture accuracy results

The loss curves in Figure 6-13 show similar results.

By following these exercises, we were able to reduce the model architecture significantly, reducing the number of parameters while improving the quality and mitigating overfitting. But there are a few more things we can do—starting with dropout.

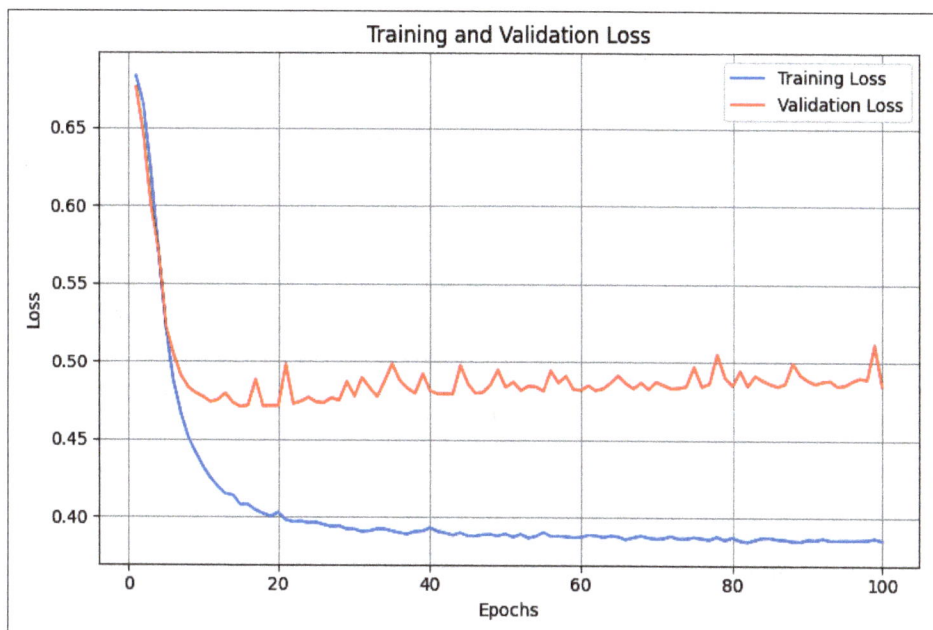

Figure 6-13. Reduced dense architecture loss results

Using dropout

A common technique for reducing overfitting is to add dropout to a dense neural network. We explored this for convolutional neural networks back in Chapter 3, so it's tempting to go straight to it here to see its effects on overfitting. But in this case, I want to wait until the vocabulary size, embedding size, and architecture complexity have been addressed. Those changes can often have a much larger impact than using dropout, and we've already seen some nice results.

Now that our architecture has been simplified to have only eight neurons in the middle dense layer, the effect of dropout may be minimized—but let's explore it anyway. Here's the updated code for the model architecture to add a dropout of 0.25 (which equates to two of our eight neurons):

```python
class TextClassificationModel(nn.Module):
    def __init__(self, vocab_size, embedding_dim, hidden_dim=8,
                 dropout_rate=0.25):
        super(TextClassificationModel, self).__init__()
        self.embedding = nn.Embedding(vocab_size, embedding_dim)
        self.global_pool = nn.AdaptiveAvgPool1d(1)
        self.fc1 = nn.Linear(embedding_dim, hidden_dim)
        self.dropout = nn.Dropout(p=dropout_rate)  # Add dropout layer
        self.fc2 = nn.Linear(hidden_dim, 1)
        self.relu = nn.ReLU()
        self.sigmoid = nn.Sigmoid()
```

```
def forward(self, x):
    x = self.embedding(x)
    x = x.transpose(1, 2)  # Change for pooling layer
    x = self.global_pool(x).squeeze(2)
    x = self.dropout(self.relu(self.fc1(x)))
    x = self.sigmoid(self.fc2(x))
    return x
```

Figure 6-14 shows the accuracy results when trained for one hundred epochs. This time, we see that the training accuracy and validation accuracy are converging, with the training accuracy now lower than before. Similarly, the loss curves in Figure 6-15 show convergence, so while dropout is making our network a little *less* accurate, it appears to generalize it better.

But do exercise caution before declaring victory! A close examination of the curves shows that the losses have nicely converged but that they *are* higher than previously. The training loss is above 0.5 with dropout but was around 0.3 without. It is also trending downward, so it's worth experimenting to see whether longer training will produce a better result.

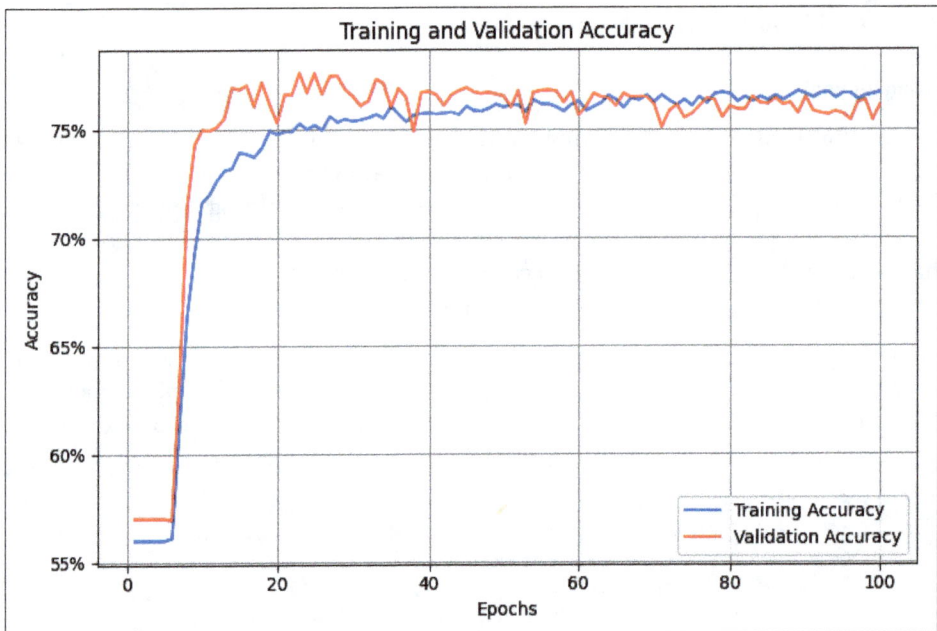

Figure 6-14. Accuracy with added dropout

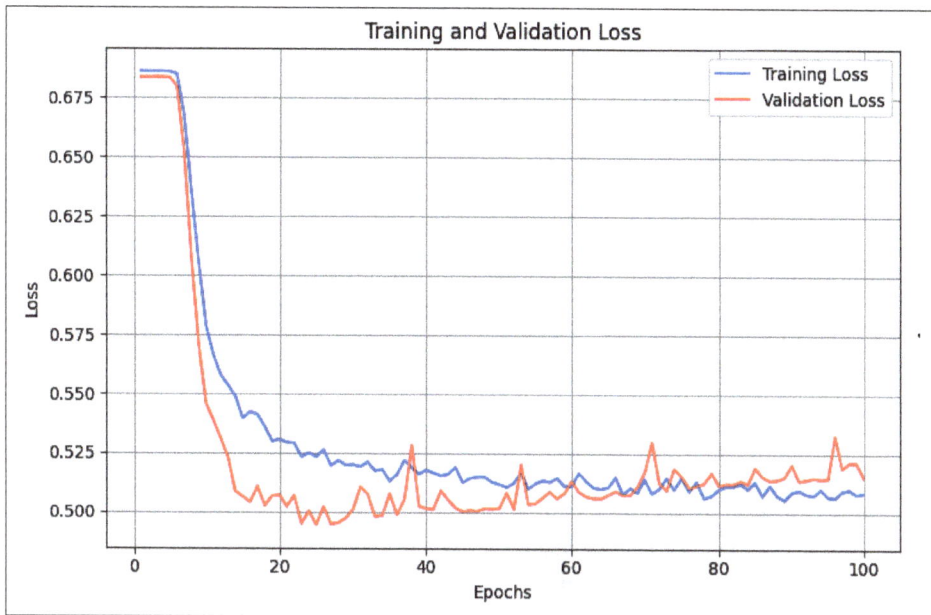

Figure 6-15. Loss with added dropout

You can also see that the model is heading back to its previous pattern of increasing validation loss over time. It's not nearly as bad as before, but it's heading in the wrong direction.

In this case, when there were very few neurons, introducing dropout probably wasn't the right idea. It's still good to have this tool in your arsenal, though, so be sure to keep it in mind for more sophisticated architectures than this one.

Using regularization

Regularization is a technique that helps prevent overfitting by reducing the polarization of weights. If the weights on some of the neurons are too heavy, regularization effectively punishes them. Broadly speaking, there are two types of regularization:

L1 regularization
This is often called *least absolute shrinkage* and *selection operator* (*lasso*) regularization. It effectively helps us ignore the zero or close-to-zero weights when calculating a result in a layer.

L2 regularization
This is often called *ridge* regression because it pushes values apart by taking their squares. This tends to amplify the differences between nonzero values and zero or close-to-zero ones, creating a ridge effect.

The two approaches can also be combined into what is sometimes called *elastic regularization*.

For NLP problems like the one we're considering, L2 is most commonly used. It can be added as the `weight_decay` attribute to the `optimizer`. Here's an example:

```
optimizer = optim.Adam(model.parameters(), lr=0.001, betas=(0.9, 0.999),
                       amsgrad=False, weight_decay=0.01)
```

This will apply the `weight_decay` of `0.01`. (Usually, you'll have a value between 0.01 and 0.001 here). Alternatively, a neat trick you can do with PyTorch is to define different weight decays for different layers by specifying them within the `Adam` declaration call, like this:

```
# Different weight decay for different layers
optimizer = torch.optim.Adam([
# L2 reg on fc1
        {'params': model.fc1.parameters(), 'weight_decay': 0.01},
    # No L2 reg on other layers
{'params': [p for name, p in model.named_parameters()
            if 'fc1' not in name]}
], lr=0.0001)
```

The impact of adding regularization in a simple model like this isn't particularly large, but it does smooth out our training loss and validation loss somewhat. It might be overkill for this scenario, but as with dropout, it's a good idea to understand how to use regularization to prevent your model from getting overspecialized.

Other optimization considerations

While the modifications we've made have given us a much-improved model with less overfitting, there are other hyperparameters that you can experiment with. For example, we chose to make the maximum sentence length one hundred words, but that was purely arbitrary and probably not optimal. It's a good idea to explore the corpus and see what a better sentence length might be. Here's a snippet of code that looks at the sentences and plots the lengths of each one, sorted from low to high:

```
xs=[]
ys=[]
current_item=1
for item in sentences:
  xs.append(current_item)
  current_item=current_item+1
  ys.append(len(item))
newys = sorted(ys)

import matplotlib.pyplot as plt
plt.plot(xs,newys)
plt.show()
```

See Figure 6-16 for the results of this.

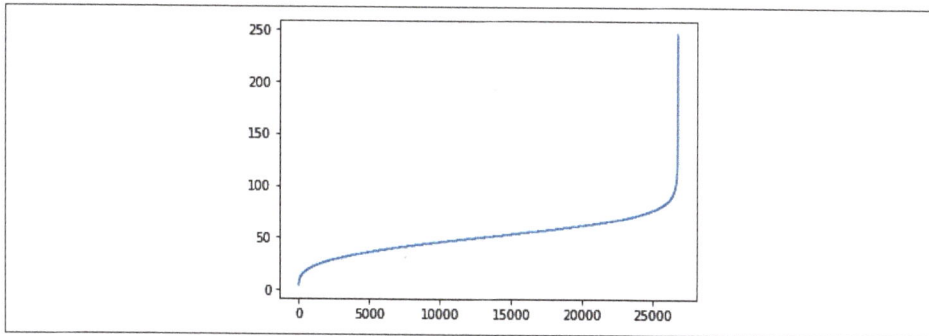

Figure 6-16. Exploring sentence length

Less than 200 sentences in the total corpus of 26,000+ have a length of 100 words or greater, so by choosing this as the maximum length, we're introducing a lot of padding that isn't necessary and thus affecting the model's performance. Reducing the maximum to 85 words would still keep 26,000 of the sentences (99%+) with greatly reduced padding.

Putting It All Together

Taking all of the preceding optimizations into effect and retraining the model for three hundred epochs gives you the results in Figure 6-17 for training and validation accuracies. Given that their curves are roughly matched, it shows that we've taken huge steps toward avoiding overfitting and that we have a network that's learning effectively.

Similarly, the training and validation loss curves over three hundred epochs are showing remarkable similarity, as depicted in Figure 6-18, which indicates that the optimizations are a step in the right direction to prevent overfitting for this model.

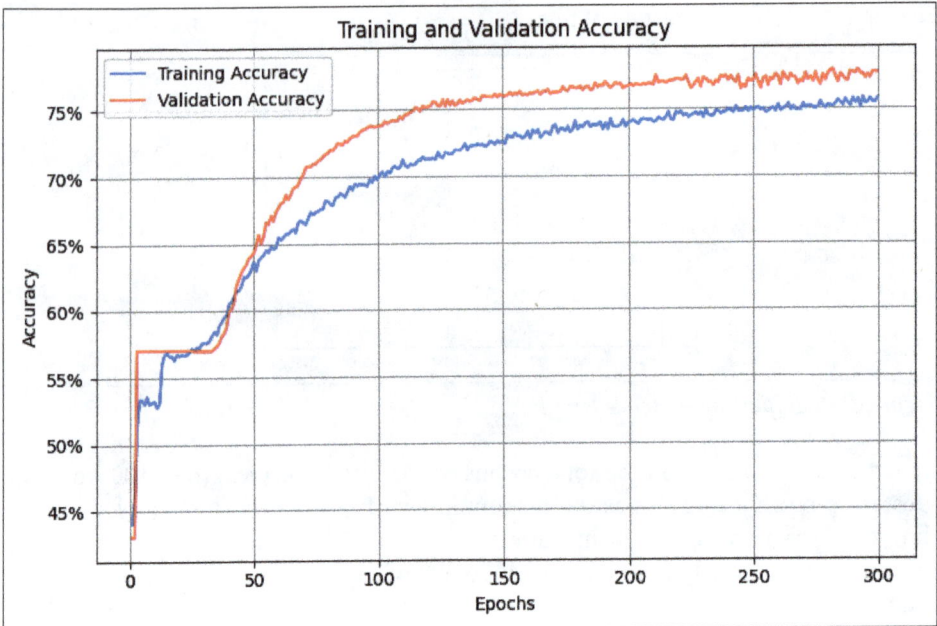

Figure 6-17. Optimized training and validation accuracy

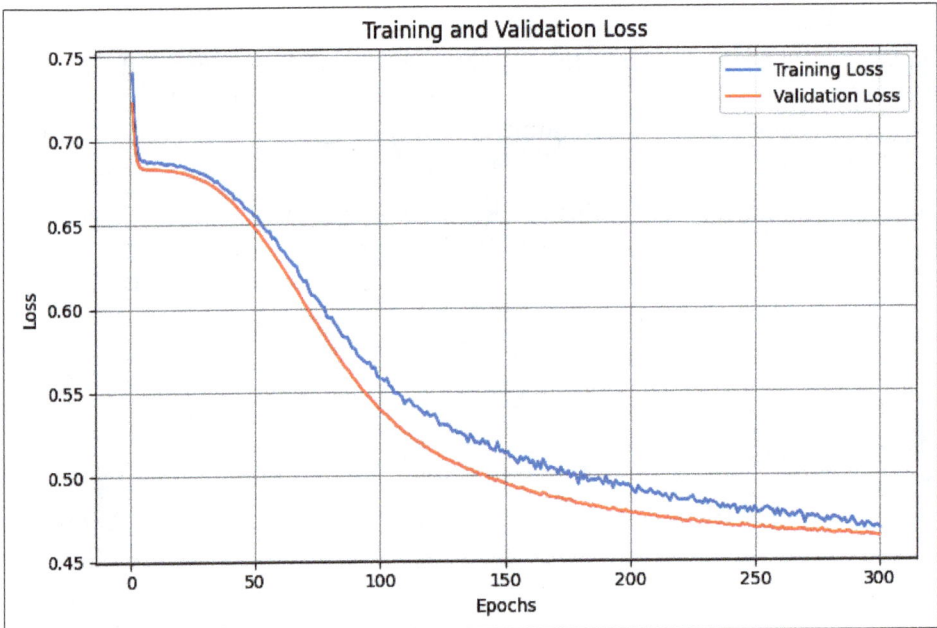

Figure 6-18. Optimized training and validation loss

Using the Model to Classify a Sentence

Now that you've created the model, trained it, and optimized it to remove a lot of the problems that caused the overfitting, the next step is to run the model and inspect its results. To do this, you'll create an array of new sentences. Consider, for example:

```
test_sentences = [
            "granny starting to fear spiders in the garden might be real",
            "game of thrones season finale showing this sunday night",
            "PyTorch book will be a best seller"]
```

You can then encode these by using the same tokenizer that you used when creating the vocabulary for training:

```
print(texts_to_sequences(test_sentences, word_index))
```

It's important to use this tokenizer because it has the tokens for the words that the network was trained on!

The output of the print statement will be the sequences for the preceding sentences:

```
[
[1, 803, 753, 1, 1, 312, 97],
[123, 1183, 160, 1, 1, 1543, 152],
[1, 235, 7, 47, 1]
]
```

There are a lot of 1 tokens here ("<OOV>"), because words like *granny* and *spiders* don't appear in the dictionary. The sequences are also shorter because the stopwords have been removed.

Next, before you can pass the sequences to the model, you'll need to put them in the shape that the model expects—that is, the desired length. You can do this with `pad_sequences` in the same way you did when training the model:

```
padded = pad_sequences(sequences, max_len)
```

This will output the sentences as sequences of length 85, so the output for the first sequence will be as follows:

```
[1, 803, 753, 1, 1, 312, 97, 0, 0, 0, 0, 0, 0, 0, 0, 0, 0, 0, 0, 0, 0, 0, 0,
 0, 0, 0, 0, 0, 0, 0, 0, 0, 0, 0, 0, 0, 0, 0, 0, 0, 0, 0, 0, 0, 0, 0, 0, 0, 0,
 0, 0, 0, 0, 0, 0, 0, 0, 0, 0, 0, 0, 0, 0, 0, 0, 0, 0, 0, 0, 0, 0, 0, 0, 0, 0,
 0, 0, 0, 0, 0, 0, 0]
```

It was a very short sentence, so it's padded up to 85 tokens with a lot of zeros!

Now that you've padded and tokenized the sentences to fit the model's expectations for the input dimensions, it's time to pass them to the model and get predictions back.

This involves multiple steps. First, convert the padded sequence into an input tensor:

```
# Convert to tensor
input_ids = torch.tensor(padded, dtype=torch.long).to(device)
```

Next, put the model into evaluation mode to get the predictions, and then simply pass the `input_ids` to it to get the outputs:

```
# Get predictions
model.eval()
with torch.no_grad():
    outputs = model(input_ids)
```

The results will be passed back as a list and printed, with high values indicating likely sarcasm. Here are the results for our sample sentences:

```
tensor([[0.5516],
        [0.0765],
        [0.0987]], device='cuda:0')
```

The high score for the first sentence ("granny starting to fear spiders in the garden might be real"), despite it having a lot of stopwords and being padded with a lot of zeros, indicates that there is a level of sarcasm there. The other two sentences scored much lower, indicating a lower likelihood of sarcasm in them.

To get the probabilities, you can call the `squeeze()` method to retrieve the tensor values. And if you want to make a comparison to a threshold to get your prediction—for example, above 0.5 indicates sarcasm and below 0.5 indicates no sarcasm—then you can use code like this:

```
probabilities = outputs.squeeze().cpu().numpy()
predictions = (probabilities >= threshold).astype(int)
```

Based on your network tuning, you could also establish what you think the appropriate threshold should be. Running this with a 0.5 threshold gives us the following:

```
Text: granny starting to fear spiders in the garden might be real
Probability: 0.5516
Classification: Sarcastic
Confidence: 0.5516
--------------------------------------------------------------------

Text: game of thrones season finale showing this sunday night
Probability: 0.0765
Classification: Not Sarcastic
Confidence: 0.9235
--------------------------------------------------------------------

Text: PyTorch book will be a best seller
Probability: 0.0987
Classification: Not Sarcastic
Confidence: 0.9013
--------------------------------------------------------------------
```

So, with these test sentences, we're beginning to get a good indication that our network is performing as desired. You should test it with other data to see if you can break it, and if you break it consistently, then it'll be time to try a different model architecture, use transfer learning from an existing working network, or explore using pretrained embeddings.

We'll learn about this in the next section, but before that, I'd like to show you how you can visualize the custom embeddings that this network learned.

Visualizing the Embeddings

To visualize embeddings, you can use an online tool called the Embedding Projector (*http://projector.tensorflow.org*). It comes preloaded with many existing datasets, but in this section, you'll see how to take the data from the model you've just trained and visualize it by using this tool.

But first, you'll need a function to reverse the word index. It currently has the word as the token and the key as the value, but you need to invert it so you'll have word values to plot on the projector. Here's the code to do this:

```
reverse_word_index = dict([(value, key)
    for (key, value) in word_index.items()])
```

You'll also need to extract the weights of the vectors in the embeddings:

```
embedding_weights = model.embedding.weight.data.cpu().numpy()
print(embedding_weights.shape)
```

If you've followed the optimizations in this chapter, the output of this will be (2000,7) because we used a 2,000 word vocabulary and 7 dimensions for the embedding. If you want to explore a word and its vector details, you can do so with code like this:

```
print(reverse_word_index[2])
print(embedding_weights[2])
```

This will produce the following output:

```
new
[-0.27116913 -1.3026129   1.6390767   0.4922502  -0.6025921   1.4584142
  0.05054485]
```

So, the word *new* is represented by a vector with those seven coefficients on its axes.

The Embedding Projector uses two tab-separated values (TSV) files, one for the vector dimensions and one for metadata. This code will generate them for you:

```python
import io
out_v = io.open('vecs.tsv', 'w', encoding='utf-8')
out_m = io.open('meta.tsv', 'w', encoding='utf-8')
for word_num in range(1, vocab_size):
  word = reverse_word_index[word_num]
  embeddings = embedding_weights[word_num]
  out_m.write(word + "\n")
  out_v.write('\t'.join([str(x) for x in embeddings]) + "\n")
out_v.close()
out_m.close()
```

Alternatively, if you are using Google Colab, you can download the TSV files with the following code or from the Files pane:

```python
try:
  from google.colab import files
except ImportError:
  pass
else:
  files.download('vecs.tsv')
  files.download('meta.tsv')
```

Once you have the files, you can press the Load button on the projector to visualize the embeddings (see Figure 6-19).

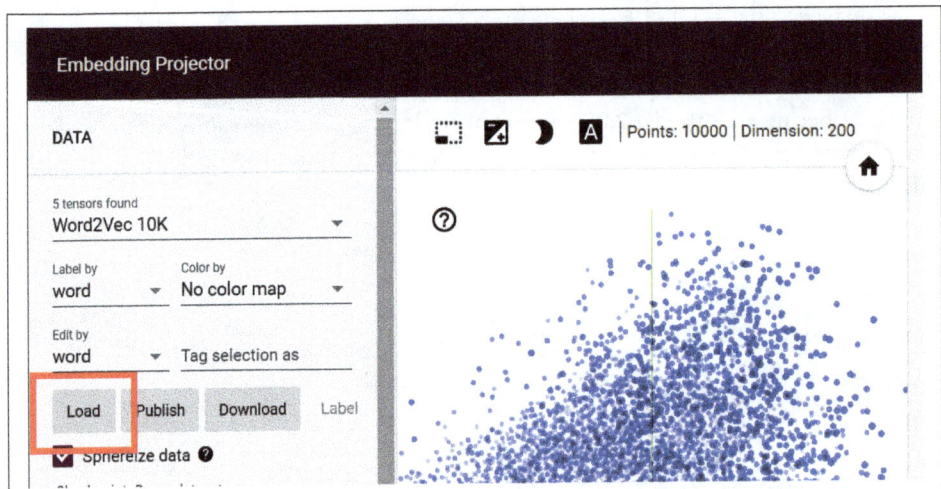

Figure 6-19. Using the Embeddings Projector

You can also use the vectors and meta TSV files where recommended in the resulting dialog and then click Sphereize Data on the projector. This will cause the words to be clustered in a sphere and will give you a clear visualization of the binary nature of this classifier. It's only been trained on sarcastic and nonsarcastic sentences, so words tend to cluster toward one label or another (see Figure 6-20).

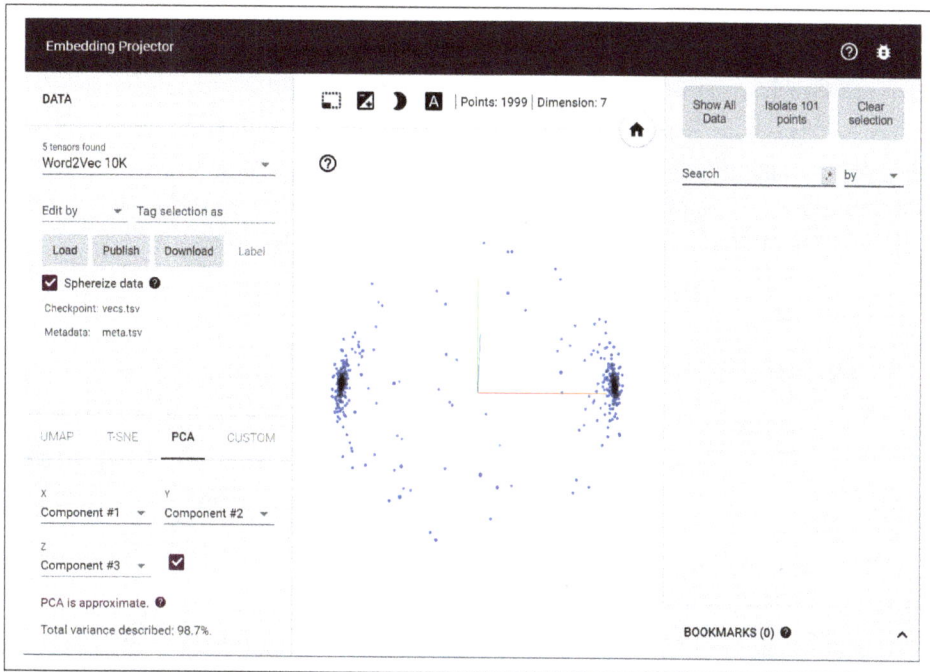

Figure 6-20. Visualizing the sarcasm embeddings

Screenshots don't do all of this justice—you should try it for yourself! You can rotate the center sphere and explore the words on each "pole" to see the impact they have on the overall classification, and you can also select words and show related ones in the righthand pane. Have a play and experiment!

Using Pretrained Embeddings

An alternative to training your own embeddings is to use ones that have been pre-trained by others on your behalf. There are many sources where you can find these, including Kaggle and Hugging Face. You can even find pretrained embeddings posted alongside research results. One such set of pretrained embeddings is the Stanford GloVe embeddings (*https://oreil.ly/s1YWw*), and we'll explore those here.

Note, however, that when using embeddings that have been pretrained, you should also consider updating and changing your tokenizer to match any rules used with the pretrained embeddings.

For example, with the GloVE pretrained embeddings—which simply comprise a large text file of words with their pretrained embedding in a number of dimensions from 50 to 300—the rules used to tokenize words are a little different from those for the handmade tokenizer we've been using for raw data. So, for GloVe, you should consider rules such as all of the words being lowercase or numbers being normalized to 0.

Once you've done this (I've provided code for GloVe in the downloads, and I discuss it in a little more detail in the next chapter), then it's simply a matter of loading the weights of the pretrained embeddings to your model definition like this:

```
# Initialize embedding layer
self.embedding = nn.Embedding(vocab_size, embedding_dim)

# Load pretrained embeddings if provided
if pretrained_embeddings is not None:
    self.embedding.weight.data.copy_(pretrained_embeddings)
    if freeze_embeddings:
        self.embedding.weight.requires_grad = False
```

If you don't want to learn from these embeddings and you want to just use them, then you should set `freeze_embeddings` to `True`. Otherwise, the network will fine-tune by using the pre-loaded embedding weights as a starting point.

This model will rapidly reach peak accuracy in training, and it will not overfit as much as we saw previously. The accuracy over three hundred epochs shows that training and validation are very much in step with each other (see Figure 6-22). The loss values are also in step, which shows that we are fitting very nicely over the first couple of hundred epochs. However, they also begin to diverge (see Figure 6-22).

On the other hand, it is worth noting that the overall accuracy (at about 70%) is quite low, considering that a coin flip would have a 50% chance of getting it right! So, while using pretrained embeddings can make for much faster training with less overfitting, you should also understand what it is that they're useful for and that they may not always be best for your scenario. You may therefore need to explore optimization methods or alternatives where appropriate.

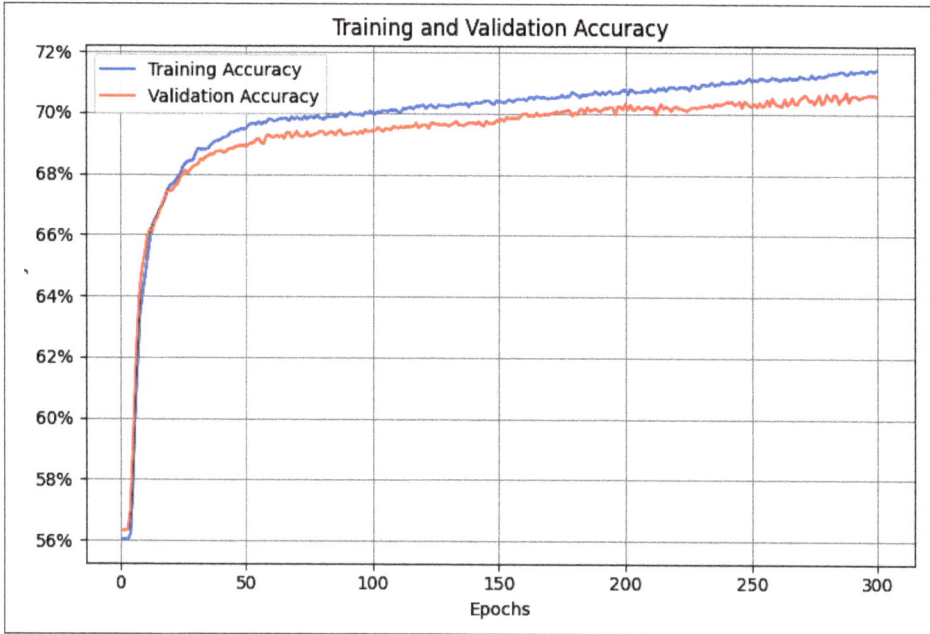

Figure 6-21. Accuracy metrics using GloVe embeddings

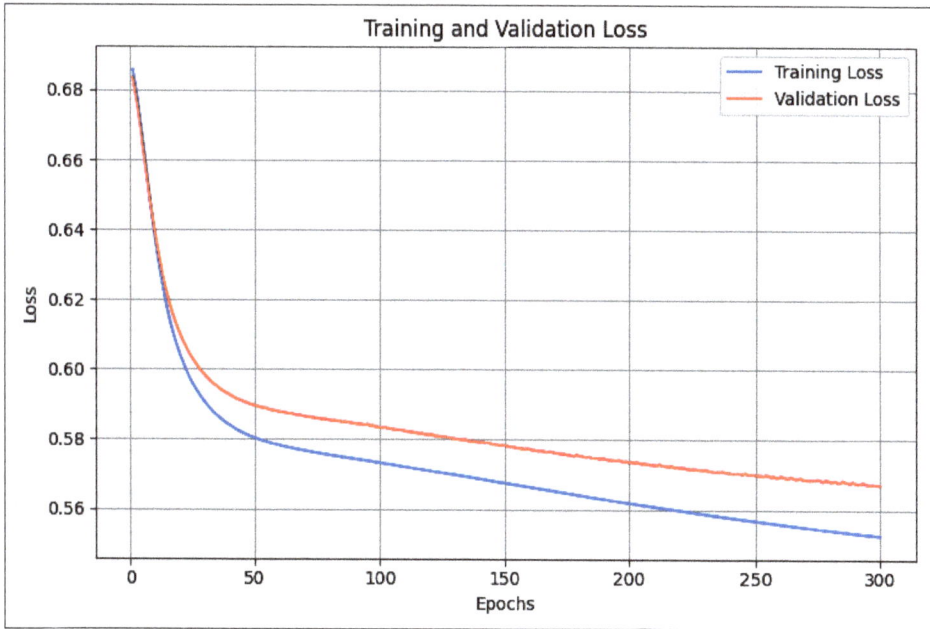

Figure 6-22. Loss metrics using GloVe embeddings

Summary

In this chapter, you built your first model that can understand sentiment in text. It did this by taking the tokenized text from Chapter 5 and mapping it to vectors. Then, using backpropagation, it learned the appropriate "direction" for each vector based on the label for the sentence containing it. Finally, it was able to use all of the vectors for a collection of words to build up an idea of the sentiment within the sentence.

You also explored ways to optimize your model to avoid overfitting, and you saw a neat visualization of the final vectors representing your words. But while this was a nice way to classify sentences, it simply treated each sentence as a bunch of words. There was no inherent sequence involved, and the order of appearance of words is very important in determining the real meaning of a sentence.

Therefore, it's a good idea to see if we can improve our models by taking sequence into account. We'll explore that in the next chapter with the introduction of a new layer type: a *recurrent* layer, which is the foundation of recurrent neural networks.

## CHAPTER 7

Recurrent Neural Networks for Natural Language Processing

In Chapter 5, you saw how to tokenize and sequence text, turning sentences into tensors of numbers that could then be fed into a neural network. You then extended that in Chapter 6 by looking at embeddings, which constitute a way to have words with similar meanings cluster together to enable the calculation of sentiment. This worked really well, as you saw by building a sarcasm classifier. But there's a limitation to that: namely, sentences aren't just collections of words—and often, the *order* in which the words appear will dictate their overall meaning. Also, adjectives can add to or change the meaning of the nouns they appear beside. For example, the word *blue* might be meaningless from a sentiment perspective, as might *sky*, but when you put them together to get *blue sky*, it indicates a clear sentiment that's usually positive. Finally, some nouns may qualify others, such as in *rain cloud*, *writing desk*, and *coffee mug*.

To take sequences like this into account, you need to take an additional approach: you need to factor *recurrence* into the model architecture. In this chapter, you'll look at different ways of doing this. We'll explore how sequence information can be learned and how you can use this information to create a type of model that is better able to understand text: the *recurrent neural network* (RNN).

The Basis of Recurrence

To understand how recurrence might work, let's first consider the limitations of the models used thus far in the book. Ultimately, creating a model looks a little bit like Figure 7-1. You provide data and labels and define a model architecture, and the model learns the rules that fit the data to the labels. Those rules then become available to you as an application programming interface (API) that will give you back predicted labels for future data.

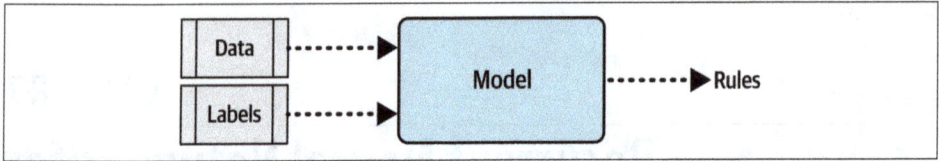

Figure 7-1. High-level view of model creation

But, as you can see, the data is lumped in wholesale. There's no granularity involved and no effort to understand the sequence in which that data occurs. This means the words *blue* and *sky* have no different meaning in sentences such as, "Today I am blue, because the sky is gray," and "Today I am happy, and there's a beautiful blue sky." To us, the difference in the use of these words is obvious, but to a model, with the architecture shown here, there really is no difference.

So, how do we fix this? Let's first explore the nature of recurrence, and from there, you'll be able to see how a basic RNN can work.

Consider the famous Fibonacci sequence of numbers. In case you aren't familiar with it, I've put some of them into Figure 7-2.

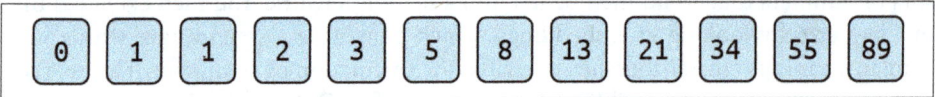

Figure 7-2. The first few numbers in the Fibonacci sequence

The idea behind this sequence is that every number is the sum of the two numbers preceding it. So if we start with 1 and 2, the next number is 1 + 2, which is 3. The one after that is 2 + 3, which is 5, and then there's 3 + 5, which is 8, and so on.

We can place this in a computational graph to get Figure 7-3.

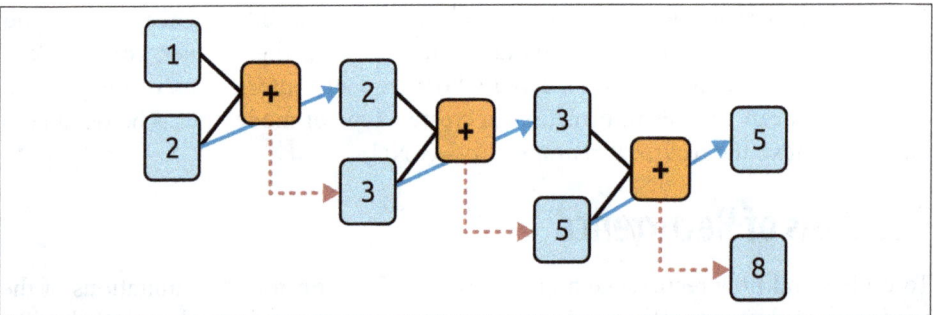

Figure 7-3. A computational graph representation of the Fibonacci sequence

Here, you can see that we feed 1 and 2 into the function and get 3 as the output. We then carry the second parameter (2) over to the next step and feed it into the function along with the output from the previous step (3). The output of this is 5, and it gets fed into the function with the second parameter from the previous step (3) to produce an output of 8. This process continues indefinitely, with every operation depending on those before it. The 1 at the top left sort of "survives" through the process—it's an element of the 3 that gets fed into the second operation, it's an element of the 5 that gets fed into the third operation, and so on. Thus, some of the essence of the 1 is preserved throughout the sequence, though its impact on the overall value is diminished.

This is analogous to how a recurrent neuron is architected. You can see the typical representation of a recurrent neuron in Figure 7-4.

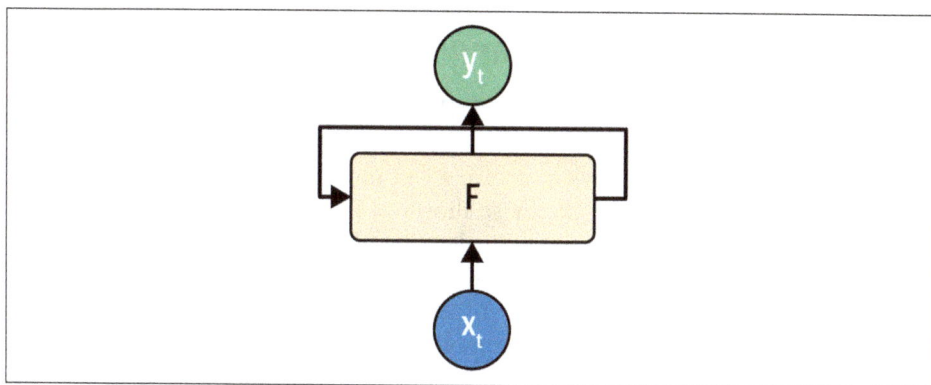

Figure 7-4. A recurrent neuron

A value x is fed into the function F at a time step, so it's typically labeled x_t. This produces an output y at that time step, which is typically labeled y_t. It also produces a value that is fed forward to the next step, which is indicated by the arrow from F to itself.

This is made a little clearer if you look at how recurrent neurons work beside one another across time steps, which you can see in Figure 7-5.

Here, x_0 is operated on to get y_0 and a value that's passed forward. The next step gets that value and x_1 and produces y_1 and a value that's passed forward. The next one gets that value and x_2 and produces y_2 and a passed-forward value, and so on down the line. This is similar to what we saw with the Fibonacci sequence, and I always find it to be a handy mnemonic when trying to remember how an RNN works.

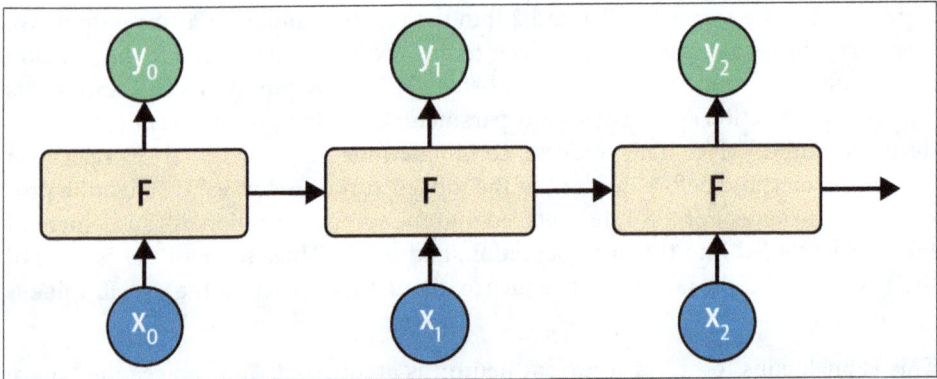

Figure 7-5. Recurrent neurons in time steps

Extending Recurrence for Language

In the previous section, you saw how an RNN operating over several time steps can help maintain context across a sequence. Indeed, we'll use RNNs for sequence modeling later in this book—but there's a nuance when it comes to language that you can miss when using a simple RNN like those shown in Figure 7-4 and Figure 7-5. As in the Fibonacci sequence example mentioned earlier, the amount of context that's carried over will diminish over time. The effect of the output of the neuron at step 1 is huge at step 2, smaller at step 3, smaller still at step 4, and so on. So, if we have a sentence like "Today has a beautiful blue <something>," the word *blue* will have a strong impact on what the next word could be: we can guess that it's likely to be *sky*. But what about context that comes from earlier in a sentence? For example, consider the sentence "I lived in Ireland, so in high school, I had to learn how to speak and write <something>."

That <something> is *Gaelic*, but the word that really gives us that context is *Ireland*, which is much earlier in the sentence. Thus, for us to be able to recognize what <something> should be, we need a way to preserve context across a longer distance. The short-term memory of an RNN needs to get longer, and in recognition of this, an enhancement to the architecture called *long short-term memory* (LSTM) was invented.

While I won't go into detail on the underlying architecture of how LSTMs work, the high-level diagram shown in Figure 7-6 gets the main point across. To learn more about the internal operations of LSTM, check out Christopher Olah's excellent blog post on the subject (*https://oreil.ly/6KcFA*).

The LSTM architecture enhances the basic RNN by adding a "cell state" that enables context to be maintained not just from step to step but across the entire sequence of steps. Remembering that these are neurons that learn in the way neurons do, you can see that this enhancement ensures that the context that is important will be learned over time.

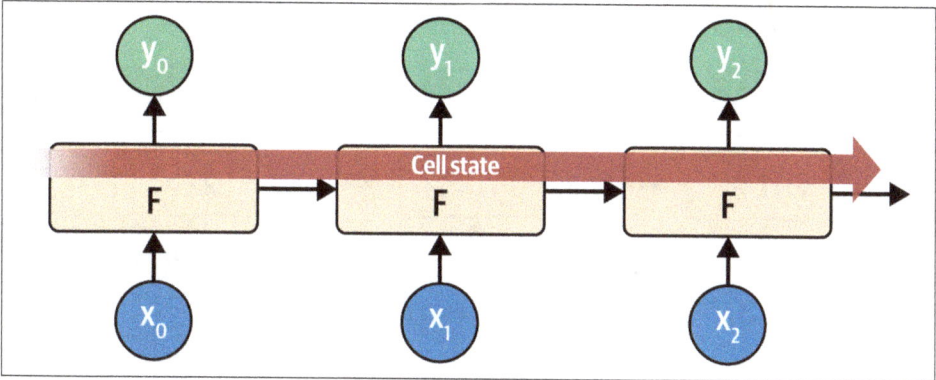

Figure 7-6. High-level view of LSTM architecture

An important part of an LSTM is that it can be *bidirectional*—the time steps can be iterated both forward and backward so that context can be learned in both directions. Often, context for a word can come *after* it in the sentence and not just before.

See Figure 7-7 for a high-level view of this.

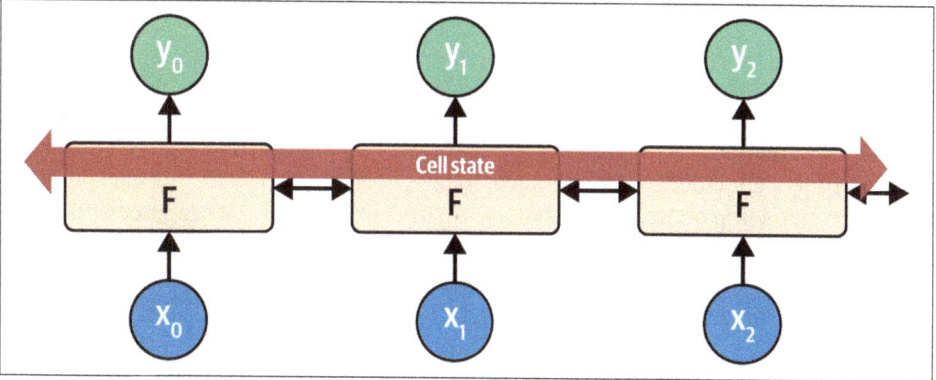

Figure 7-7. High-level view of LSTM bidirectional architecture

This is how evaluation in the direction from 0 to `number_of_steps` is done, and it's also how evaluation from `number_of_steps` to 0 is done. At each step, the *y* result is an aggregation of the "forward" pass and the "backward" pass. You can see this in Figure 7-8.

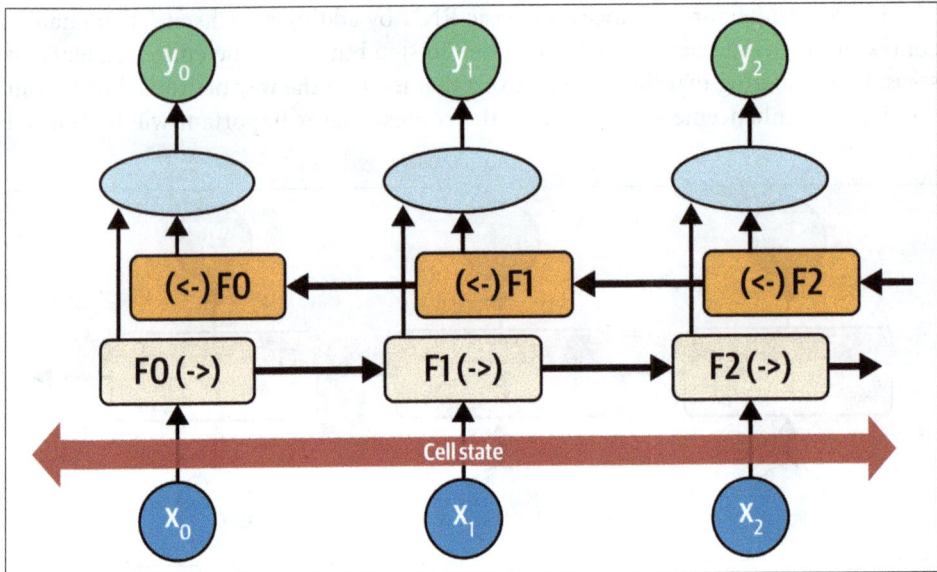

Figure 7-8. Bidirectional LSTM

It's easy to confuse the bidirectional nature of the LSTM with the terms *forward* and *backward* when it comes to the training of the network, but they're very different. When I refer to the forward and backward pass, I'm referring to the setting of the parameters of the neurons and their updating from the learning process, respectively. Don't confuse this with the values that the LSTM is paying attention to as being the next or previous tokens in the sequence.

Also, consider each neuron at each time step to be F0, F1, F2, etc. The direction of the time step is shown, so the calculation at F1 in the forward direction is F1(->), and in the reverse direction, it's (<-)F1. The values of these are aggregated to give the *y* value for that time step. Additionally, the cell state is bidirectional, and this can be really useful for managing context in sentences. Again, considering the sentence "I lived in Ireland, so in high school, I had to learn how to speak and write <something>," you can see how the <something> was qualified to be *Gaelic* by the context word *Ireland*. But what if it were the other way around: "I lived in <this country>, so in high school, I had to learn how to speak and write Gaelic"? You can see that by going *backward* through the sentence, we can learn about what <this country> should be. Thus, using bidirectional LSTMs can be very powerful for understanding sentiment in text. (And as you'll see in Chapter 8, they're really powerful for generating text, too!)

Of course, there's a lot going on with LSTMs, in particular bidirectional ones, so expect training to be slow. Here's where it's worth investing in a GPU or at the very least using a hosted one in Google Colab if you can.

Creating a Text Classifier with RNNs

In Chapter 6, you experimented with creating a classifier for the Sarcasm dataset by using embeddings. In that case, you turn words into vectors before aggregating them and then feeding them into dense layers for classification. But when you're using an RNN layer such as an LSTM, you don't do the aggregation, and you can feed the output of the embedding layer directly into the recurrent layer. When it comes to the dimensionality of the recurrent layer, a rule of thumb you'll often see is that it's the same size as the embedding dimension. This isn't necessary, but it can be a good starting point. Also note that while in Chapter 6 I mentioned that the embedding dimension is often the fourth root of the size of the vocabulary, when using RNNs, you'll often see that that rule may be ignored because it would make the size of the recurrent layer too small.

For this example, I have used the number of neurons in the hidden layer as a starting point, and you can experiment from there.

So, for example, you could update the simple model architecture for the sarcasm classifier you developed in Chapter 6 to the following to use a bidirectional LSTM:

```python
class TextClassificationModel(nn.Module):
    def __init__(self, vocab_size, embedding_dim,
                 hidden_dim=24, lstm_layers=1):
        super(TextClassificationModel, self).__init__()

        # Embedding layer
        self.embedding = nn.Embedding(vocab_size, embedding_dim)

        # LSTM layer
        self.lstm = nn.LSTM(
            input_size=embedding_dim,
            hidden_size=hidden_dim,
            num_layers=lstm_layers,
            batch_first=True,
            bidirectional=True
        )

        # Global pooling
        self.global_pool = nn.AdaptiveAvgPool1d(1)

        # Fully connected layers
        self.fc1 = nn.Linear(hidden_dim * 2, hidden_dim)
        self.fc2 = nn.Linear(hidden_dim, 1)

        # Activation functions
```

```python
        self.relu = nn.ReLU()
        self.sigmoid = nn.Sigmoid()

    def forward(self, x):
        # x shape: (batch_size, sequence_length)

        # Get embeddings
        embedded = self.embedding(x)
        # Shape: (batch_size, sequence_length, embedding_dim)

        # Pass through LSTM
        lstm_out, _ = self.lstm(embedded)
        # Shape: (batch_size, sequence_length, hidden_dim)

        # Transpose for global pooling
        # (expecting: batch, channels, sequence_length)
        lstm_out = lstm_out.transpose(1, 2)
        # Shape: (batch_size, hidden_dim, sequence_length)

        # Apply global pooling
        pooled = self.global_pool(lstm_out)
        # Shape: (batch_size, hidden_dim, 1)
        pooled = pooled.squeeze(-1)  # Shape: (batch_size, hidden_dim)

        # Pass through fully connected layers
        x = self.relu(self.fc1(pooled))
        x = self.sigmoid(self.fc2(x))

        return x
```

You can then set the loss function and optimizer to this. (Note that the LR is 0.001, or 1e–3.):

```python
# Define loss function and optimizer
criterion = nn.BCELoss()
optimizer = optim.Adam(model.parameters(), lr=0.001,
                       betas=(0.9, 0.999), amsgrad=False)
```

When you print out the model architecture summary, you'll see something like the following:

```
===============================================================================
Layer (type:depth-idx)                    Output Shape              Param #
===============================================================================
TextClassificationModel                   [32, 1]                   --
├─Embedding: 1-1                          [32, 85, 7]               14,000
├─LSTM: 1-2                               [32, 85, 48]              6,336
├─AdaptiveAvgPool1d: 1-3                  [32, 48, 1]               --
├─Linear: 1-4                             [32, 24]                  1,176
├─ReLU: 1-5                               [32, 24]                  --
├─Linear: 1-6                             [32, 1]                   25
├─Sigmoid: 1-7                            [32, 1]                   --
===============================================================================
```

```
Total params: 21,537
Trainable params: 21,537
Non-trainable params: 0
Total mult-adds (M): 17.72
================================================================
Input size (MB): 0.02
Forward/backward pass size (MB): 1.20
Params size (MB): 0.09
Estimated Total Size (MB): 1.31
================================================================
```

Note that the vocab size is 2,000 and the embedding dimension is 7. This gives 14,000 parameters in the embedding layer, and the bidirectional layer will have 48 neurons (24 out, 24 back) with a sequence length of 85 characters

Figure 7-9 shows the results of training with this over three hundred epochs.

This gives us a network with only 21,537 parameters. As you can see, the accuracy of the network on training data rapidly climbs toward 85%, but the validation data plateaus at around 75%. This is similar to the figures we got earlier, but inspecting the loss chart in Figure 7-10 shows that while the loss for the training set diverged after 15 epochs, the validation loss turned to increase, indicating we have overfitting.

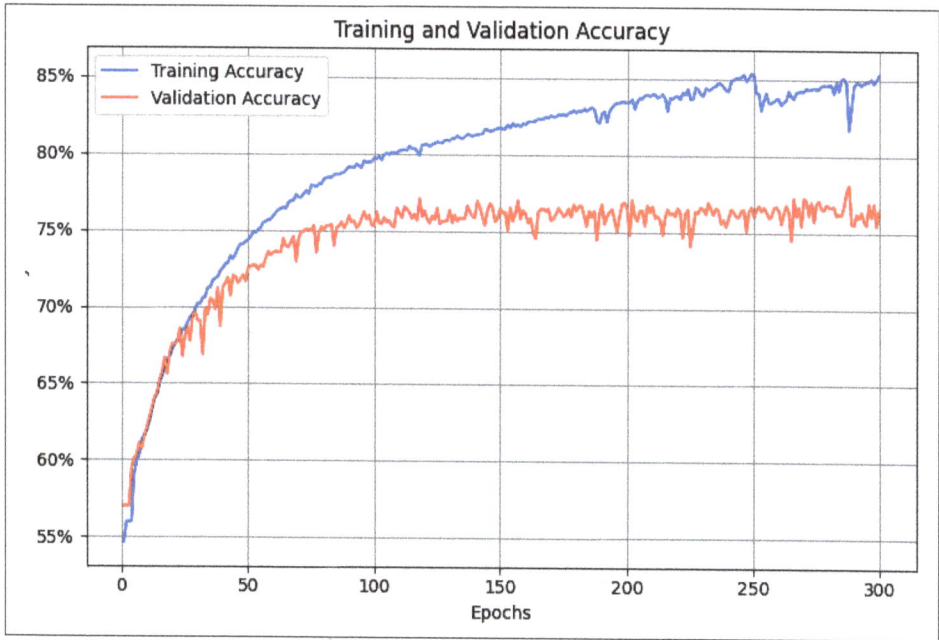

Figure 7-9. Accuracy for LSTM over 300 epochs

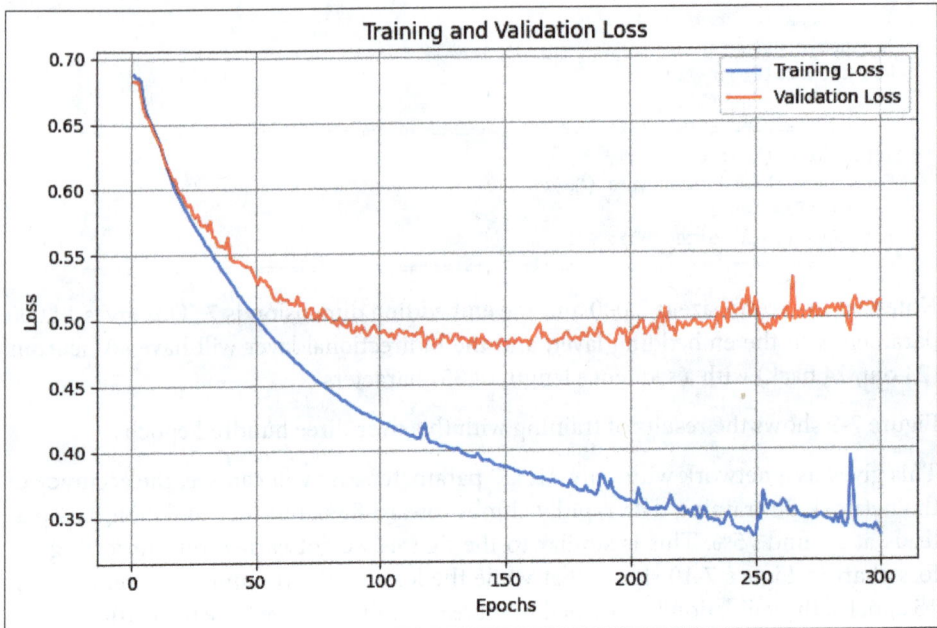

Figure 7-10. Loss with LSTM over 300 epochs

However, this was just using a single LSTM layer with a hidden layer of 24 neurons. In the next section, you'll see how to use stacked LSTMs and explore the impact on the accuracy of classifying this dataset.

Stacking LSTMs

In the previous section, you saw how to use an LSTM layer after the embedding layer to help classify the contents of the sarcasm dataset. But LSTMs can be stacked on top of one another, and this approach is used in many state-of-the-art NLP models.

Stacking LSTMs with PyTorch is pretty straightforward. You add them as extra layers just like you would with any other layer, but you will need to be careful in specifying the dimensions. So, for example, if the first LSTM has *x* number of hidden layers, then the next LSTM will have *x* number of inputs. If the LST is bidirectional, then the next will need to double the size. Here's an example:

```python
# First LSTM layer
self.lstm1 = nn.LSTM(
    input_size=embedding_dim,
    hidden_size=hidden_dim,
    num_layers=lstm_layers,
    batch_first=True,
    bidirectional=True
)
```

```
# Second LSTM layer
# Note: Input size is hidden_dim*2 because first LSTM is bidirectional.
self.lstm2 = nn.LSTM(
    input_size=hidden_dim * 2,
    hidden_size=hidden_dim,
    num_layers=lstm_layers,
    batch_first=True,
    bidirectional=True
)
```

Note that the `input_size` for the first layer is the embedding dimension because it's preceded by the embedding layer. The second LSTM then has its input size as (`hidden_dim * 2`) because the output from the first LSTM is that size, given that it's bidirectional.

The model architecture will look like this:

```
=================================================================
Layer (type:depth-idx)            Output Shape          Param #
=================================================================
TextClassificationModel           [32, 1]               --
├─Embedding: 1-1                   [32, 85, 7]           14,000
├─LSTM: 1-2                        [32, 85, 48]          6,336
├─LSTM: 1-3                        [32, 85, 48]          14,208
├─AdaptiveAvgPool1d: 1-4           [32, 48, 1]           --
├─Linear: 1-5                      [32, 24]              1,176
├─ReLU: 1-6                        [32, 24]              --
├─Linear: 1-7                      [32, 1]               25
├─Sigmoid: 1-8                     [32, 1]               --
=================================================================
Total params: 35,745
Trainable params: 35,745
Non-trainable params: 0
Total mult-adds (M): 56.37
=================================================================
Input size (MB): 0.02
Forward/backward pass size (MB): 2.25
Params size (MB): 0.14
Estimated Total Size (MB): 2.41
=================================================================
```

Adding the extra layer will give us roughly 14,000 extra parameters that need to be learned, which is an increase of about 75%. So, it might slow the network down, but the cost is relatively low if there's a reasonable benefit.

After training for three hundred epochs, the result looks like Figure 7-11. While the accuracy on the validation set is flat, examining the loss (shown in Figure 7-12) tells a different story. As you can see in Figure 7-12, while the accuracy for both training and validation looked good, the validation loss quickly took off upward, which is a clear sign of overfitting.

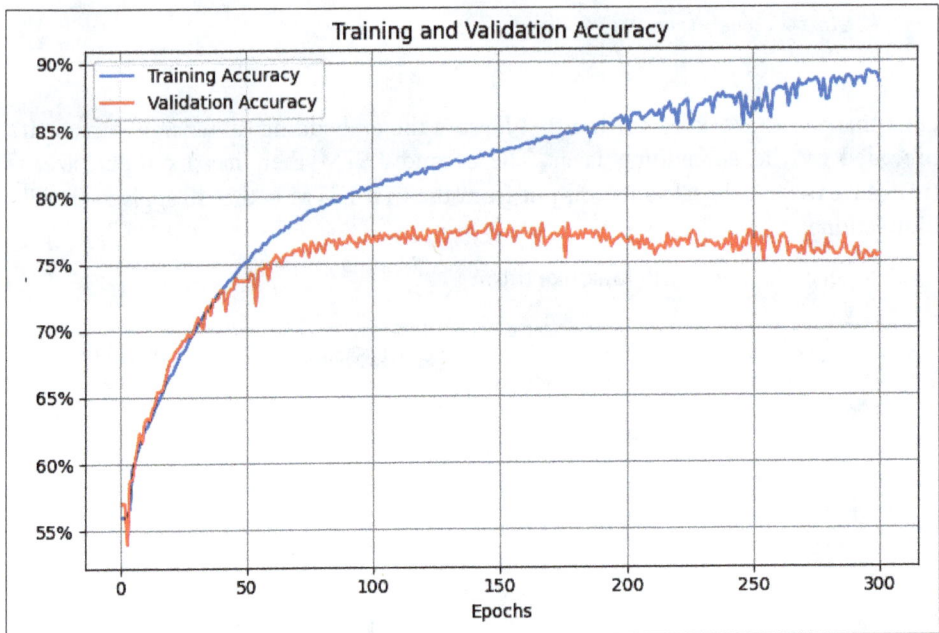

Figure 7-11. Accuracy for stacked LSTM architecture

This overfitting (which is indicated by the training accuracy climbing toward 100% as the loss falls smoothly while the validation accuracy is relatively steady and the loss increases drastically) is a result of the model getting overspecialized for the training set. As with the examples in Chapter 6, this shows that it's easy to be lulled into a false sense of security if you just look at the accuracy metrics without examining the loss.

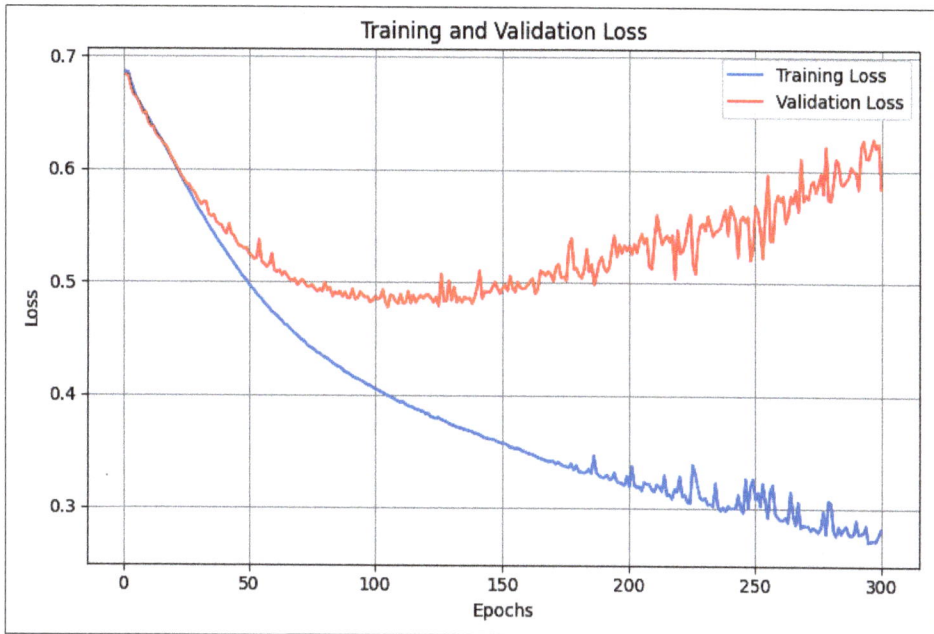

Figure 7-12. Loss for stacked LSTM architecture

Optimizing stacked LSTMs

In Chapter 6, you saw that a very effective method of reducing overfitting was to reduce the LR. It's worth exploring here whether that will have a positive effect on an RNN, too.

For example, the following code reduces the LR by 50%, from 0.0001 to 0.00005:

```python
# Define loss function and optimizer
criterion = nn.BCELoss()
optimizer = optim.Adam(model.parameters(), lr=0.00005,
                       betas=(0.9, 0.999), amsgrad=False)
```

Figure 7-13 demonstrates the impact of this on training. As you can see, there's a small difference in the validation accuracy, indicating that we're overfitting a bit less.

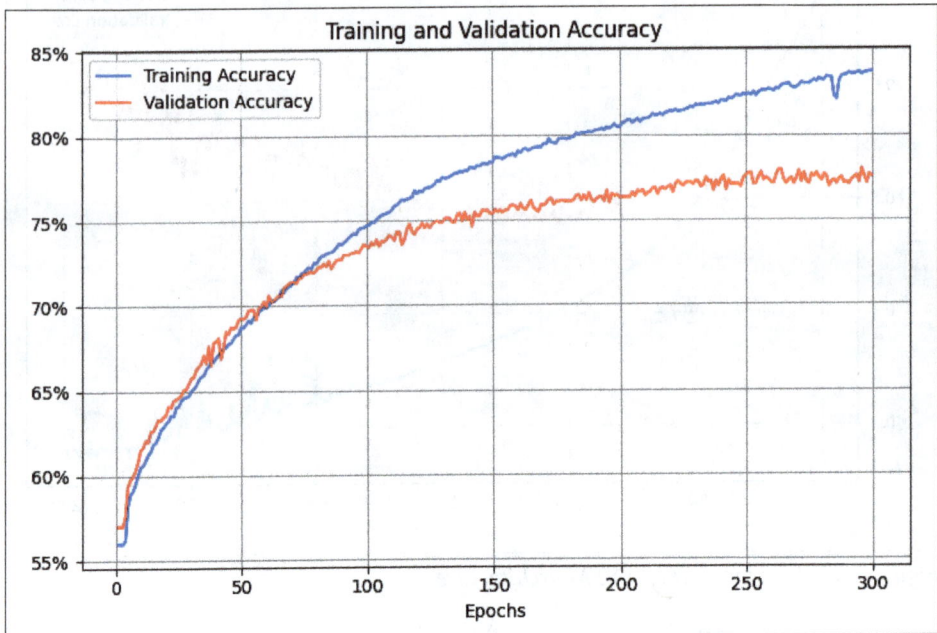

Figure 7-13. Impact of reduced LR on accuracy with stacked LSTMs

While an initial look at Figure 7-14 similarly suggests a decent impact on loss due to the reduced LR, with the curve not moving up so sharply, it's worth looking a little closer. We see that the loss on the training set is actually a little higher (~0.35 versus ~0.27) than the previous example, while the loss on the validation set is lower (~0.5 versus 0.6).

Adjusting the LR hyperparameter certainly seems worth investigation.

Indeed, further experimentation with the LR showed a marked improvement in getting training and validation curves to converge, indicating that while the network was less accurate after training, we could tell that it was generalizing better. Figures 7-15 and 7-16 show the impact of using a lower LR (.0003 rather than .0005).

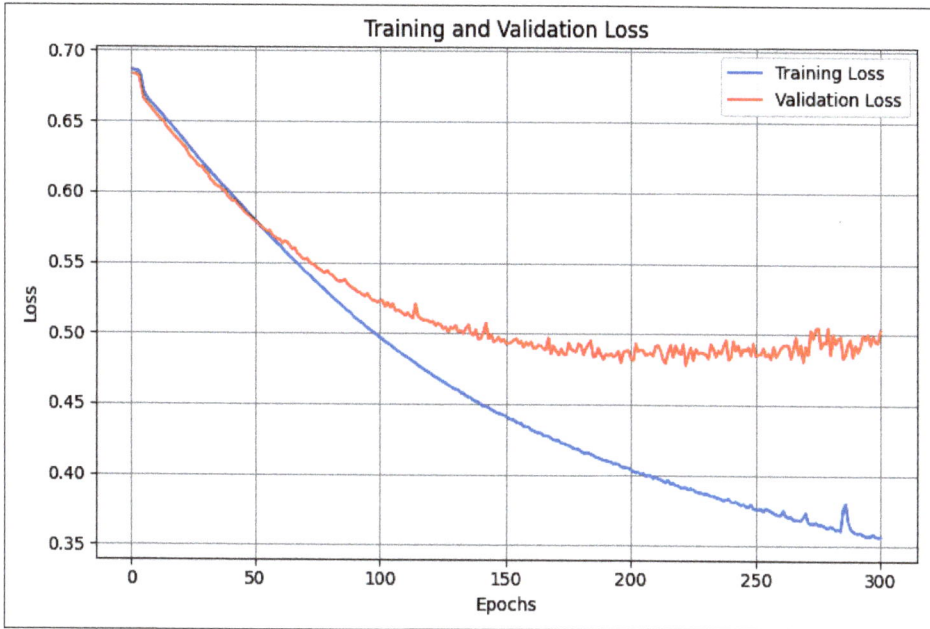

Figure 7-14. Impact of reduced LR on loss with stacked LSTMs

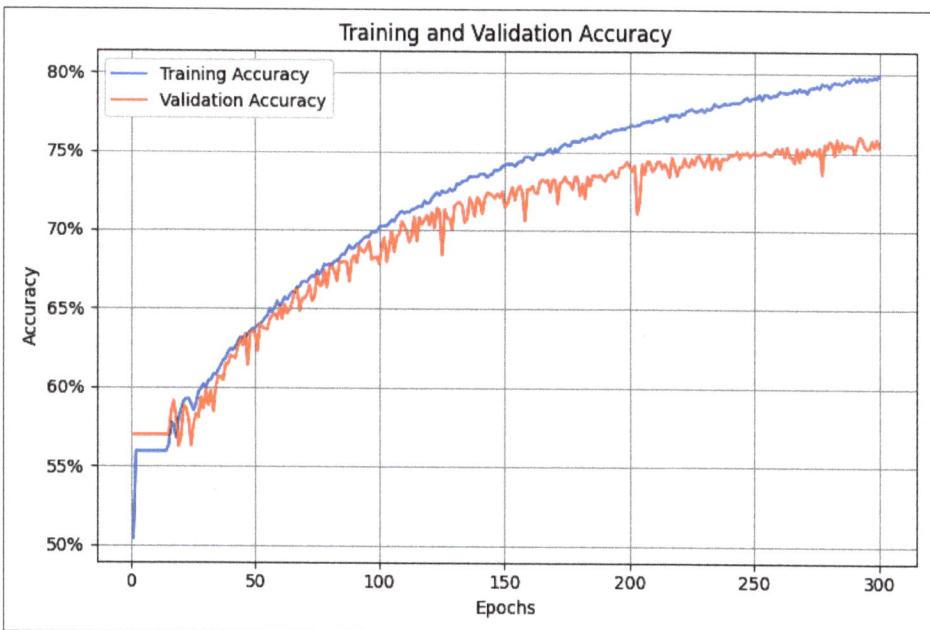

Figure 7-15. Accuracy with further-reduced LR with stacked LSTM

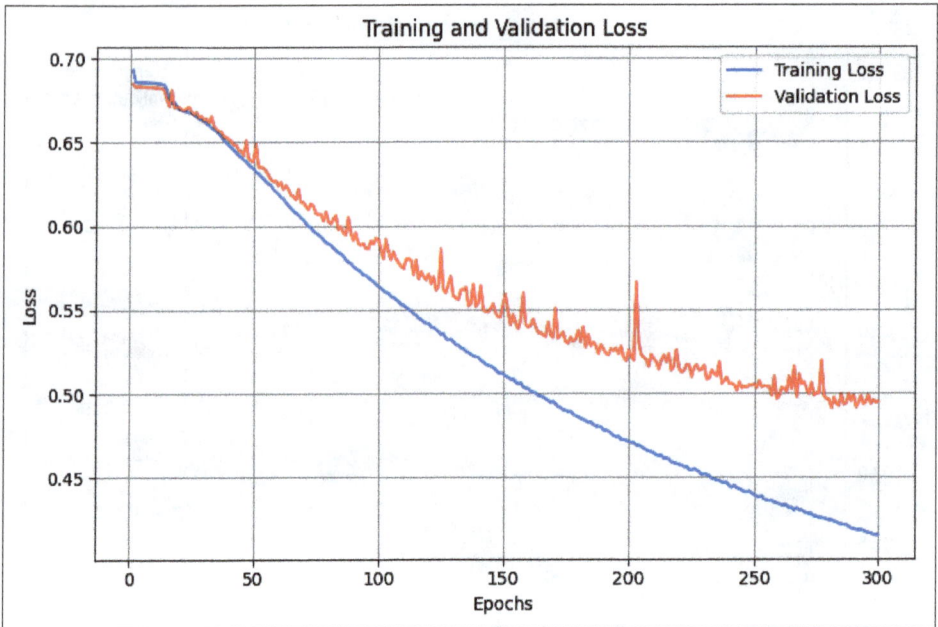

Figure 7-16. Loss with further-reduced LR and stacked LSTM

Indeed, reducing the LR even further, to .00001, gave potentially even better results, as shown in Figures 7-17 and 7-18. As with the previous diagrams, while the overall accuracy isn't as good and the loss is higher, that's an indication that we're getting closer to a "realistic" result for this network architecture and not being led into having a false sense of security by overfitting on the training data.

In addition to changing the LR parameter, you should also consider using dropout in the LSTM layers. It works exactly the same as for dense layers, as discussed in Chapter 3, where random neurons are dropped to prevent a proximity bias from impacting the learning. That being said, you should be careful about setting it *too* low, because when you start tweaking with different architectures, you might freeze the ability of the network to learn.

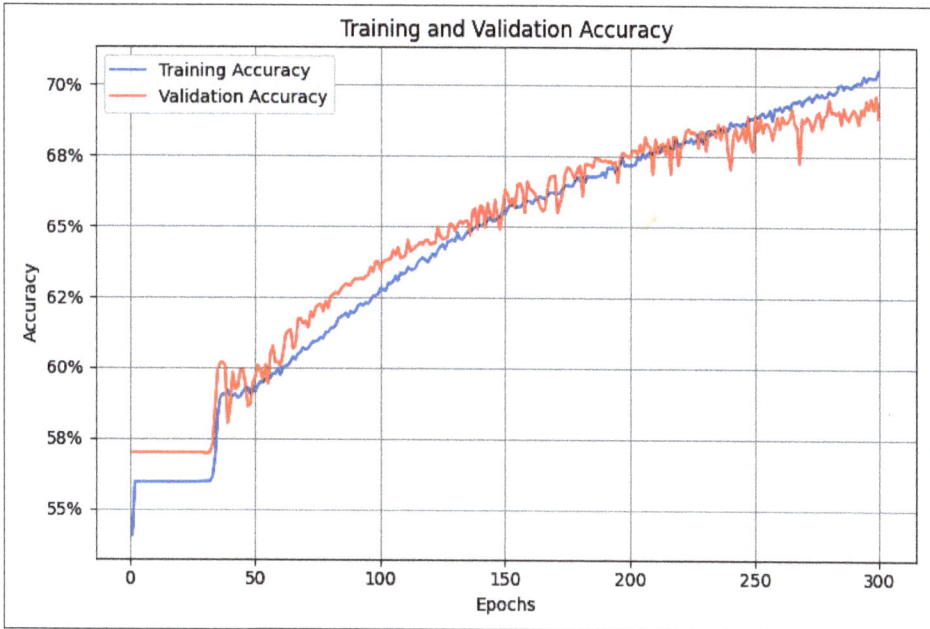

Figure 7-17. Accuracy with lower LR

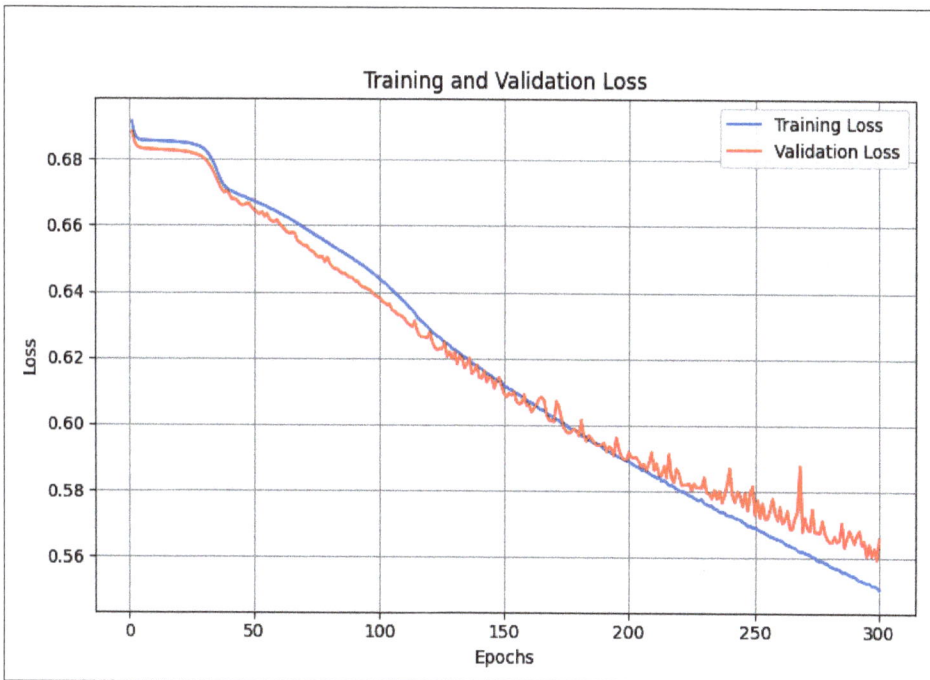

Figure 7-18. Loss with lower LR

Using dropout

In addition to changing the LR parameter, you should also consider using dropout in the LSTM layers. It works exactly the same as for dense layers, as discussed in Chapter 3, where random neurons are dropped to prevent a proximity bias from impacting the learning.

You can implement dropout by using nn.Dropout. Here's an example:

```
self.embedding_dropout = nn.Dropout(dropout_rate)
self.lstm_dropout = nn.Dropout(dropout_rate)
self.final_dropout = nn.Dropout(dropout_rate)
```

Then, in your forward pass, you can apply the dropouts at the appropriate levels, like this:

```
def forward(self, x):
    # Get embeddings
    embedded = self.embedding(x)

    # Apply first dropout after embedding layer
    embedded = self.embedding_dropout(embedded)

    lstm1_out, _ = self.lstm1(embedded)

    # Apply dropout between LSTM layers
    lstm1_out = self.lstm_dropout(lstm1_out)

    lstm2_out, _ = self.lstm2(lstm1_out)

    # Apply final dropout
    lstm2_out = self.final_dropout(lstm2_out)

    lstm_out = lstm2_out.transpose(1, 2)

    pooled = self.global_pool(lstm_out)
    pooled = pooled.squeeze(-1)

    x = self.relu(self.fc1(pooled))
    x = self.sigmoid(self.fc2(x))

    return x
```

When I ran this with the lowest LR I had tested prior to dropout, the network didn't learn. So, I moved the LR back up to 0.0003 and ran for 300 epochs using this dropout (note that the dropout rate is 0.2, so about 20% of neurons are dropped at random). The accuracy results can be seen in Figure 7-19. The curves for training and validation are still close to each other, but they're hitting greater than 75% accuracy, whereas without dropout, it was hard to get above 70%.

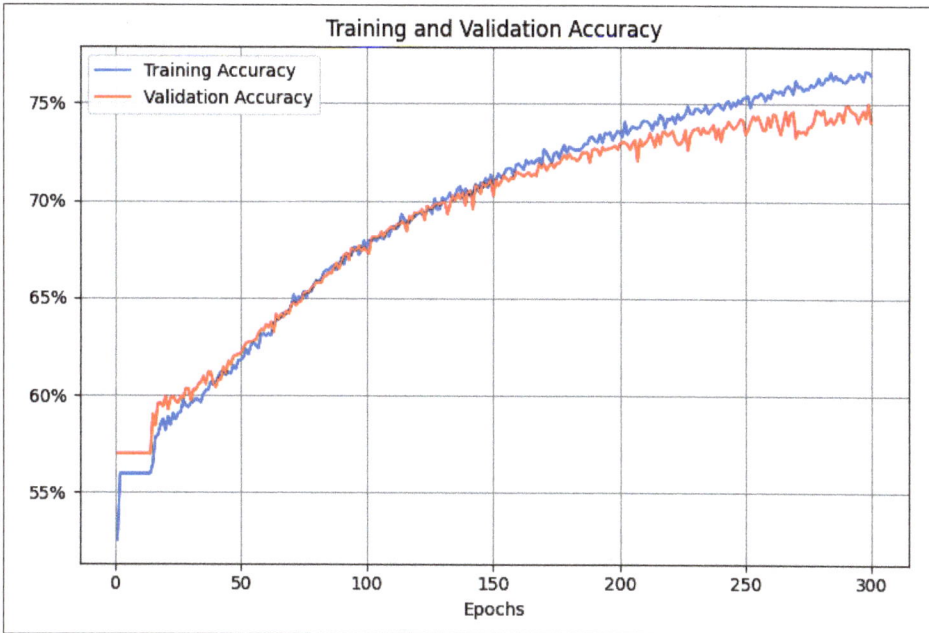

Figure 7-19. Accuracy of stacked LSTMs using dropout

As you can see, using dropout can have a positive impact on the accuracy of the network, which is good! There's always a worry that losing neurons will make your model perform worse, but as we can see here, that's not the case. But do be careful when using dropout because it can lead to underfitting or overfitting if not used appropriately.

There's also a positive impact on loss, as you can see in Figure 7-20. While the curves are clearly diverging, they are closer than they were previously, and the validation set is flattening out at a loss of about 0.45, which also demonstrates an improvement! As this example shows, dropout is another handy technique that you can use to improve the performance of LSTM-based RNNs.

It's worth exploring these techniques for avoiding overfitting in your data, and it's also worth exploring the techniques for preprocessing your data that we covered in Chapter 6. But there's one thing that we haven't yet tried: a form of transfer learning in which you can use pre-learned embeddings for words instead of trying to learn your own. We'll explore that next.

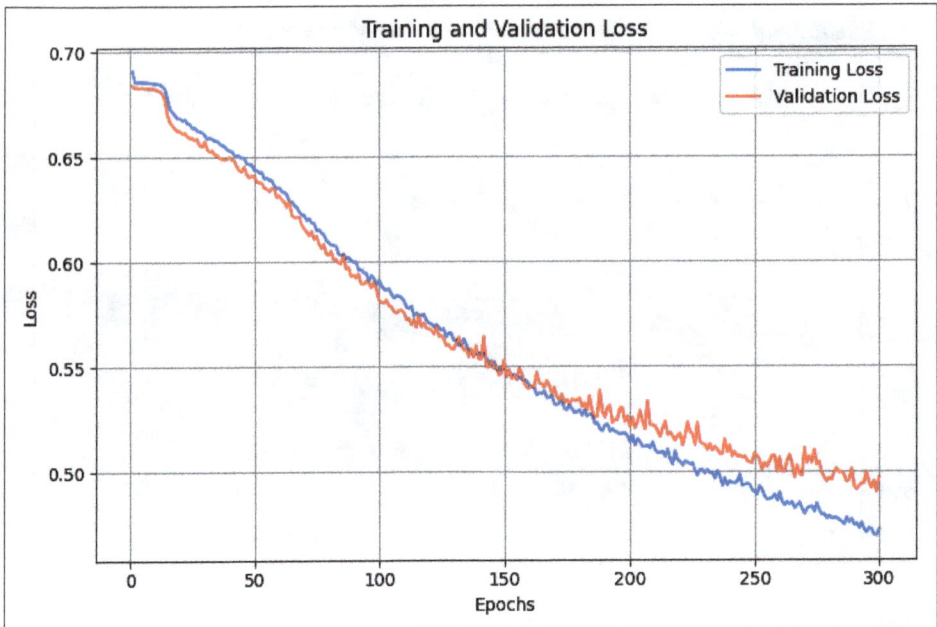

Figure 7-20. Loss curves for dropout-enabled LSTMs

Using Pretrained Embeddings with RNNs

In all the previous examples, you gathered the full set of words to be used in the training set and then trained embeddings with them. You initially aggregated them before feeding them into a dense network, and in this chapter, you explored how to improve the results using an RNN. While doing this, you were restricted to the words in your dataset and how their embeddings could be learned by using the labels from that dataset.

Now, think back to Chapter 4, where we discussed transfer learning. What if instead of learning the embeddings for yourself, you could use pre-learned embeddings, where researchers have already done the hard work of turning words into vectors and those vectors are proven? One example of this, as we saw in Chapter 6, is the GloVe (Global Vectors for Word Representation) model (*https://oreil.ly/4ENdQ*) developed by Jeffrey Pennington, Richard Socher, and Christopher Manning at Stanford.

In this case, the researchers have shared their pretrained word vectors for a variety of datasets:

- A 6-billion-token, 400,000-word vocabulary set in 50, 100, 200, and 300 dimensions with words taken from Wikipedia and Gigaword

- A 42-billion-token, 1.9-million-word vocabulary in 300 dimensions from a common crawl

- An 840-billion-token, 2.2-million-word vocabulary in 300 dimensions from a common crawl

- A 27-billion-token, 1.2-million-word vocabulary in 25, 50, 100, and 200 dimensions from a Twitter crawl of 2 billion tweets

Given that the vectors are already pretrained, it's simple for you to reuse them in your PyTorch code, instead of learning them from scratch. First, you'll have to download the GloVe data. I've opted to use the 6-billion-word version, in 50 dimensions, using this code to download and unzip it:

```python
import urllib.request
import zipfile

# Download GloVe embeddings
url = "https://nlp.stanford.edu/data/glove.6B.zip"
urllib.request.urlretrieve(url, "glove.6B.zip")

# Unzip
with zipfile.ZipFile("glove.6B.zip", 'r') as zip_ref:
    zip_ref.extractall()

# You can use glove.6B.50d.txt (50 dimensions)
# or glove.6B.100d.txt (100 dimensions)
```

Each entry in the file is a word, followed by the dimensional coefficients that were learned for it. The easiest way to use this is to create a dictionary where the key is the word and the values are the embeddings. You can set up this dictionary like this:

```python
import numpy as np
glove_embeddings = dict()
f = open('glove.6B.50d.txt')
for line in f:
    values = line.split()
    word = values[0]
    coefs = np.asarray(values[1:], dtype='float32')
    glove_embeddings[word] = coefs
f.close()
```

At this point, you'll be able to look up the set of coefficients for any word simply by using it as the key. So, for example, to see the embeddings for the word *frog*, you could use this:

```python
glove_embeddings['frog']
```

With these pretrained embeddings in hand, you can now load them into the embeddings layer in your neural architecture and use them as pretrained embeddings instead of learning them from scratch. See the following model architecture definition. If the `pretrained_embeddings` value is not null, then the weights for the embedding layer will be loaded from that. If `freeze_embeddings` is `True`, then they'll be frozen; otherwise, they'll be used as the starting point for learning (i.e., you'll fine-tune the embeddings based on your corpus):

```python
class TextClassificationModel(nn.Module):
    def __init__(self, vocab_size, embedding_dim=100, hidden_dim=16,
                 dropout_rate=0.25, pretrained_embeddings=None,
                 freeze_embeddings=True, lstm_layers=2):
        super(TextClassificationModel, self).__init__()

        # Initialize embedding layer
        self.embedding = nn.Embedding(vocab_size, embedding_dim)

        # Load pretrained embeddings if provided
        if pretrained_embeddings is not None:
            self.embedding.weight.data.copy_(pretrained_embeddings)
            if freeze_embeddings:
                self.embedding.weight.requires_grad = False

        # LSTM layer
        self.lstm = nn.LSTM(
            input_size=embedding_dim,
            hidden_size=hidden_dim,
            num_layers=lstm_layers,
            batch_first=True
        )

        # Global pooling
        self.global_pool = nn.AdaptiveAvgPool1d(1)

        # Fully connected layers
        self.fc1 = nn.Linear(hidden_dim, hidden_dim)
        self.fc2 = nn.Linear(hidden_dim, 1)

        # Activation functions
        self.relu = nn.ReLU()
        self.sigmoid = nn.Sigmoid()
```

This model shows a total of 406,817 parameters of which only 6,817 are trainable, so training will be fast!

```
========================================================================
Layer (type:depth-idx)                 Output Shape            Param #
========================================================================
TextClassificationModel                [32, 1]                --
├─Embedding: 1-1                       [32, 60, 50]           (400,000)
├─Dropout: 1-2                         [32, 60, 50]           --
├─LSTM: 1-3                            [32, 60, 16]           6,528
├─Dropout: 1-4                         [32, 60, 16]           --
├─AdaptiveAvgPool1d: 1-5              [32, 16, 1]            --
├─Linear: 1-6                          [32, 16]               272
```

```
├─ReLU: 1-7                          [32, 16]                --
├─Dropout: 1-8                       [32, 16]                --
├─Linear: 1-9                        [32, 1]                 17
├─Sigmoid: 1-10                      [32, 1]                 --
================================================================================
Total params: 406,817
Trainable params: 6,817
Non-trainable params: 400,000
Total mult-adds (M): 25.34
================================================================================
Input size (MB): 0.02
Forward/backward pass size (MB): 1.02
Params size (MB): 1.63
Estimated Total Size (MB): 2.66
================================================================================
```

You can now train as before, and you can see how this architecture, with the pretrained embeddings and stacked LSTMs, reduces overfitting really nicely! Figure 7-21 shows the Training versus Validation accuracy on the sarcasm dataset using LSTMs and pretrained GloVe embeddings, while Figure 7-22 shows the loss on training versus validation, where the closeness of the curves demonstrates that we're not overfitting.

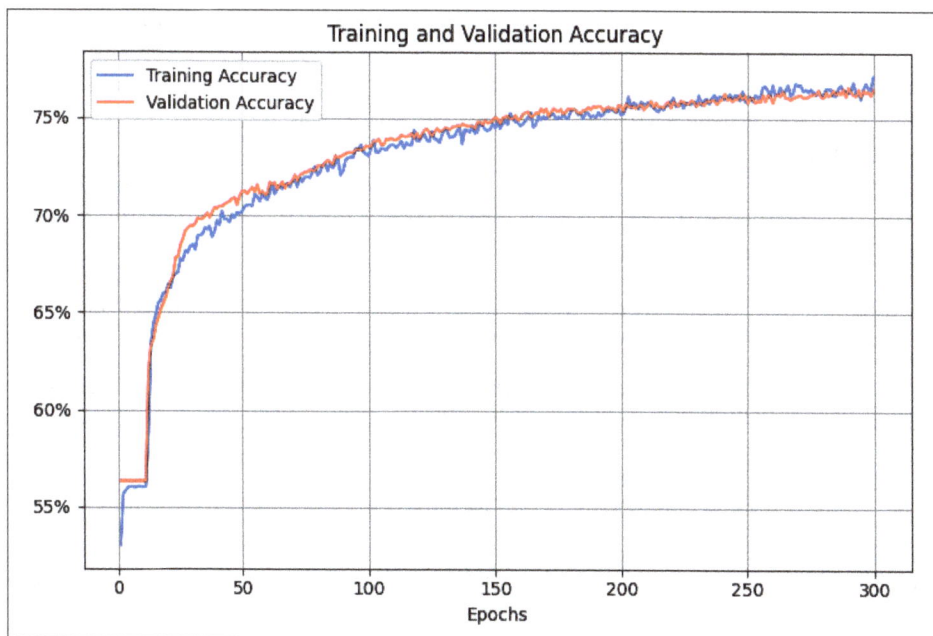

Figure 7-21. Training versus validation accuracy on the sarcasm dataset with LSTMs and GloVe

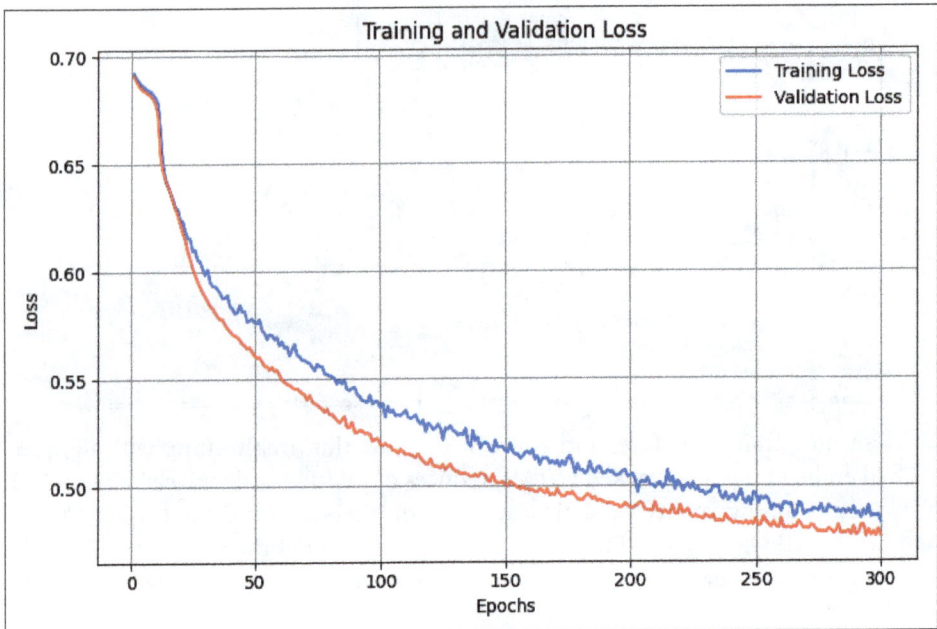

Figure 7-22. Training and validation loss on the sarcasm dataset with LSTMs and GloVe

For further analysis, you'll want to consider your vocab size. One of the optimizations you did in the previous chapter to avoid overfitting was intended to prevent the embeddings becoming overburdened with learning low-frequency words: you avoided overfitting by using a smaller vocabulary of frequently used words. In this case, as the word embeddings have already been learned for you with GloVe, you could expand the vocabulary—but by how much?

The first thing to explore is how many of the words in your corpus are actually in the GloVe set. It has 1.2 million words, but there's no guarantee it has *all* of your words.

When building the `word_index`, you can call `build_vocab_glove` with a *really* large number and it will ignore any words over the total amount. So, for example, say you call this:

```
word_index = build_vocab_glove(training_sentences, max_vocab_size=100,000)
```

With the sarcasm dataset, you'll get a vocab_size of 22,457 returned. If you like, you can then explore the GloVe embeddings to see just how many of these words are present in GloVE. Start by creating a dictionary for the embeddings and reading the GloVE file into it:

```
embeddings_dict = {}
embedding_dim = 50
glove_file = f'glove.6B.{embedding_dim}d.txt'
```

```
# Read GloVe embeddings
print(f"Loading GloVe embeddings from {glove_file}...")
with open(glove_file, 'r', encoding='utf-8') as f:
    for line in f:
        values = line.split()
        word = values[0]
        vector = np.asarray(values[1:], dtype='float32')
        embeddings_dict[word] = vector
```

Then, you can compare this with your word_index that you created from the entire corpus with the preceding line:

```
found_words = 0
for word, idx in word_index.items():
    if word in embeddings_dict:
        found_words += 1
print(found_words)
```

In the case of sarcasm, 21,291 of the words were found in GloVE, which is the vast majority, so the principles you used in Chapter 6 to choose how many you should train on (i.e., picking those with sufficient frequency to have a signal) will still apply!

Using this method, I chose to use a vocabulary size of 8,000 (instead of the 2,000 that was previously used to avoid overfitting) to get the results you saw just now. I then tested it with headlines from *The Onion*, the source of the sarcastic headlines in the sarcasm dataset, against other sentences, as shown here:

```
test_sentences = ["It Was, For, Uh, Medical Reasons, Says Doctor To
                  Boris Johnson, Explaining Why They Had To Give Him Haircut",
                  "It's a beautiful sunny day",
                  "I lived in Ireland, so in high school they made me
                  learn to speak and write in Gaelic",
                  "Census Foot Soldiers Swarm Neighborhoods, Kick Down
                  Doors To Tally Household Sizes"]
```

The results for these headlines are as follows—remember that values close to 50% (0.5) are considered neutral, those close to 0 are considered nonsarcastic, and those close to 1 are considered sarcastic:

```
tensor([[0.9316],
        [0.1603],
        [0.6959],
        [0.9594]], device='cuda:0')

Text: It Was, For, Uh, Medical Reasons, Says Doctor To Boris Johnson,
      Explaining Why They Had To Give Him Haircut
Probability: 0.9316
Classification: Sarcastic
------------------------------------------------------------------------

Text: It's a beautiful sunny day
```

```
Probability: 0.1603
Classification: Not Sarcastic
------------------------------------------------------------

Text: I lived in Ireland, so in high school they made me learn to speak
      and write in Gaelic
Probability: 0.6959
Classification: Sarcastic
------------------------------------------------------------

Text: Census Foot Soldiers Swarm Neighborhoods, Kick Down Doors To Tally
      Household Sizes
Probability: 0.9594
Classification: Sarcastic
------------------------------------------------------------
```

The first and fourth sentences, which are taken from *The Onion*, showed 93%+ likelihood of sarcasm. The statement about the weather was strongly nonsarcastic (16%), and the sentence about going to high school in Ireland was deemed to be potentially sarcastic but not with high confidence (69%).

Summary

This chapter introduced you to recurrent neural networks, which use sequence-oriented logic in their design and can help you understand the sentiment in sentences based not only on the words they contain but also on the order in which they appear. You saw how a basic RNN works, as well as how an LSTM can build on this to enable context to be preserved over the long term. These models are the precursors to the popular and famous "transformers" models used to underpin generative AI.

You also used LSTMs to improve the sentiment analysis model you've been working on, and you then looked into overfitting issues with RNNs and techniques to improve them, including by using transfer learning from pretrained embeddings.

In Chapter 8, you'll use what you've learned so far to explore how to predict words, and from there, you'll be able to create a model that creates text and writes poetry for you!

Using ML to Create Text

With the release of ChatGPT in 2022, the words *generative AI* entered the common lexicon. This simple application that allowed you to chat with a cloud-based AI seemed almost miraculous in how it could answer your queries with knowledge of almost everything in human experience. It worked by using a very advanced evolution beyond the recurrent neural networks you saw in the last chapter, by using a technique called *transformers*.

A *transformer* learns the patterns that turn one piece of text into another. With a large enough transformer architecture and a large enough set of text to learn from, the GPT model (GPT stands for generative pretrained transformers) could predict the next tokens to follow a piece of text. When GPT was wrapped in an application that made it more user friendly, a whole new industry was born.

While creating models with transformers is beyond the scope of this book, we will look at their architecture in detail in Chapter 15.

The principles involved in training models with transformers can be replicated with smaller, simpler, architectures like RNNs or LSTM. We'll explore that in this chapter and with a much smaller corpus of text—traditional Irish songs.

So, for example, consider this line of text from a famous TV show:

> You know nothing, Jon Snow.

A next-token-predictor model, created with RNNs, came up with these song lyrics in response:

> You know nothing, Jon Snow
> the place where he's stationed
> be it Cork or in the blue bird's son
> sailed out to summer

old sweet long and gladness rings
so I'll wait for the wild Colleen dying

This text was generated by a very simple model that was trained on a small corpus. I've enhanced it a little by adding line breaks and punctuation, but other than the first line, all of the lyrics were generated by the model you'll learn how to build in this chapter. It's kind of cool that it mentions a *wild Colleen dying*—if you've watched the show that Jon Snow comes from, you'll understand why!

In the last few chapters, you saw how you can use PyTorch with text-based data—first tokenizing it into numbers and sequences that can be processed by a neural network, then using embeddings to simulate sentiment using vectors, and finally using deep and recurrent neural networks to classify text. We used the sarcasm dataset, a small and simple one, to illustrate how all this works.

In this chapter we're going to shift gears: instead of classifying existing text, you'll create a neural network that can *predict* text and thus *generate* text.

Given a corpus of text, the network will attempt to learn and understand the *patterns* of words within the text so that it can, given a new piece of text called a *seed*, predict what word should come next. Once the network has that, the seed and the predicted word become the new seed, and it can predict the next word. Thus, when trained on a corpus of text, a neural network can attempt to write new text in a similar style. To create the preceding piece of poetry, I collected lyrics from a number of traditional Irish songs, trained a neural network with them, and used it to predict words.

We'll start simple, using a small amount of text to illustrate how to build up to a predictive model, and we'll end by creating a full model with a lot more text. After that, you can try it out to see what kind of poetry it can create!

To get started, you'll have to treat the text a little differently from how you've been treating it thus far. In the previous chapters, you took sentences and turned them into sequences that were then classified based on the embeddings for the tokens within them. But when it comes to creating data that can be used to train a predictive model like this one, there's an additional step in which you need to transform the sequences into *input sequences* and *labels*, where the input sequence is a group of words and the label is the next word in the sentence. You can then train a model to match the input sequences to their labels, so that future predictions can pick a label that's close to the input sequence.

Turning Sequences into Input Sequences

When predicting text, you need to train a neural network with an input sequence (feature) that has an associated label. Matching sequences to labels is the key to predicting text. In this case, you won't have explicit labels like you do when you are classifying, but instead, you'll split the sentence, and for a block of *n* words, the next word in the sentence will be the label.

So, for example, if in your corpus you had the sentence "Today has a beautiful blue sky," then you could split it into "Today has a beautiful blue" as the feature and "sky" as the label. Then, if you were to get a prediction for the text "Today has a beautiful blue," it would likely be "sky." If, in the training data, you also had "Yesterday had a beautiful blue sky," you would split it in the same way, and if you were to get a prediction for the text "Tomorrow will have a beautiful blue," then there's a high probability that the next word would be "sky."

If you train a network with lots of sentences, where you remove the last word and make it the label, you can quickly build up a predictive model in which the most likely next word in the sentence can be predicted from an existing body of text.

We'll start with a very small corpus of text—an excerpt from a traditional Irish song from the 1860s, some of the lyrics of which are as follows:

> In the town of Athy one Jeremy Lanigan
> Battered away 'til he hadn't a pound.
> His father died and made him a man again
> Left him a farm and ten acres of ground.
> He gave a grand party for friends and relations
> Who didn't forget him when come to the wall,
> And if you'll but listen I'll make your eyes glisten
> Of the rows and the ructions of Lanigan's Ball.
> Myself to be sure got free invitation,
> For all the nice girls and boys I might ask,
> And just in a minute both friends and relations
> Were dancing round merry as bees round a cask.
> Judy O'Daly, that nice little milliner,
> She tipped me a wink for to give her a call,
> And I soon arrived with Peggy McGilligan
> Just in time for Lanigan's Ball.

You'll want to create a single string with all the text and set that to be your data. Use \n for the line breaks. Then, this corpus can be easily loaded and tokenized. First, the `tokenize` function will split the text into individual words, and then the `create_word_dictionary` will create a dictionary with an index for each individual word in the text:

```python
def tokenize(text):
    tokens = text.lower().split()
    return tokens

def create_word_dictionary(word_list):
```

```
# Create an empty dictionary
word_dict = {}
word_dict["UNK"] = 0
# Counter for unique values
counter = 1

# Iterate through the list and assign numbers to unique words
for word in word_list:
    if word not in word_dict:
        word_dict[word] = counter
        counter += 1

return word_dict
```

Note that this is a very simplistic approach for the purpose of learning how these work. In production systems, you'd likely either use off-the-shelf components that have been built for scale or greatly enhance them for scale and exception checking.

With these functions, you can then create a word_index of the simple corpus, like this:

```
data="In the town of Athy one Jeremy Lanigan \n
     Battered away til he hadnt a pound. ..."

tokens = tokenize(data)
word_index = create_word_dictionary(tokens)

total_words = len(tokenizer.word_index) + 1
```

The result of this process is to replace the words with their token values (see Figure 8-1).

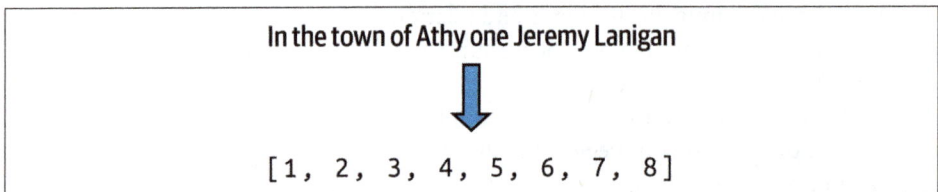

In the town of Athy one Jeremy Lanigan

⬇

[1, 2, 3, 4, 5, 6, 7, 8]

Figure 8-1. Tokenizing a sentence

To train a predictive model, we should take a further step here: splitting the sentence into multiple smaller sequences so, for example, we can have one sequence consisting of the first two tokens, another consisting of the first three, etc. We would then pad these out to be the same length as the input sequence by prepending zeros (see Figure 8-2).

Line:	Input sequences:
`[1, 2, 3, 4, 5, 6, 7, 8]`	`[0, 0, 0, 0, 0, 0, 1, 2]`
	`[0, 0, 0, 0, 0, 1, 2, 3]`
	`[0, 0, 0, 0, 1, 2, 3, 4]`
	`[0, 0, 0, 1, 2, 3, 4, 5]`
	`[0, 0, 1, 2, 3, 4, 5, 6]`
	`[0, 1, 2, 3, 4, 5, 6, 7]`
	`[1, 2, 3, 4, 5, 6, 7, 8]`

Figure 8-2. Turning a sequence into a number of input sequences

To do this, you'll need to go through each line in the corpus and turn it into a list of tokens, using functions to convert the text words into an array of their lookup values in the word dictionary, and then to create the padded versions of the subsequences. To assist you with this task, I've provided these functions: text_to_sequence and pad_sequence.

```python
def text_to_sequence(sentence, word_dict):
    # Convert sentence to lowercase and split into words
    words = sentence.lower().strip().split()

    # Convert each word to its corresponding number
    number_sequence = [word_dict[word] for word in words]

    return number_sequence

def pad_sequences(sequences, max_length=None):
    # If max_length is not specified, find the length of the longest sequence
    if max_length is None:
        max_length = max(len(seq) for seq in sequences)

    # Pad each sequence with zeros at the beginning
    padded_sequences = []
    for seq in sequences:
        # Calculate number of zeros needed
        num_zeros = max_length - len(seq)
        # Create padded sequence
        padded_seq = [0] * num_zeros + list(seq)
        padded_sequences.append(padded_seq)

    return padded_sequences
```

You can then create the input sequences using these helper functions like this:

```python
corpus = data.lower().split("\n")
```

```
input_sequences = []
for line in corpus:
    token_list = text_to_sequence(line, word_index)
    for i in range(1, len(token_list)):
        n_gram_sequence = token_list[:i+1]
        input_sequences.append(n_gram_sequence)

max_sequence_len = max([len(x) for x in input_sequences])
input_sequences = pad_sequences(input_sequences, max_sequence_len)
```

Finally, once you have a set of padded input sequences, you can split them into features and labels, where each label is simply the last token in each input sequence (see Figure 8-3).

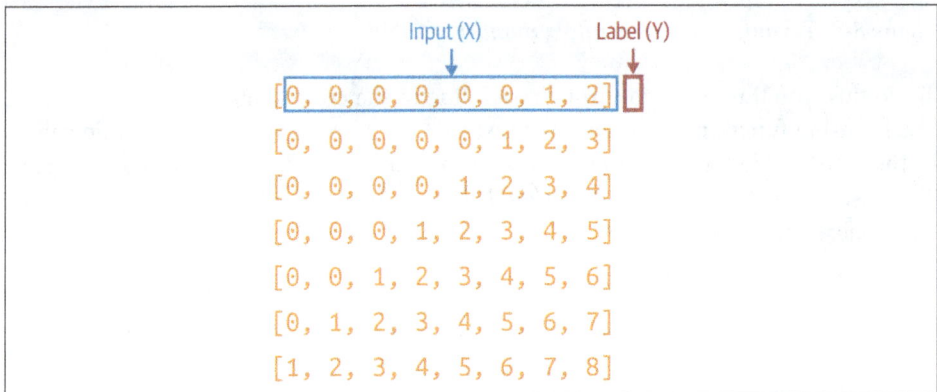

Figure 8-3. *Turning the padded sequences into features (x) and labels (y)*

When you're training a neural network, you're going to match each feature to its corresponding label. So, for example, the label for [0 0 0 0 0 0 1] will be [2].

Here's the code you use to separate the labels from the input sequences:

```
def split_sequences(sequences):
    # Create xs by removing the last element from each sequence
    xs = [seq[:-1] for seq in sequences]

    # Create labels by taking just the last element from each sequence
    labels = [seq[-1:] for seq in sequences]
    # Using [-1:] to keep it as a single-element list
    # Alternative if you want labels as single numbers instead of lists:
    # labels = [seq[-1] for seq in sequences]

    return xs, labels
xs, labels = split_sequences(input_sequences)
```

Next, you need to encode the labels. Right now, they're just tokens—for example, the number 2 at the top of Figure 8-3 is a token. But if you want to use a token as a label in a classifier, you'll have to map it to an output neuron. Thus, if you're going to classify *n* words, with each word being a class, you'll need to have *n* neurons. Here's where it's important to control the size of the vocabulary, because the more words you have, the more classes you'll need. Remember back in Chapter 2 and 3, when you were classifying fashion items with the Fashion MNIST dataset and you had 10 types of items of clothing? That required you to have 10 neurons in the output layer—but in this case, what if you want to predict up to 10,000 vocabulary words? You'll need an output layer with 10,000 neurons!

Additionally, you need to one-hot encode your labels so that they match the desired output from a neural network. Consider Figure 8-3 again. If a neural network is fed the input *x* consisting of a series of 0s followed by a 1, you'll want the prediction to be 2—but how the network delivers that is by having an output layer of vocabulary_size neurons, where the second one has the highest probability.

To encode your labels into a set of Ys that you can then use to train, you can use this code:

```python
def one_hot_encode_with_checks(value, corpus_size):
    # Check if value is within valid range
    if not 0 <= value < corpus_size:
        raise ValueError(f"Value {value} is out of range for corpus size
                          {corpus_size}")
    # Create and return one-hot encoded list
    encoded = [0] * corpus_size
    encoded[value] = 1
    return encoded
```

Note that there are many libraries and helper functions out there that can do this for you, so feel free to use them instead of the simple code in this chapter. But I want to put these methodologies in here, so you can see how it works under the hood.

You can see this visually in Figure 8-4.

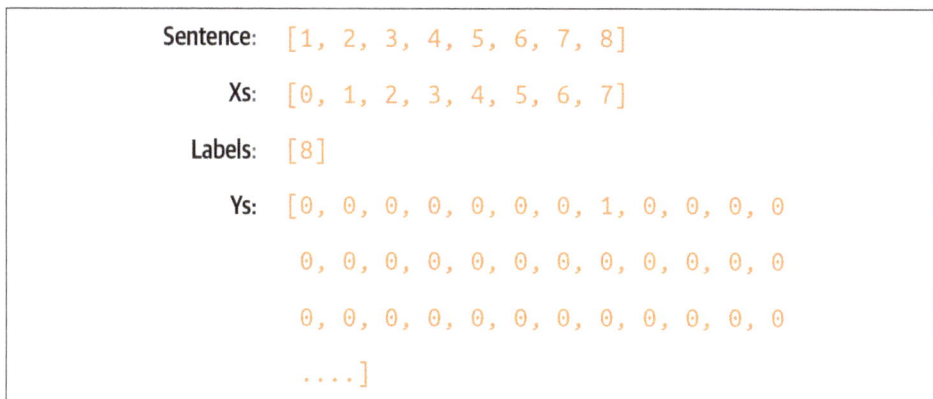

```
Sentence:  [1, 2, 3, 4, 5, 6, 7, 8]

      Xs:  [0, 1, 2, 3, 4, 5, 6, 7]

  Labels:  [8]

      Ys:  [0, 0, 0, 0, 0, 0, 0, 1, 0, 0, 0, 0

            0, 0, 0, 0, 0, 0, 0, 0, 0, 0, 0, 0

            0, 0, 0, 0, 0, 0, 0, 0, 0, 0, 0, 0

            ....]
```

Figure 8-4. One-hot encoding labels

This is a very sparse representation that, if you have a lot of training data and a lot of potential words, will eat memory very quickly! Suppose you had 100,000 training sentences, with a vocabulary of 10,000 words—you'd need 1,000,000,000 bytes just to hold the labels! But that's the way we have to design our network if we're going to classify and predict words.

Creating the Model

Let's now create a simple model that can be trained with this input data. It will consist of just an embedding layer, followed by an LSTM, followed by a dense layer.

For the embedding, you'll need one vector per word, so the parameters will be the total number of words and the number of dimensions you want to embed on. In this case, we don't have many words, so eight dimensions should be enough.

You can make the LSTM bidirectional, and the number of steps can be the length of a sequence, which is our max length minus 1 (because we took one token off the end to make the label).

Finally, the output layer will be a dense layer with the total number of words as a parameter, activated by Softmax. Each neuron in this layer will be the probability that the next word matches the word for that index value:

```python
class LSTMPredictor(nn.Module):
    def __init__(self, total_words, embedding_dim=8, hidden_dim=None):
        super(LSTMPredictor, self).__init__()

        # If hidden_dim not specified, use max_sequence_len-1 as in TF version
        if hidden_dim is None:
            hidden_dim = max_sequence_len-1

        # Embedding layer
        self.embedding = nn.Embedding(total_words, embedding_dim)

        # Bidirectional LSTM
        self.lstm = nn.LSTM(
            input_size=embedding_dim,
            hidden_size=hidden_dim,
            bidirectional=True,
            batch_first=True
        )

        # Final dense layer (accounting for bidirectional LSTM)
        self.fc = nn.Linear(hidden_dim * 2, total_words)

        # Softmax activation
        self.softmax = nn.Softmax(dim=1)

    def forward(self, x):
        # Embedding layer
```

```
x = self.embedding(x)

# LSTM layer
lstm_out, _ = self.lstm(x)

# Take the output from the last time step
lstm_out = lstm_out[:, -1, :]

# Dense layer
out = self.fc(lstm_out)

# Softmax activation
out = self.softmax(out)

return out
```

Next, you compile the model with a categorical loss function such as categorical cross entropy and an optimizer like Adam. You can also specify that you want to capture metrics:

```
# Training setup
total_words = len(word_index)
model = LSTMPredictor(total_words)
criterion = nn.CrossEntropyLoss()
optimizer = torch.optim.Adam(model.parameters())
```

It's a very simple model without a lot of data, and it trains quickly. Here are the results of training for about 15,000 epochs, which takes maybe 10 minutes. In the real world, you'll likely be training with a lot more data and thus taking a lot more time, so you'll have to consider some of the techniques we saw in Chapter 7 to ensure greater accuracy and potentially less time to train.

You'll see that it has reached very high accuracy, and there may be room for more (see Figure 8-5).

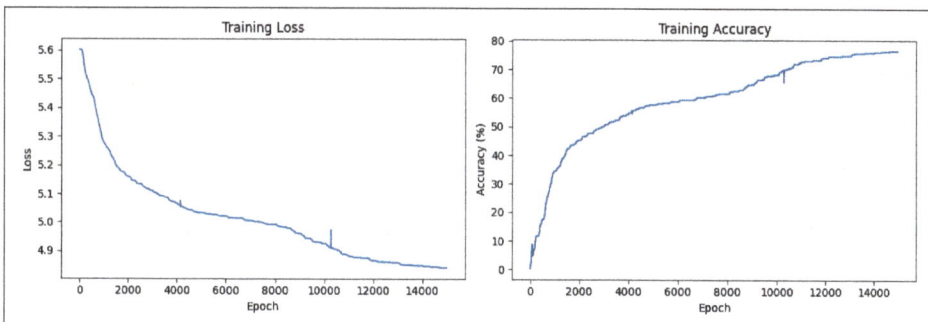

Figure 8-5. Training loss and accuracy

With the model at 80%+ accuracy, we can be assured that if we have a string of text that it has already seen, it will predict the next word accurately about 80% of the time.

Note, however, that when generating text, it will continually see words that it hasn't previously seen, so despite this good number, you'll find that the network will rapidly end up producing nonsensical text. We'll explore this in the next section.

Generating Text

Now that you've trained a network that can predict the next word in a sequence, the next step is to give it a sequence of text and have it predict the next word. Let's take a look at how to do that.

Predicting the Next Word

You'll start by creating a phrase called the *seed text*. This is the initial expression on which the network will base all the content it generates, and it will do this by predicting the next word.

Start with a phrase that the network has *already* seen, such as "in the town of Athy":

```
input_text = "In the town of Athy"
```

Next, you need to tokenize this and turn it into a sequence of tokens of the same length as used for training:

```
# Convert text to lowercase and split into words
words = input_text.lower().strip().split()

# Convert words to numbers using the word dictionary, use 0 for unknown words
number_sequence = [word_dict.get(word, 0) for word in words]

# Pad the sequence
padded_sequence = [0] * (sequence_length - len(number_sequence))
                    + number_sequence
```

Then, you need to pad that sequence to get it into the same shape as the data used for training by converting it to a tensor:

```
input_tensor = torch.LongTensor([padded_sequence])
```

Now, you can predict the next word for this token list by passing this input tensor to the model to get the output. This will be a tensor that contains the probabilities for each of the words in the dictionary and the likelihood that it will be the next token.

```
# Get prediction
with torch.no_grad():  # No need to track gradients for prediction
    output = model(input_tensor)
```

This will return the probabilities for each word in the corpus, so you should pass the results to torch.argmax to get the most likely one:

```
# Get the predicted word index (highest probability)
predicted_idx = torch.argmax(output[0]).item()
```

This should give you the value 6. If you look at the word index, you'll see that it's the word "one":

```
{'UNK': 0, 'in': 1, 'the': 2, 'town': 3, 'of': 4, 'Athy': 5,
 'one': 6, 'Jeremy': 7, 'Lanigan': 8,
```

You can also look it up in code by searching through the word index items until you find the predicted word and printing it out. You can do this by creating a reverse dictionary (mapping the value to the word, instead of vice-versa):

```
# Create reverse dictionary to convert number back to word
reverse_dict = {v: k for k, v in word_dict.items()}

# Convert predicted index to word
predicted_word = reverse_dict[predicted_idx]
```

So, starting from the text "in the town of Athy," the network predicted that the next word should be "one"—which, if you look at the training data, is correct because the song begins with this line:

> In the town of Athy one Jeremy Lanigan
> Battered away til he hadn't a pound

Indeed, if you check the top five predictions based on their values in the index, you'll get something like this:

```
Top 5 predictions:
one: 0.9999
youll: 0.0000
didnt: 0.0000
creature: 0.0000
nelly: 0.0000
```

Now that you've confirmed that the model is working, you can get creative and use different seed text. For example, when I used the seed text "sweet Jeremy saw Dublin," the next word it predicted was "his."

```
Top 5 predictions:
his: 0.7782
go: 0.1393
bellows,: 0.0605
accident: 0.0090
til: 0.0048
```

This text was chosen because all of those words are in the corpus. In such cases, you should expect more accurate results, at least at the beginning, for the predicted words.

Compounding Predictions to Generate Text

In the previous section, you saw how to use the model to predict the next word given a seed text. Now, to have the neural network create new text, you simply repeat the prediction, adding new words each time.

For example, earlier, when I used the phrase "sweet Jeremy saw Dublin," it predicted that the next word would be "his." You can build on this by appending "his" to the seed text to get "sweet Jeremy saw Dublin his" and getting another prediction. Repeating this process will give you an AI-created string of text.

Here's the updated code from the previous section that performs this loop a number of times, with the number set by the num_words parameter:

```python
def generate_sequence(model, initial_text, word_dict,
                      sequence_length, num_words=10):
    # Set model to evaluation mode
    model.eval()

    # Start with the initial text
    current_text = initial_text
    generated_sequence = initial_text

    # Create reverse dictionary for converting numbers back to words
    reverse_dict = {v: k for k, v in word_dict.items()}

    print(f"Initial text: {initial_text}")

    for i in range(num_words):
        # Convert current text to lowercase and split into words
        words = current_text.lower().strip().split()

        # Take the last 'sequence_length' words if we exceed it
        if len(words) > sequence_length:
            words = words[-sequence_length:]

        # Convert words to numbers using the word dictionary, use 0 for unknown
        number_sequence = [word_dict.get(word, 0) for word in words]

        # Pad the sequence
        padded_sequence = [0] * (sequence_length - len(number_sequence)) \
                                                + number_sequence

        # Convert to PyTorch tensor and add batch dimension
        input_tensor = torch.LongTensor([padded_sequence])

        # Get prediction
        with torch.no_grad():
            output = model(input_tensor)

        # Get the predicted word index (highest probability)
```

```python
    predicted_idx = torch.argmax(output[0]).item()

    # Convert predicted index to word
    predicted_word = reverse_dict[predicted_idx]

    # Add the predicted word to the sequence
    generated_sequence += " " + predicted_word

    # Update current text for next prediction
    current_text = generated_sequence

    # Print progress
    print(f"Generated word {i+1}: {predicted_word}")

    # Optionally print top 5 predictions for each step
    _, top_indices = torch.topk(output[0], 5)
    print(f"\nTop 5 predictions for step {i+1}:")
    for idx in top_indices:
        word = reverse_dict[idx.item()]
        probability = output[0][idx].item()
        print(f"{word}: {probability:.4f}")
    print("\n" + "-"*50 + "\n")

    return generated_sequence

# Example usage:
initial_text = "sweet Jeremy saw Dublin"
generated_text = generate_sequence(
    model=model,
    initial_text=initial_text,
    word_dict=word_index,
    sequence_length=max_sequence_len,
    num_words=10
)

print("\nFinal generated sequence:")
print(generated_text)
```

This will end up creating a string something like this:

```
sweet jeremy saw dublin his right leg acres of the nolans, dolans, daughter, of
```

It rapidly descends into gibberish. Why? The first reason is that the body of training text is really small, so it has very little context to work with. The second is that the prediction of the next word in the sequence depends on the previous words in the sequence, and if there is a poor match on the previous ones, even the best "next" match will have a low probability of being accurate. When you add this to the sequence and predict the next word after that, the likelihood of it having a low probability of accuracy is even higher—thus, the predicted words will seem semirandom.

So, for example, while all of the words in the phrase "sweet Jeremy saw Dublin" exist in the corpus, they never exist in that order. When the model made the first prediction, it chose the word "his" as the most likely candidate, and it had quite a high probability of accuracy (78%):

```
Initial text: sweet Jeremy saw Dublin
Generated word 1: his

Top 5 predictions for step 1:
his: 0.7782
go: 0.1393
bellows,: 0.0605
accident: 0.0090
til: 0.0048
```

When the model added that word to the seed to get "sweet Jeremy saw Dublin his," we had another phrase not seen in the training data, so the prediction gave the highest probability to the word "right," at 44%:

```
Generated word 2: right

Top 5 predictions for step 2:
right: 0.7678
pipes,: 0.1376
creature: 0.0458
didnt: 0.0136
youll: 0.0113
```

While occasionally there will be high certainty of a token following another (like "leg" following "right"), over time, you'll see that continuing to add words to the sentence reduces the likelihood of a match in the training data, and as such, the prediction accuracy will suffer—leading to there being a more random "feel" to the words being predicted.

This leads to the phenomenon of AI-generated content getting increasingly nonsensical over time.

For an example, check out the excellent sci-fi short *Sunspring* (*https://oreil.ly/hTBtJ*), which was written entirely by an LSTM-based network (like the one you're building here) that was trained on science fiction movie scripts. The model was given seed content and tasked with generating a new script. The results were hilarious, and you'll see that while the initial content makes sense, as the movie progresses, it becomes less and less comprehensible.

This is also the basis of *hallucination* in LLMs, a common phenomenon that reduces trust in them.

Extending the Dataset

You can easily extend the same pattern that you used for the hardcoded dataset to use a text file. I've hosted a text file containing about 1,700 lines of text gathered from a number of songs that you can use for experimentation. With a little modification, you can use this instead of the single hardcoded song.

To download the data in Colab, use the following code:

```
!wget --no-check-certificate \
    https://storage.googleapis.com/learning-datasets/ \
    irish-lyrics-eof.txt -O /tmp/irish-lyrics-eof.txt
```

Then, you can simply load the text from it into your corpus like this:

```
data = open('/tmp/irish-lyrics-eof.txt').read()
corpus = data.lower().split("\n")
```

The rest of your code will then work with very little modification!

Training this for 50,000 epochs—which takes about 30 minutes on a T4 Colab instance—brings you to about 30% accuracy, with the curve flattening out (see Figure 8-6).

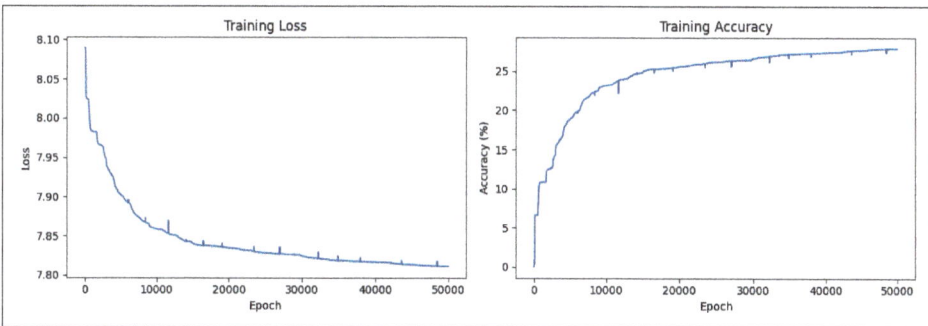

Figure 8-6. Training on a larger dataset

When using Google Colab, you can choose various accelerators on the backend, which can make your training go much faster. In this case, as noted, I used a T4. To do this for yourself, when in Colab, under the Connect menu, you'll see a Change Runtime Type option. Select that, and you'll see the accelerators available for you to use.

Trying the phrase "in the town of Athy" again yields a prediction of "one" but this time with just over 83% probability:

```
Initial text: in the town of Athy
Using device: cuda
```

```
Generated word 1: one

Top 5 predictions for step 1:
one: 0.8318
is: 0.1648
she: 0.0016
thee: 0.0013
that: 0.0003


--------------------------------------------------

Generated word 2: my

Top 5 predictions for step 2:
my: 0.9377
of: 0.0622
is: 0.0001
that: 0.0000
one: 0.0000
```

Running for a few more tokens, we can see output like the following. It's beginning to create songs, though it's quite quickly descending into gibberish!

```
in the town of athy one my heart
was they were the a reflections
on me all the frivolity;
of me and me and me and the
there was my heart was
on the a over the frivolity;
```

For "sweet Jeremy saw Dublin," the predicted next word is "she," with a probability of 80%. Predicting the next few words yields this:

```
sweet jeremy saw dublin she of his on the frivolity; of a heart is the ground
```

It's looking a little better! But can we improve it further?

Improving the Model Architecture

One way that you can improve the model is to change its architecture, using multiple stacked LSTMs and some other optimization techniques. Given that there's no clear benchmark for accuracy with a dataset like this one—there's no right or wrong classification, nor is there a target regression—it's difficult to establish when a model is good or bad. Therefore, accuracy results can be very subjective.

That being said, they're still a good yardstick, so in this section, I'm going to explore some additions you can make to the architecture to improve the accuracy metric.

Embedding Dimensions

In Chapter 6, we discussed the fact that the optimal dimension for embeddings is the fourth root of the number of words. In this scenario, the vocabulary has 3,259 words, and the fourth root of this is approximately 8. Another rule of thumb is the log of this number—and log(3259) is a little over 32. So, if the network is learning slowly, you have the option to pick a number between these two values. That gives you enough "room" to capture the relationships between words.

Initializing the LSTMs

Often, parameters in a neural network are initialized to zero. You can give learning a little bit of a kickstart by initializing different layers to different types supported by various research findings. We briefly cover these types of layers in the following subsections.

Embedding layers

Embedding layers can be initialized with a normal distribution, with a standard deviation scaled by `1/sqrt(embedding_dim)` for better gradient flow. That's similar to `word2vec`-style initialization.

LSTM layers

LSTM has four internal neural types—input, forget, cell, and output—and their weight matrix is a bunch of them stacked together. The different types benefit from different initializations. Two great papers that discuss this are "An Empirical Exploration of Recurrent Network Architectures" by Rafal Jozefowicz et al. (*https://oreil.ly/UuvQO*) and "On the Difficulty of Training Recurrent Neural Networks" (*https://oreil.ly/Ttvll*) by Razvan Pascanu et al. The specifics of LSTM are beyond the scope of this chapter, but check the associated code for one methodology to initialize them.

Final linear layer

In their 2015 paper "Delving Deep into Rectifiers," Kaiming He et al. (*https://oreil.ly/_MM6A*) explored initialization of linear layers and proposed "Kaiming" initialization (aka "He" initialization). A detailed explanation of this is beyond the scope of this book, but the code is available in the notebooks in the GitHub repository (*https://github.com/lmoroney/PyTorch-Book-FIles*).

Variable Learning Rate

In every example we've seen so far, we've explored different learning rates and their impact on the network—but you can actually *vary* the learning rate as the network learns. Values that work well in early epochs may not work so well in later ones, so putting together a scheduler that adjusts this learning rate epoch by epoch can help you create networks that learn more effectively.

For this, PyTorch provides a `torch.optim.lr_scheduler` that you can program to change over the course of the training:

```
scheduler = torch.optim.lr_scheduler.OneCycleLR(
    optimizer,
    max_lr=0.01,                # Peak learning rate
    epochs=20000,               # Total epochs
    steps_per_epoch=1,          # Steps per epoch
    pct_start=0.1,              # Percentage of training spent increasing lr
    div_factor=10.0,            # Initial lr = max_lr/10
    final_div_factor=1000.0     # Final lr = initial_lr/1000
)
```

In Chapter 7, we looked at the idea of a *learning rate* (LR), which is a hyperparameter that, if set to be too large, will cause the network to overlearn, and if set to be too small will prevent the network from learning effectively. The nice thing about this is that you can set it as a variable rate, which we do here. In the early epochs, we want the network to learn fast, so we have a large LR—and in the later epochs, we don't want it to overfit, so we gradually reduce the LR.

The `pct_start` parameter defines a warm-up period as the first 10% of the training, during which the learning rate gradually increases up to the maximum (in this case, 0.01) and then decreases to 1/1000 of the initial learning rate (determined by `final_div_factor`).

You can see the impact this has on training in Figure 8-7, where it reached ~90% accuracy in 6,800 epochs before triggering early stopping.

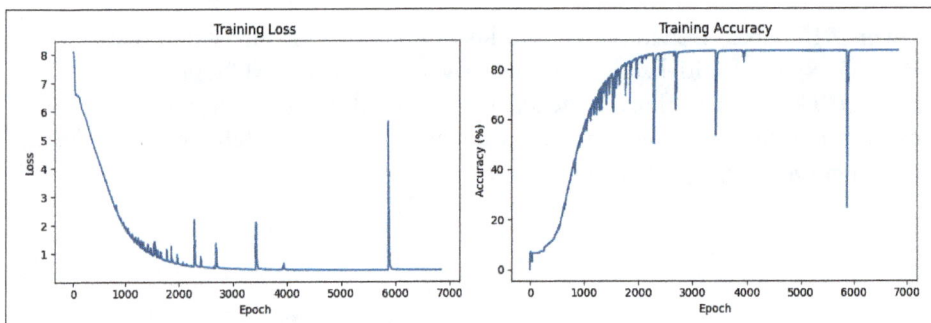

Figure 8-7. Accuracy and loss with variable learning rate

This time, when testing with the same phrases as before, I got "little" as the next word after "in the town of Athy" with a 26% probability, and I got "one" as the next word after "sweet Jeremy saw Dublin" with a 32% probability. Again, when predicting more words, the output quickly descended into gibberish.

Here are some examples:

```
sweet jeremy saw dublin one evening two white ever we once to raise you,
tis young i was told my heart as found has

you know nothing jon snow you should laugh all the while at me curious style,
twould set your heart a bubblin will lámh. you that

in the town of athy one jeremy lanigan do lámh. pretty generation her soul,
fell on the stony ground red we were feeble was down
```

The word lámh shows up in this text, and it's Gaelic for *hand*. And do lámh means *your hand*.

If you get different results, don't worry—you didn't do anything wrong, but the random initialization of the neurons will impact the final scores.

Improving the Data

There's a little trick that you can use to extend the size of this dataset without adding any new songs. It's called *windowing* the data. Right now, every line in every song is read as a single line and then turned into input sequences, as you saw in Figure 8-2. While humans read songs line by line to hear rhyme and meter, the model doesn't have to, in particular when using bidirectional LSTMs.

So, instead of taking the line "In the town of Athy, one Jeremy Lanigan," processing that, and then moving to the next line ("Battered away 'til he hadn't a pound") and processing that, we could treat all the lines as one long, continuous text. We could then create a "window" into that text of *n* words, process that, and then move the window forward one word to get the next input sequence (see Figure 8-8).

In this case, far more training data can be yielded in the form of an increased number of input sequences. Moving the window across the entire corpus of text would give us ((*number_of_words* – *window_size*) × *window_size*) input sequences that we could train with.

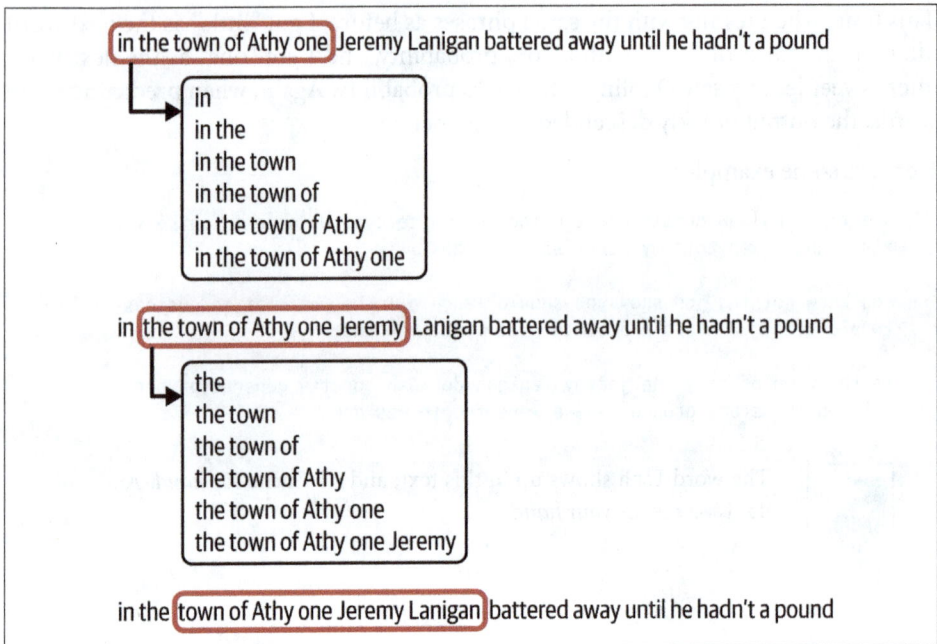

Figure 8-8. A moving word window

The code is pretty simple—when loading the data, instead of splitting each song line into a "sentence," we can create them on the fly from the words in the corpus:

```
window_size=10
sentences=[]
alltext=[]
data = open('/tmp/irish-lyrics-eof.txt').read()
corpus = data.lower()
words = corpus.split(" ")
range_size = len(words)-max_sequence_len
for i in range(0, range_size):
    thissentence=""
    for word in range(0, window_size-1):
        word = words[i+word]
        thissentence = thissentence + word
        thissentence = thissentence + " "
    sentences.append(thissentence)
```

In this case, because we no longer have sentences and we're creating sequences that are the same size as the moving window, `max_sequence_len` is the size of the window. The full file is read, converted to lowercase, and split into an array of words using string splitting. The code then loops through the words and makes sentences of each word from the current index up to the current index plus the window size, adding each of those newly constructed sentences to the sentences array.

To train this, you'll likely need a higher-memory GPU. I used the 40Gb A100 that's available in Colab.

When training, you'll notice that the extra data makes training much slower per epoch, but the results are greatly improved and the generated text descends into gibberish much more slowly.

Here's an example that caught my eye—particularly the last line:

```
you know nothing jon snow
tell the loved ones and the friends
we would neer see again.
and the way of their guff again
and high tower might ask,
not see night unseen
```

There are many hyperparameters you can try to tune. Changing the window size will change the amount of training data—a smaller window size can yield more data, but there will be fewer words to give to a label, so if you set it too small, you'll end up with nonsensical poetry. You can also change the dimensions in the embedding, the number of LSTMs, or the size of the vocabulary to use for training. Given that percentage accuracy isn't the best measurement—you'll want to make a more subjective examination of how much "sense" the poetry makes—there's no hard-and-fast rule to follow to determine whether your model is "good" or not. Of course, you *will* be limited by the amount of data you have and the compute that's available to you—the bigger the model, the more power you will need.

Ultimately, the important thing is to experiment and have fun!

Character-Based Encoding

For the last few chapters, we've been looking at NLP using word-based encoding. I find that much easier to get started with, but when it comes to generating text, you might also want to consider *character-based encoding* because the number of unique *characters* in a corpus tends to be a lot less than the number of unique *words*. If you use this approach, you can have a lot fewer neurons in your output layer, and your output predictions can be spread across fewer probabilities. For example, if you look at the dataset of the complete works of Shakespeare (*https://oreil.ly/XW_ab*), you'll see that there are only 65 unique characters in the entire set. Basically, Shakespeare only really used uppercase and lowercase letters and some punctuation to give a unique set of 65 characters!

So, when you are making predictions with this dataset, instead of looking at the probabilities for the next word across 2,700 words as in the Irish songs dataset, you're only looking at 65. This makes your model a bit simpler!

What's also nice about character-based encoding is that punctuation characters are also included, so line breaks, etc., can be predicted. As an example, when I used an RNN that was trained on the Shakespeare corpus to predict the text following on from my favorite *Game of Thrones* line, I got the following:

YGRITTE:

You know nothing, Jon Snow.
Good night, we'll prove those body's servants to
The traitor be these mine:
So diswarl his body in hope in this resceins,
I cannot judg appeal't.

MENENIUS:

Why, 'tis pompetsion.

KING RICHARD II:

I think he make her thought on mine;
She will not: suffer up thy bonds:
How doched it, I pray the gott,
We'll no fame to this your love, and you were ends

It's kind of cool that she identifies him as a traitor and wants to tie him up ("diswarl his body"), but I have no idea what "resceins" means! If you watch the show, this is part of the plot, so maybe Shakespeare was on to something without realizing it!

Of course, I do think we tend to be a little more forgiving when using something like Shakespeare's texts as our training data, because the language is already a little unfamiliar.

As with the Irish songs model, the output from the Shakespeare dataset does quickly degenerate into nonsensical text, but it's still fun to play with. To try it for yourself, you can check out the Colab (*https://oreil.ly/cbz9c*). This Colab is TensorFlow based, not PyTorch based. See the GitHub repository (*https://oreil.ly/kQ7aa*) for a similar example that gives different results but is PyTorch based.

Summary

In this chapter, we explored how to do basic text generation using a trained LSTM-based model. You learned how you can split text into training features and labels by using words as labels, and you also learned how create a model that, when given seed text, can predict the next likely word. Then, you iterated on this to improve the model for better results by exploring a dataset of traditional Irish songs. Hopefully, this was a fun introduction to how ML models can synthesize text and also gave you the knowledge you need to understand the foundational principles of generative AI. This approach was massively improved with the transformer architecture that under-pins how LLM models like GPT and Gemini work!

## CHAPTER 9

Understanding Sequence and Time Series Data

Time series are everywhere. You've probably seen them in things like weather forecasts, stock prices, and historic trends like Moore's law. If you're not familiar with *Moore's law*, it predicts that the number of transistors on a microchip will roughly double every two years—and for almost 50 years, it has proven to be an accurate predictor of the future of computing power and cost (see Figure 9-1).

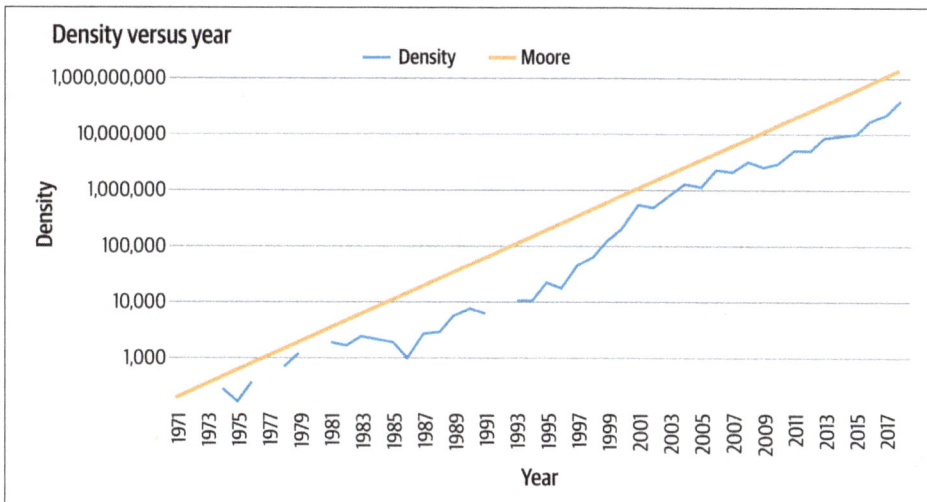

Figure 9-1. Moore's law

The gaps in Figure 9-1 are missing data for that period of time, but the general trend still holds.

Time series data is a set of values that are spaced over time, usually in a particular order or denoting values of a thing at a timestamped point in time. When a time series is plotted on a graph, the *x*-axis is usually temporal in nature. Often, there are a number of values plotted on the time axis, such as in the example shown in Figure 9-1, where the number of transistors is one plot and the predicted value from Moore's law is the other. This is called a *multivariate* time series. If there's just a single value—for example, the volume of rainfall over time—then it's called a *univariate* time series.

With Moore's law, predictions are simple because there's a fixed and simple rule that allows us to roughly predict the future—a rule that has held for about 50 years.

But what about a time series like the one shown in Figure 9-2?

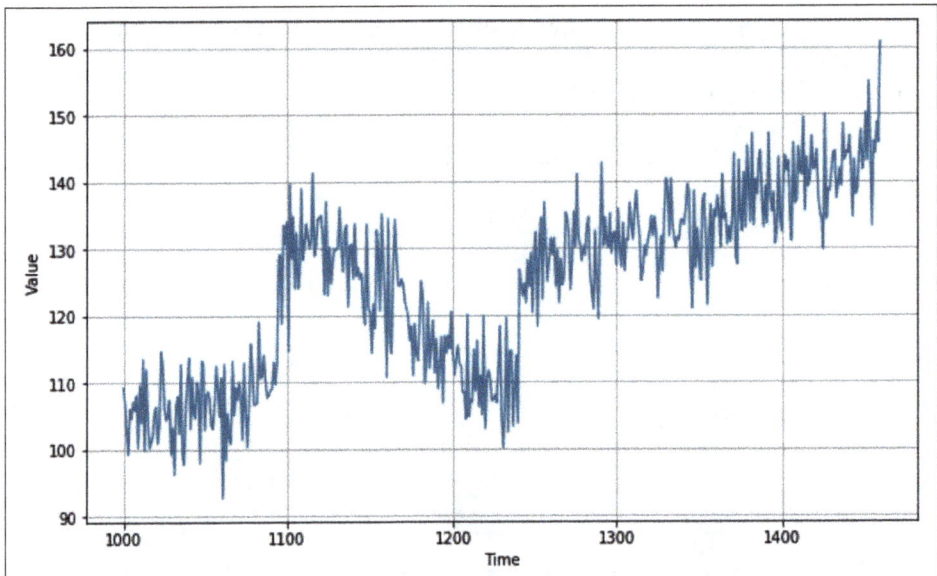

Figure 9-2. A real-world time series

While this time series was artificially created (and you'll see how to do that later in this chapter), it has all the attributes of a complex real-world time series, like a stock chart or a chart depicting seasonal rainfall. Despite their seeming randomness, time series have some common attributes that are helpful in designing ML models that can predict them, as described in the next section.

Common Attributes of Time Series

While time series might appear random and noisy, they often have common attributes that are predictable. In this section, we'll explore some of them.

Trend

Time series typically move in a specific direction. In the case of Moore's law, it's easy to see that over time, the values on the *y*-axis increase and there's an upward trend. There's also an upward trend in the time series in Figure 9-2. Of course, this won't always be the case: some time series may be roughly level over time, despite seasonal changes, and others may have a downward trend. For example, this is the case in the inverse version of Moore's law that predicts the price per transistor.

Seasonality

Many time series have a repeating pattern over time, with the repeats happening at regular intervals called *seasons*. Consider, for example, temperature in weather. We typically have four seasons per year, with the temperature being highest in summer. So, if you plotted weather over several years, you'd see peaks happening every four seasons, giving us the concept of seasonality. But this phenomenon isn't limited to weather—consider, for example, Figure 9-3, which is a plot of traffic to a website.

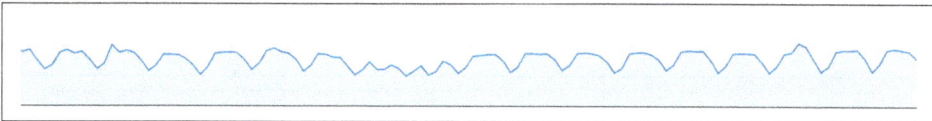

Figure 9-3. Website traffic

It's plotted week by week, and you can see regular dips. Can you guess what they are? The site in this case is one that provides information for software developers, and as you would expect, it gets less traffic on weekends! Thus, the time series has a seasonality of five high days and two low days. The data is plotted over several months, with the Christmas and New Year's holidays roughly in the middle, so you can see an additional seasonality there. If I had plotted it over some years, you'd clearly see the additional end-of-year dip.

There are many ways that seasonality can manifest in a time series. Traffic to a retail website, for instance, might peak on the weekends.

Autocorrelation

Another feature that you may see in time series is predictable behavior after an event. You can see this in Figure 9-4, in which there are clear spikes, but after each spike, there's a deterministic decay. This is called *autocorrelation*.

In this case, we can see a particular set of behavior that is repeated. Autocorrelations may be hidden in a time series pattern, but they have inherent predictability, so a time series containing many autocorrelations may be predictable.

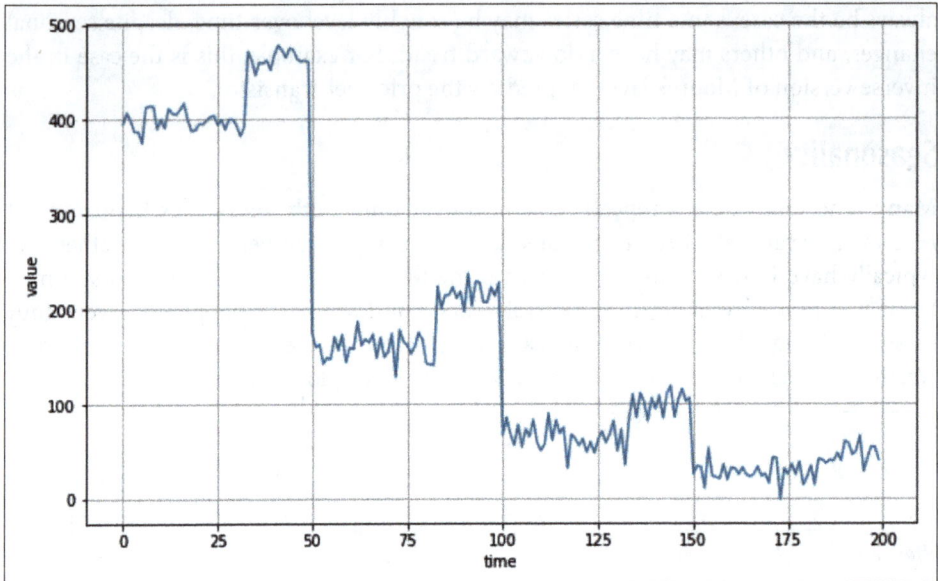

Figure 9-4. Autocorrelation

Noise

As its name suggests, *noise* is a set of seemingly random perturbations in a time series. These perturbations lead to a high level of unpredictability and can mask trends,
seasonal behavior, and autocorrelation. For example, Figure 9-5 shows the same autocorrelation from Figure 9-4 but with a little noise added. Suddenly, it's much harder
to see the autocorrelation and predict values.

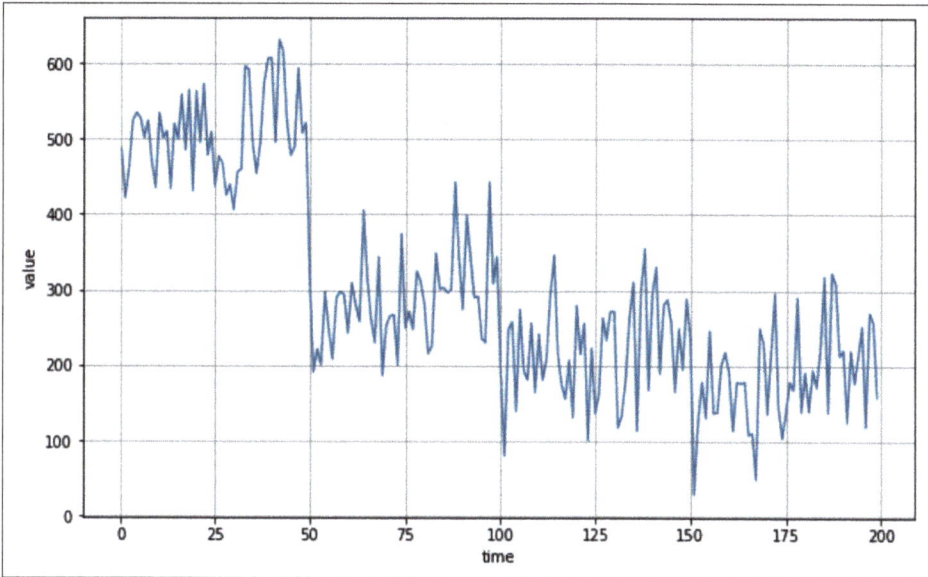

Figure 9-5. Autocorrelated series with added noise

Given all of these factors, let's explore how you can make predictions on time series
that contain these attributes.

Techniques for Predicting Time Series

Before we get into ML-based prediction—which is the topic of the next few chapters—we'll explore some more naive prediction methods. These will enable you to establish a baseline that you can use to measure the accuracy of your ML predictions.

Naive Prediction to Create a Baseline

The most basic method to predict a time series is to say that the predicted value at time $t + 1$ is the same as the value from time t, effectively shifting the time series by a single period.

Let's begin by creating a time series that has trend, seasonality, and noise:

```python
def plot_series(time, series, format="-", start=0, end=None):
    plt.plot(time[start:end], series[start:end], format)
    plt.xlabel("Time")
    plt.ylabel("Value")
    plt.grid(True)

def trend(time, slope=0):
    return slope * time

def seasonal_pattern(season_time):
    """Just an arbitrary pattern, you can change it if you wish"""
    return np.where(season_time < 0.4,
                    np.cos(season_time * 2 * np.pi),
                    1 / np.exp(3 * season_time))

def seasonality(time, period, amplitude=1, phase=0):
    """Repeats the same pattern at each period"""
    season_time = ((time + phase) % period) / period
    return amplitude * seasonal_pattern(season_time)

def noise(time, noise_level=1, seed=None):
    rnd = np.random.RandomState(seed)
    return rnd.randn(len(time)) * noise_level

time = np.arange(4 * 365 + 1, dtype="float32")
baseline = 10
series = trend(time, .05)
baseline = 10
amplitude = 15
slope = 0.09
noise_level = 6

# Create the series
series = baseline + trend(time, slope)
                  + seasonality(time, period=365, amplitude=amplitude)
# Update with noise
series += noise(time, noise_level, seed=42)
```

After plotting this, you'll see something like Figure 9-6.

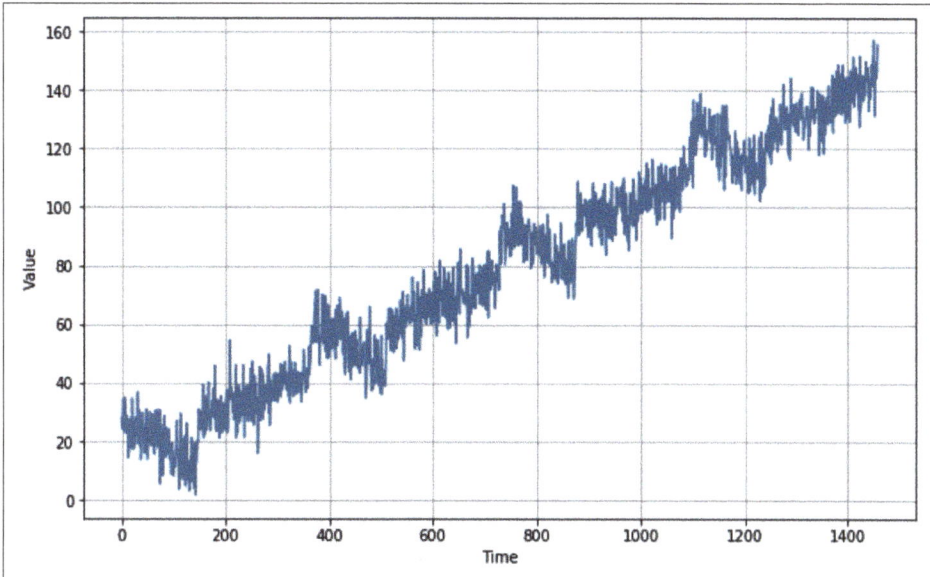

Figure 9-6. A time series showing trend, seasonality, and noise

Now that you have the data, you can split it like any data source into a training set, a validation set, and a test set. When there's some seasonality in the data, as you can see in this case, it's a good idea when splitting the series to ensure that there are whole seasons in each split. So, for example, if you wanted to split the data in Figure 9-6 into training and validation sets, a good place to do this might be at time step 1,000, which would give you training data up to step 1,000 and validation data after step 1,000.

However, you don't actually need to do the split here because you're just doing a naive forecast where each value t is simply the value at step $t - 1$, but for the purposes of illustration in the next few figures, we'll zoom in on the data from time step 1,000 onward.

To predict the series from a split time period onward, where the period that you want to split from is in the variable split_time, you can use code like this:

```
naive_forecast = series[split_time - 1:-1]
```

Figure 9-7 shows the validation set (from time step 1,000 onward, which you get by setting split_time to 1000) with the naive prediction overlaid.

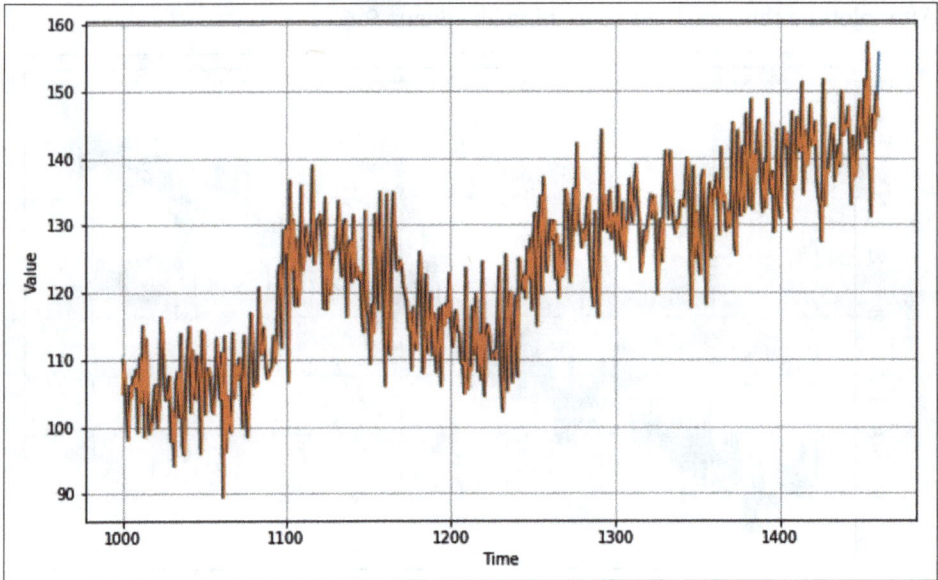

Figure 9-7. Naive forecast on time series

It looks pretty good—there is a relationship between the values—and, when charted over time, the predictions appear to closely match the original values. But how would you measure the accuracy?

Measuring Prediction Accuracy

There are a number of ways to measure prediction accuracy, but we'll concentrate on two of them: the *mean squared error* (MSE) and *mean absolute error* (MAE).

With MSE, you simply take the difference between the predicted value and the actual value at time *t*, square it (to remove negatives), and then find the average of all of them.

With MAE, you calculate the difference between the predicted value and the actual value at time *t*, take its absolute value to remove negatives (instead of squaring), and find the average of all of them.

For the naive forecast you just created based on our synthetic time series, you can get the MSE and MAE like this:

```python
import torch
import torch.nn.functional as F

# Mean Squared Error
mse = F.mse_loss(torch.tensor(x_valid), torch.tensor(naive_forecast)).item()
print(mse)

# Mean Absolute Error
mae = F.l1_loss(torch.tensor(x_valid), torch.tensor(naive_forecast)).item()
print(mae)
```

I got an MSE of 76.47 and an MAE of 6.89. As with any prediction, if you can reduce the error, you can increase the accuracy of your predictions. We'll look at how to do that next.

Less Naive Predictions: Using a Moving Average for Prediction

The previous naive prediction took the value at time $t - 1$ to be the forecasted value at time t. Using a *moving average* is similar, but instead of just taking the value from $t - 1$, it takes a group of values (say, 30), averages them out, and sets that average value to be the predicted value at time t. Here's the code:

```python
def moving_average_forecast(series, window_size):
    """Forecasts the mean of the last few values.
       If window_size=1, then this is equivalent to naive forecast"""
    forecast = []
    for time in range(len(series) - window_size):
        forecast.append(series[time:time + window_size].mean())
    return np.array(forecast)

moving_avg = moving_average_forecast(series, 30)[split_time - 30:]

plt.figure(figsize=(10, 6))
plot_series(time_valid, x_valid)
plot_series(time_valid, moving_avg)
```

Figure 9-8 shows the plot of the moving average against the data.

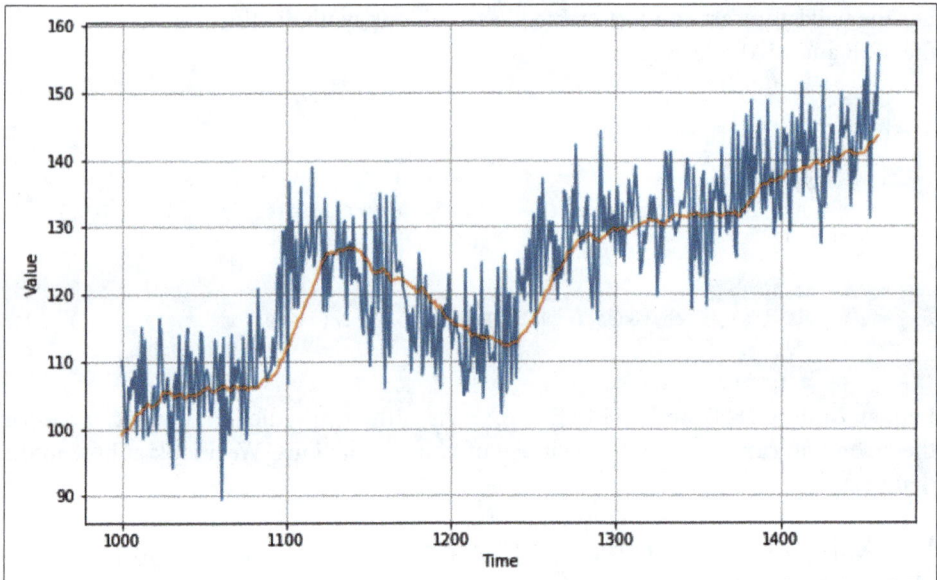

Figure 9-8. Plotting the moving average

When I plotted this time series, I got an MSE and MAE of 49 and 5.5, respectively, so it definitely improved the prediction a little. But this approach doesn't take into account the trend or the seasonality, so we may be able to improve it further with a little analysis.

Improving the Moving-Average Analysis

Given that the seasonality in this time series is 365 days, you can smooth out the trend and seasonality by using a technique called *differencing*, which just subtracts the value at *t* – 365 from the value at *t*. This will flatten out the diagram. Here's the code:

```
diff_series = (series[365:] - series[:-365])
diff_time = time[365:]
```

You can now calculate a moving average of *these* values and add back in the past values:

```
diff_moving_avg =
    moving_average_forecast(diff_series, 50)[split_time - 365 - 50:]

diff_moving_avg_plus_smooth_past =
    moving_average_forecast(series[split_time - 370:-360], 10) +
    diff_moving_avg
```

When you plot this (see Figure 9-9), you can already see an improvement in the predicted values: the trend line is very close to the actual values, albeit with the noise smoothed out. Seasonality seems to work, as does the trend.

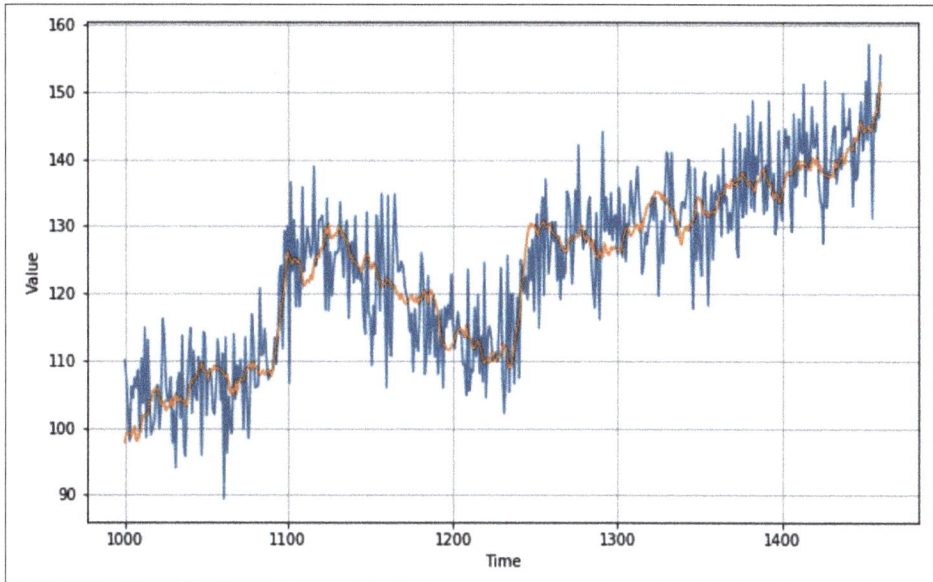

Figure 9-9. Improved moving average

This impression is confirmed when you calculate the MSE and MAE—in this case, I got 40.9 and 5.13, respectively, showing a clear improvement in the predictions.

Summary

This chapter introduced time series data and some of the common attributes of time series. You created a synthetic time series and saw how you can start making naive predictions on it, and from these predictions, you established baseline measurements using MSE and MAE. Synthetic data is also a really cool area for exploration—and hopefully, some of the techniques you explored in this chapter will be useful on your learning journey.

This chapter was a nice break from PyTorch and ML, but in the next chapter, you'll go back to using ML to see if you can improve on your predictions!

## CHAPTER 10

Creating ML Models to Predict Sequences

Chapter 9 introduced sequence data and the attributes of a time series, including seasonality, trend, autocorrelation, and noise. You created a synthetic series to use for predictions, and you explored how to do basic statistical forecasting.

Over the next couple of chapters, you'll learn how to use ML for forecasting. But before you start creating models, you need to understand how to structure the time series data for training predictive models by creating what we'll call a *windowed dataset*.

To understand why you need to do this, consider the time series you created in Chapter 9. You can see a plot of it in Figure 10-1.

If at any point, you want to predict a value at time *t*, you'll want to predict it as a function of the values preceding time *t*. For example, say you want to predict the value of the time series at time step 1,200 as a function of the 30 values preceding it. In this case, the values from time steps 1,170 to 1,199 would determine the value at time step 1,200 (see Figure 10-2).

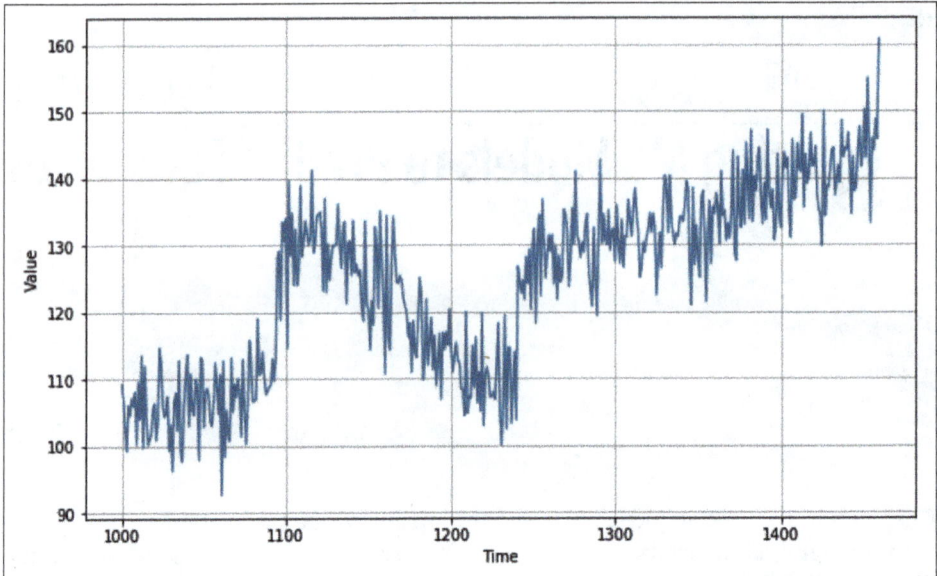

Figure 10-1. Synthetic time series

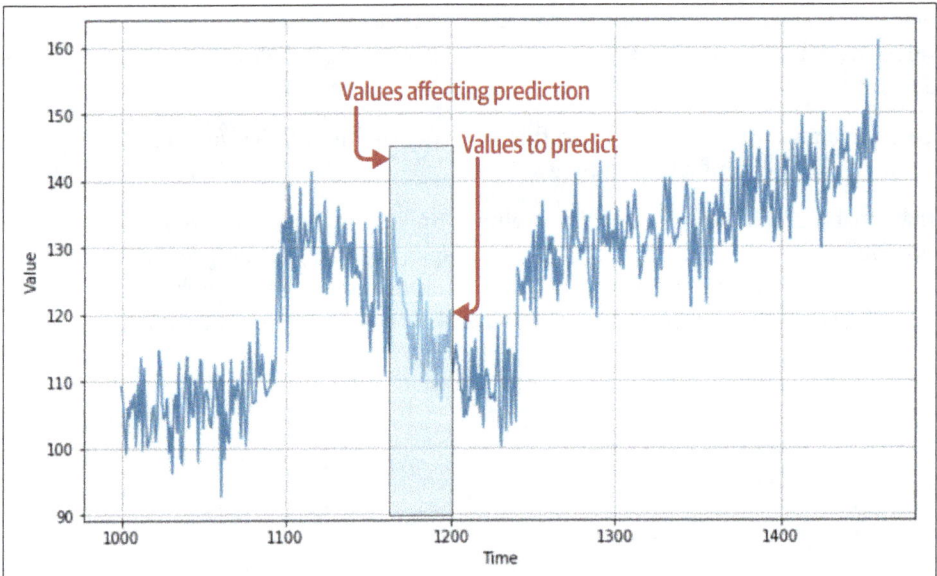

Figure 10-2. Previous values impacting prediction

Now, this begins to look familiar: you can consider the values from 1,170 to 1,199 to be your *features* and the value at 1,200 to be your *target label*. If you can get your data-set into a condition where you have a certain number of values as features and the following value as the *target label*, and if you do this for every known value in the dataset, then you'll end up with a pretty decent set of features and labels that you can use to train a model.

Before doing this for the time series dataset from Chapter 9, let's create a very simple dataset that has all the same attributes but a much smaller amount of data.

Creating a Windowed Dataset

PyTorch has a lot of APIs that are useful for manipulating data. For example, you can use `torch.arange(10)` to create a basic dataset containing the numbers 0–9, thus emulating a time series. You can then turn that dataset into the beginnings of a windowed dataset. Here's the code:

```python
import torch

def create_sliding_windows(data, window_size, shift=1):
    # Convert input to tensor if it isn't already
    if not isinstance(data, torch.Tensor):
        data = torch.tensor(data)

    # Calculate number of valid windows
    n = len(data)
    num_windows = max(0, (n - window_size) // shift + 1)

    # Create strided view of data
    windows = data.unfold(0, window_size, shift)

    return windows

# Example usage:
data = torch.arange(10)
windows = create_sliding_windows(data, window_size=5, shift=1)

# Print each window
for window in windows:
    print(window.numpy())
```

First, it creates the dataset by using a range, which simply makes the dataset contain the values 0 to $n - 1$, where n is, in this case, 10.

Next, calling `create_sliding_windows` and passing a parameter of 5 specifies that the network should split the dataset into windows of five items. Specifying `shift=1` causes each window to then be shifted one spot from the previous one: the first window will contain the five items beginning at 0, the next window will contain the five items beginning at 1, etc.

Running this code will give you the following result:

```
[0 1 2 3 4]
[1 2 3 4 5]
[2 3 4 5 6]
[3 4 5 6 7]
[4 5 6 7 8]
[5 6 7 8 9]
```

Earlier, you saw that we want to make training data out of this, where there are *n* values defining a feature and there is a subsequent value giving a label. You can do this with some simple Python list slicing that splits each window into two things: everything before the last value and the last value only.

This uses the unfold technique on tensor data, which creates a sliding window over your data to turn it into sets like those outlined previously.

It also takes three parameters:

Dimension (0 in this case)
    This is the dimension along which to unfold.

window_size
    This is the size of each sliding window.

shift
    This is the stride/step size between windows.

This process gives us an *x* and a *y* dataset, as shown here:

```python
import torch

def create_sliding_windows_with_target(data, window_size, shift=1):
    # Convert input to tensor if it isn't already
    if not isinstance(data, torch.Tensor):
        data = torch.tensor(data, dtype=torch.float32)

    # Create windows using unfold
    windows = data.unfold(0, window_size, shift)

    # Split each window into features
    features = windows[:, :-1]  # All elements except the last
    targets = windows[:, -1:]   # Just the last element

    return features, targets

# Example usage:
data = torch.arange(10)
features, targets = create_sliding_windows_with_target(data, window_size=5,
                                                        shift=1)

# Print each window's features and target
```

```
for x, y in zip(features, targets):
    print(f"Features: {x.numpy()}, Target: {y.numpy()}")
```

The results are now in line with what you'd expect. The first four values in the window can be thought of as the features, with the subsequent value being the label:

```
[0 1 2 3] [4]
[1 2 3 4] [5]
[2 3 4 5] [6]
[3 4 5 6] [7]
[4 5 6 7] [8]
[5 6 7 8] [9]
```

Now, with the PyTorch TensorDataset type, you can turn this into a dataset and do things like shuffling and batching natively.

Note that when shuffling data, it's good practice to ensure that the validation and test datasets are separated first. In time series data, shuffling before the split can cause information from the training set to bleed into the test set, which would compromise the evaluation process.

Here, it's been shuffled and batched with a batch size of 2:

```
from torch.utils.data import TensorDataset, DataLoader

# Create dataset
data = torch.arange(10)
features, targets = create_sliding_windows_with_target(data, window_size=5,
                                                        shift=1)

# Combine features and targets into a dataset
dataset = TensorDataset(features, targets)

# Create DataLoader with shuffling and batching
batch_size = 2
dataloader = DataLoader(dataset, batch_size=batch_size, shuffle=True)

# Example iteration
for batch_features, batch_targets in dataloader:
    print(f"Batch features shape: {batch_features.shape}")
    print(f"Features:\n{batch_features}")
    print(f"Targets:\n{batch_targets}\n")
```

The results show that the first batch has two sets of *x* (starting at 5 and 0, respectively) with their target labels, the second batch has two sets of *x* (starting at 1 and 3, respectively) with their target labels, and so on:

```
tensor([[5, 6, 7, 8],
        [0, 1, 2, 3]])
Targets:
tensor([[9],
```

```
        [4]])


Features:
tensor([[1, 2, 3, 4],
        [3, 4, 5, 6]])
Targets:
tensor([[5],
        [7]])


Features:
tensor([[4, 5, 6, 7],
        [2, 3, 4, 5]])
Targets:
tensor([[8],
        [6]])
```

With this technique, you can now turn any time series dataset into a set of training data for a neural network. In the next section, you'll explore how to take the synthetic data from Chapter 9 and create a training set from it. From there, you'll move on to creating a simple DNN that is trained on this data and can be used to predict future values.

Creating a Windowed Version of the Time Series Dataset

As a recap, here's the code we used in the previous chapter to create a synthetic time series dataset:

```python
import numpy as np
def trend(time, slope=0):
    return slope * time

def seasonal_pattern(season_time):
    return np.where(season_time < 0.4,
                    np.cos(season_time * 2 * np.pi),
                    1 / np.exp(3 * season_time))

def seasonality(time, period, amplitude=1, phase=0):
    season_time = ((time + phase) % period) / period
    return amplitude * seasonal_pattern(season_time)

def noise(time, noise_level=1, seed=None):
    rnd = np.random.RandomState(seed)
    return rnd.randn(len(time)) * noise_level

time = np.arange(4 * 365 + 1, dtype="float32")
series = trend(time, 0.1)
baseline = 10
amplitude = 20
```

```
slope = 0.09
noise_level = 5

series = baseline + trend(time, slope)
series += seasonality(time, period=365, amplitude=amplitude)
series += noise(time, noise_level, seed=42)
```

This will create a time series that looks like the one in Figure 10-3. If you want to change it, feel free to tweak the values of the various constants.

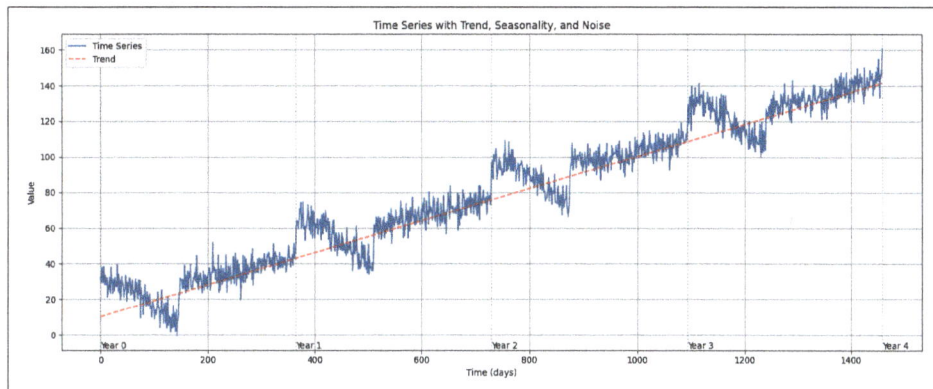

Figure 10-3. Plotting the time series with trend, seasonality, and noise

Once you have the series, you can turn it into a windowed dataset with code similar to that in the previous section:

```python
import torch
from torch.utils.data import TensorDataset, DataLoader

# Convert the numpy series to a PyTorch tensor
series_tensor = torch.tensor(series, dtype=torch.float32)

# Create windowed dataset with 30-day windows (predicting next day)
window_size = 30
features, targets = create_sliding_windows_with_target(
    series_tensor, window_size=window_size, shift=1)

# Create PyTorch Dataset and DataLoader
dataset = TensorDataset(features, targets)
batch_size = 32
train_loader = DataLoader(dataset, batch_size=batch_size, shuffle=True)

# Print some information about the dataset
print(f"Series length: {len(series)}")
print(f"Number of windows: {len(features)}")
print(f"Feature shape: {features.shape}")  # Should be (num_windows,
                                           #             window_size-1)
print(f"Target shape: {targets.shape}")    # Should be (num_windows, 1)
```

```
# Show a few examples
print("\nFirst few windows:")
for i in range(3):
    print(f"\nWindow {i+1}:")
    print(f"Features (previous {window_size-1} days): {features[i].numpy()}")
    print(f"Target (next day): {targets[i].item():.2f}")
```

You can see the output here:

```
Series length: 1461
Number of windows: 1432
Feature shape: torch.Size([1432, 29])
Target shape: torch.Size([1432, 1])

First few windows:

Window 1:
Features (previous 29 days): [32.48357  29.395714 33.40659  37.858486
 29.14184  29.20528  38.32948  34.322147 28.183279 33.283253 28.287313
 28.303862 31.864614 21.104889 22.057411 27.875519 25.622026 32.25094
 26.127428 23.588236 37.95459  29.468477 30.900469 23.39905  27.755371
 30.980967 24.615065 32.186863 27.23822 ]
Target (next day): 28.71

Window 2:
Features (previous 29 days): [29.395714 33.40659  37.858486 29.14184
 29.20528  38.32948  34.322147 28.183279 33.283253 28.287313 28.303862
 31.864614 21.104889 22.057411 27.875519 25.622026 32.25094  26.127428
 23.588236 37.95459  29.468477 30.900469 23.39905  27.755371 30.980967
 24.615065 32.186863 27.23822  28.710733]
Target (next day): 27.08

Window 3:
Features (previous 29 days): [33.40659  37.858486 29.14184  29.20528
 38.32948  34.322147 28.183279 33.283253 28.287313 28.303862 31.864614
 21.104889 22.057411 27.875519 25.622026 32.25094  26.127428 23.588236
 37.95459  29.468477 30.900469 23.39905  27.755371 30.980967 24.615065
 32.186863 27.23822  28.710733 27.083256]
Target (next day): 39.27
```

To train a model with this data, you'll split the series into training and validation datasets. In this case, we'll train on 1,000 records by splitting the list and turning it into `train_dataset` and `val_dataset` subsets. This uses the `Subset` class from `torch.utils.data`.

You can then load these with a `DataLoader`, as we saw in Chapter 3:

```
train_size = 1000
total_windows = len(full_dataset)
train_indices = list(range(train_size))
val_indices = list(range(train_size, total_windows))
```

```
# Create training and validation datasets using Subset
train_dataset = Subset(full_dataset, train_indices)
val_dataset = Subset(full_dataset, val_indices)

# Create DataLoaders
batch_size = 32
train_loader = DataLoader(train_dataset, batch_size=batch_size, shuffle=True)
val_loader = DataLoader(val_dataset, batch_size=batch_size, shuffle=False)
```

The important thing to remember now is that your data is a dataset, so you can easily use it in model training without further coding.

If you want to inspect what the data looks like, you can do so with code like this:

```
features, target = train_dataset[0]
print("First window:")
print(f"Features shape: {features.shape}")
print(f"Features: {features.numpy()}")
print(f"Target: {target.item()}\n")
```

Here, the `batch_size` is set to 1, just to make the results more readable. You'll therefore end up with output like this, in which a single set of data is in the batch:

```
First window:
Features shape: torch.Size([29])
Features: [32.48357  29.395714 33.40659  37.858486 29.14184  29.20528  38.32948
 34.322147 28.183279 33.283253 28.287313 28.303862 31.864614 21.104889
 22.057411 27.875519 25.622026 32.25094  26.127428 23.588236 37.95459
 29.468477 30.900469 23.39905  27.755371 30.980967 24.615065 32.186863
 27.23822 ]
Target: 28.71073341369629
```

The first batch of numbers are the features. We've set the window size to 30, so it's a 1×30 tensor. The second number is the label (28.710 in this case), which the model will try to fit the features to. You'll see how that works in the next section.

Creating and Training a DNN to Fit the Sequence Data

Now that you have the data, creating a neural network model becomes very straightforward. Let's first explore a simple DNN. Here's the model definition:

```
# Define the model
class TimeSeriesModel(nn.Module):
    def __init__(self, window_size):
        super(TimeSeriesModel, self).__init__()
        # window_size-1 because our features are window_size-1
        self.network = nn.Sequential(
            nn.Linear(window_size-1, 10),
            nn.ReLU(),
            nn.Linear(10, 10),
```

```
                nn.ReLU(),
                nn.Linear(10, 1)
        )

    def forward(self, x):
        return self.network(x)
```

It's a super-simple model with three linear layers, the first of which accepts the input shape of `window_size` and then a hidden layer of 10 neurons, before an output layer that will contain the predicted value.

The model is initialized with a loss function and optimizer, as before:

```
# Initialize model, loss function, and optimizer
model = TimeSeriesModel(window_size)
criterion = nn.MSELoss()
optimizer = optim.Adam(model.parameters())
```

In this case, the loss function is specified as `MSELoss`, which stands for "mean squared error" and is commonly used in regression problems (which is what this ultimately boils down to). For the optimizer, `Adam` is a good fit. I won't go into detail on these types of functions in this book, but any good resource on ML will teach you about them—Andrew Ng's seminal "Deep Learning Specialization (*https://oreil.ly/A8QzN*)" on Coursera is a great place to start.

Training is then pretty standard. It's composed of loading the batches from the training loader and performing a forward pass with them, followed by a backward pass with optimization:

```
for batch_features, batch_targets in train_loader:
    batch_features = batch_features.to(device)
    batch_targets = batch_targets.to(device)

    # Forward pass
    outputs = model(batch_features)
    loss = criterion(outputs, batch_targets)

    # Backward pass and optimize
    optimizer.zero_grad()
    loss.backward()
    optimizer.step()

    train_loss += loss.item()
```

Given that we also have a validation dataset, we can also perform a validation pass for each epoch:

```
# Validation phase
model.eval()
val_loss = 0
with torch.no_grad():
```

```
    for batch_features, batch_targets in val_loader:
        batch_features = batch_features.to(device)
        batch_targets = batch_targets.to(device)

        outputs = model(batch_features)
        val_loss += criterion(outputs, batch_targets).item()

# Calculate average losses
train_loss /= len(train_loader)
val_loss /= len(val_loader)

train_losses.append(train_loss)
val_losses.append(val_loss)
```

As you train, you'll see the loss function report a number that will start high but decline steadily. Figure 10-4 shows the loss over 100 epochs.

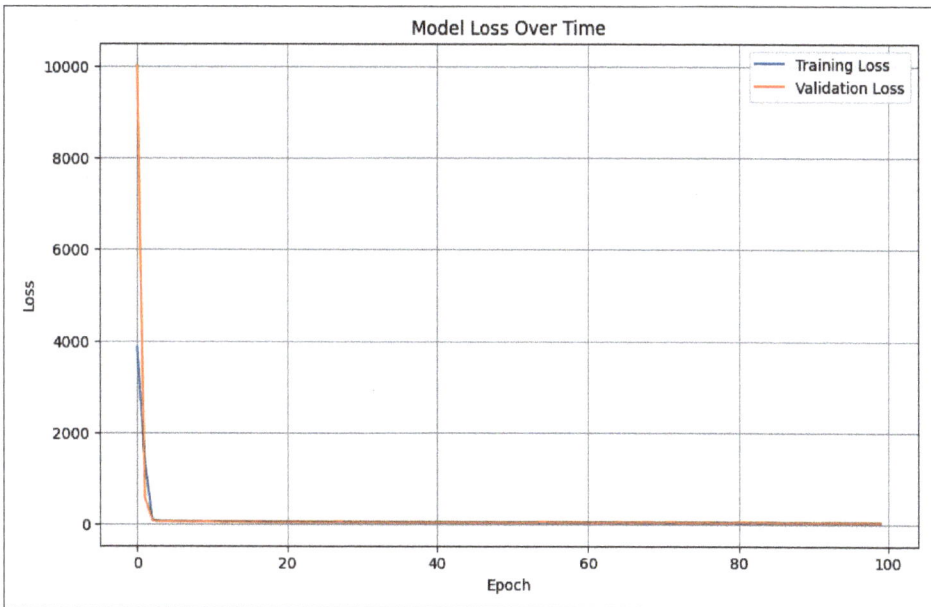

Figure 10-4. *The DNN predictor of model loss over time for the time series data*

Evaluating the Results of the DNN

Once you have a trained DNN, you can start predicting with it. But remember, you have a windowed dataset, so the prediction for a given point is based on the values of a certain number of time steps before it.

Also, given that the dataset is batched, we can easily use the loaders to access batches and explore what it looks like to predict on them.

Here's the code. We iterate through each batch in the loader and get the features and targets, and then we can get the predictions for the batch by sending the batch to the model:

```
for batch_features, batch_targets in val_loader:
    batch_features = batch_features.to(device)
    predictions = model(batch_features)
    val_predictions.extend(predictions.cpu().numpy())
    val_targets.extend(batch_targets.numpy())
```

The predictions are then converted from PyTorch tensors into NumPy arrays using .numpy(), and then the batches are turned into a single list with the extend() call.

You might have also noticed the .cpu() in this code. PyTorch allows you to designate *where* your code runs, and if you have a GPU or other accelerator available, you can push the intense calculations of ML to it. You can also use a CPU to do other things like processing and preprocessing data to save accelerator time. This code allows you to explicitly express that.

So, you can then compare your predictions with the actual values quite easily. Here's the code you use to plot the predicted values against the actual ones using matplotlib:

```
# Make predictions
model.eval()
with torch.no_grad():
    # Get predictions for validation set
    val_predictions = []
    val_targets = []
    for batch_features, batch_targets in val_loader:
        batch_features = batch_features.to(device)
        outputs = model(batch_features)
        val_predictions.extend(outputs.cpu().numpy())
        val_targets.extend(batch_targets.numpy())

# Plot predictions vs actual for validation set
plt.figure(figsize=(15, 6))
plt.plot(val_targets, label='Actual', color="lightgrey")
plt.plot(val_predictions, label='Predicted', color="red")
plt.title('Predictions vs Actual Values (Validation Set)')
plt.xlabel('Time')
plt.ylabel('Value')
plt.legend()
plt.grid(True)
plt.show()
```

You can see the results of this in Figure 10-5. The line for the predicted values (in red) closely matches the overall pattern of the original data, but it's less noisy, with a lot less variance.

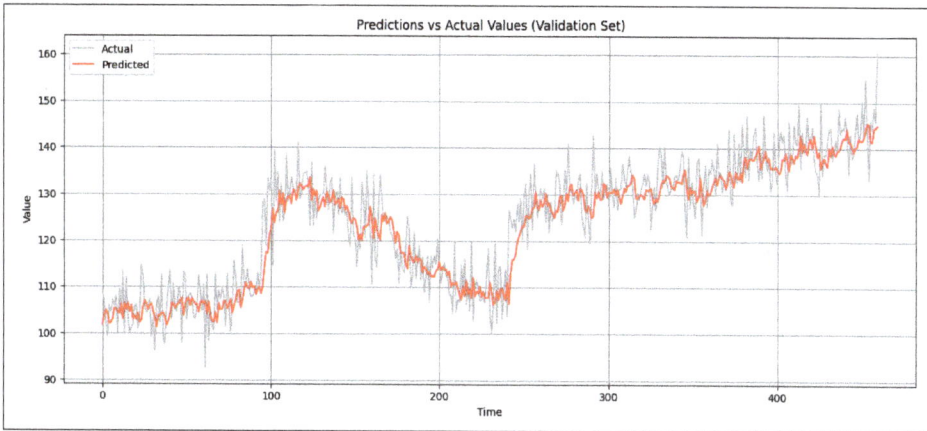

Figure 10-5. Predicted versus actual values in the validation set

From a quick visual inspection, you can see that the prediction isn't bad because it's generally following the curve of the original data. When there are rapid changes in the data, the prediction takes a little time to catch up, but on the whole, it isn't bad.

However, it's hard to be precise when eyeballing the curve. It's best to have a good metric, and in Chapter 9 you learned about one—the MAE. Now that you have the valid data and the results, you can measure the MAE with this code by using `torch.mean` and `torch.abs`:

```
val_predictions_tensor = torch.tensor(val_predictions)
val_targets_tensor = torch.tensor(val_targets)
mae_torch = torch.mean(torch.abs(
                    val_predictions_tensor - val_targets_tensor))
print(f"Validation MAE (PyTorch): {mae_torch:.4f}")
```

Randomness has been introduced into the data, so your results may vary, but when I tried it, I got a value of 4.57 as the MAE.

You could also argue that at this point, the process of getting the predictions as accurate has become the process of minimizing that MAE. There are some techniques that you can use to do this, including the obvious changing of the window size. I'll leave you to experiment with that, but in the next section, you'll do some basic hyperparameter tuning on the optimizer to improve how your neural network learns, and see what impact that will have on the MAE.

Tuning the Learning Rate

In the previous example, you might recall that you compiled the model with an optimizer that looked like this:

```
optimizer = optim.Adam(model.parameters())
```

In that case, you didn't specify an LR, so the model used the default LR of 1×10^{-3}. But that seemed to be a really arbitrary number. What if you changed it, and how should you go about changing it? It would take a lot of experimentation to find the best rate.

One way to experiment with this is by using a `torch.optim.lr_scheduler`, which can change the LR on the fly, epoch by epoch, as the model trains.

A good practice is to start with a higher LR and gradually reduce it as the network learns. Here's an example:

```
optimizer = optim.Adam(model.parameters(), lr=0.01)
```

In this case, you're going to start the LR at 1e – 2, which is really high. However, you can set a scheduler to multiply the LR rate by a "gamma" amount every n epochs. So, for example, the following code will change it every 30 epochs. We'll start at 0.01, and then, after the thirtieth epoch, it will multiply the LR by .1 to get 0.001. After 60, it will be 0.0001, etc.:

```
scheduler = lr_scheduler.StepLR(optimizer, step_size=30, gamma=0.1)
```

When we run the scheduler like this, the performance improves a little—giving MAE of 4.36.

You can continue to explore like this by tweaking the LR and also the size of the window—30 days of data to predict 1 day may not be enough, so you might want to try a window of 40 days. Also, try training for more epochs. With a bit of experimentation, you could get an MAE of close to 4, which isn't bad.

Summary

In this chapter, you took the statistical analysis of the time series from Chapter 9 and applied ML to try to do a better job of prediction. ML really is all about pattern matching, and, as expected, you were able to quickly create a deep neural network to spot the patterns with low error before exploring some hyperparameter tuning to improve the accuracy further.

In Chapter 11, you'll go beyond a simple DNN and examine the implications of using an RNN to predict sequential values.

Using Convolutional and Recurrent Methods for Sequence Models

The last few chapters introduced you to sequence data. You saw how to predict it, first by using statistical methods and then by using basic ML methods with a deep neural network. You also explored how to tune the model's hyperparameters for better performance.

In this chapter, you'll look at additional techniques that may further enhance your ability to predict sequence data by using convolutional neural networks as well as recurrent neural networks.

Convolutions for Sequence Data

In Chapter 3, you were introduced to convolutions in which a two-dimensional (2D) filter was passed over an image to modify it and potentially extract features. Over time, the neural network learned which filter values were effective at matching the modifications that had been made to the pixels to their labels, thus effectively extracting features from the image. The same technique can be applied to numeric time series data, but with one modification: the convolution will be one dimensional (1D) instead of two dimensional.

Consider, for example, the series of numbers in Figure 11-1.

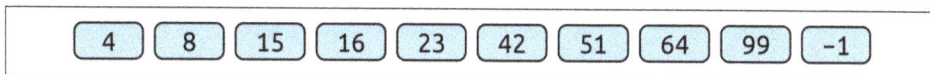

Figure 11-1. A sequence of numbers

A 1D convolution could operate on these as follows. Consider the convolution to be a 1×3 filter with filter values of –0.5, 1, and –0.5, respectively. In this case, the first value in the sequence will be lost and the second value will be transformed from 8 to −1.5 (see Figure 11-2).

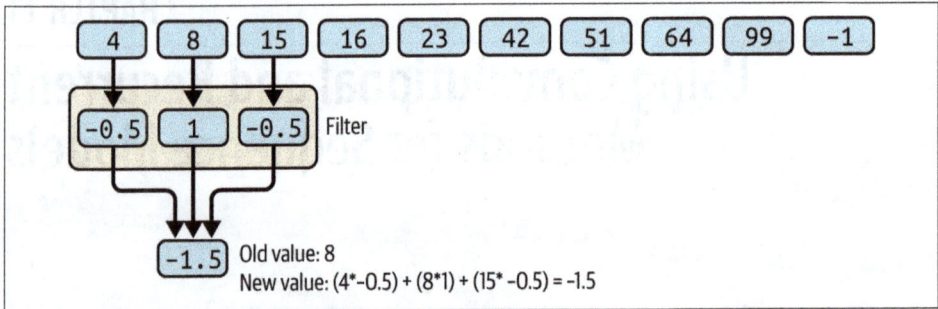

Figure 11-2. Using a convolution with the number sequence

The filter will then stride across the values, calculating new ones as it goes. So, for example, in the next stride, 15 will be transformed into 3 (see Figure 11-3).

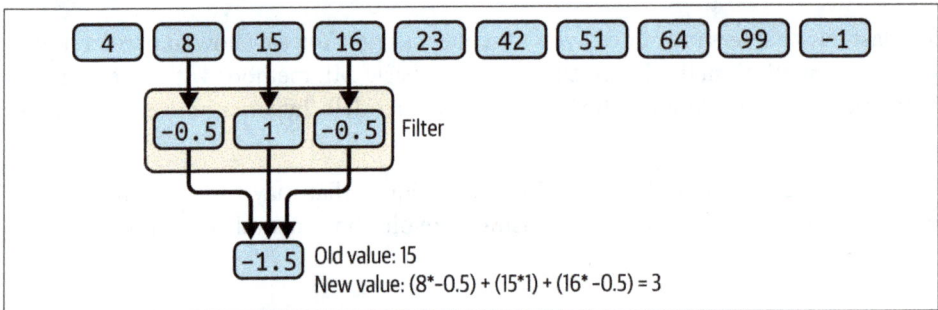

Figure 11-3. An additional stride in the 1D convolution

Using this method, it's possible to extract the patterns between values and learn the filters that extract them successfully, in much the same way that convolutions on the pixels in images can extract features. In this instance, there are no labels, but the convolutions that minimize overall loss can be learned.

Coding Convolutions

Before coding convolutions, you'll need to use the *sliding windows* technique to create a dataset, as shown in Chapter 10. The code is available on this book's GitHub page (*https://oreil.ly/pytorch_ch11*).

Once you have that dataset, you can add a convolutional layer before the dense layers that you had previously. Here's the code, which we'll look at line by line:

```
class CNN1D(nn.Module):
    def __init__(self, input_size):
        super(CNN1D, self).__init__()
```

```
self.conv1 = nn.Conv1d(in_channels=1,
                       out_channels=128,
                       kernel_size=3,
                       padding=1)

conv_output_size = input_size  # Same padding maintains input size

self.relu = nn.ReLU()
self.flatten = nn.Flatten()
self.dense1 = nn.Linear(128 * conv_output_size, 28)
self.dense2 = nn.Linear(28, 10)
self.dense3 = nn.Linear(10, 1)

def forward(self, x):
    # Transpose input from [batch_size, sequence_length]
    # to [batch_size, 1, sequence_length]
    if len(x.shape) == 2:
        x = x.unsqueeze(1)
    elif len(x.shape) == 3 and x.shape[1] != 1:
        x = x.transpose(1, 2)

    x = self.relu(self.conv1(x))
    x = self.flatten(x)
    x = self.relu(self.dense1(x))
    x = self.relu(self.dense2(x))
    x = self.dense3(x)
    return x
```

First, notice the new line that defines a 1D convolutional layer:

```
self.conv1 = nn.Conv1d(in_channels=1,
                       out_channels=128,
                       kernel_size=3,
                       padding=1)
```

The in_channels parameter defines the dimensionality of the input data. As we have a single sequence of numbers with a single value per data point, this is 1. If we were using multiple features per time step, such as perhaps an RGB color value, this would be 3.

The out_channels parameter is the number of filters (aka convolutions) that the network will learn.

The kernel_size parameter determines the size of the convolution (i.e., the number of data points on the line that a convolution will filter). Refer back to Figure 11-2 and Figure 11-3 and you'll see a convolution there with a kernel size of 3.

The padding parameter adds elements to the beginning and end of your list of data. So, for example, the list of numbers in Figure 11-1 is [4 8 15 16 23 42 51 64 99 –1]. When the filter of kernel size 3 looks at this list, it begins with [4 8 15], and 4 never gets to be the "middle" number. The filter effectively ignores the numbers at the beginning and end of the list. With padding, a 0 will be added at the front and back of the list to make it [0 4 8 15 16 23 42 51 64 99 –1 0], and you can now see that the filter will look first at [0 4 8].

Next, we see a line that looks like this:

```
conv_output_size = input_size   # Same padding maintains input size
```

This helps us to know the size of the output from the convolutional layer to inform the "next" layer in the sequence.

Why is it the input size, you might wonder. This is the idea of "same padding" that comes about from setting the `padding=1` parameter.

If you consider what would happen if you slid a kernel of size 3 across a list of values as in Figures 11-2 and 11-3, you'd see an odd effect. Because the kernel starts with its left side aligned with the first value and its center at the second value, and because it slides across to the end of the list where the center of the kernel will be aligned to the second-to-last value, the result of the calculations against the values in the list will give us n – 2 answers, where n is the length of the list. But if we pad the list with `padding=1`, then the kernel sliding across the list will give us n answers, so the output size from the layer will be the *same* as the input size.

So now, after ReLUing and flattening the results, we can see the next line:

```
self.dense1 = nn.Linear(128 * conv_output_size, 28)
```

The input to this will be a number of values: the size of the list, multiplied by 128, where 128 is the number of kernels. It will then output 28 values, which will be fed into the next linear layer.

Now, when you get to the forward function, it begins with this line:

```
if len(x.shape) == 2:
    x = x.unsqueeze(1)
elif len(x.shape) == 3 and x.shape[1] != 1:
    x = x.transpose(1, 2)
```

This looks quite unusual, but it's necessary for dealing with the convolutions. First of all, consider what the input to a convolution should look like. There'll be batches of them being fed in, each batch will have 1 dimension, and each item in the batch will have a number of items in it. If we are learning from sequences of 20 items, for example, and if we batch them 32 at a time, then the dimensionality of data being fed into the neural network will be [32, 1, 20].

But if our dataset isn't giving us that—and if, for example, there are only two dimensions [32, 20]—then we want to use unsqueeze to slip in another dimension. When we pass a 1 into it, it will be put at position 1, so we'll get [32, 1, 20] as desired.

The other case might be if we haven't put our dimension in correctly and added it on the end, like in [32, 20, 1], so the x = x.transpose(1, 2) will flip these around and make the dimension [32, 1, 20] again.

Now, these are two specific cases I hardcoded for. You may encounter others, so watch out for issues with your data when feeding it into the neural network. This is likely a place where you can fix them.

The rest of the forward pass is pretty straightforward; it's just passing the data through the different layers.

The loss function and optimizer are going to be pretty straightforward, too, using a mean-squared-error loss and an Adam optimizer:

```
criterion = nn.MSELoss()
optimizer = optim.Adam(model.parameters(), lr=learning_rate)
```

Training with this will give you a model as before, and to get predictions from the model, you can just use the loader in the same way as you did for training the model. So, for example, you can do this:

```
# Create DataLoaders
batch_size = 32
train_loader = DataLoader(train_dataset, batch_size=batch_size,
                          shuffle=True)
val_loader = DataLoader(val_dataset, batch_size=batch_size, shuffle=False)
```

And here's a helper function that can predict an entire series, batch by batch:

```
def predict(model, loader):
    device = torch.device("cuda" if torch.cuda.is_available() else "cpu")
    model.eval()
    predictions = []

    with torch.no_grad():
        for inputs, _ in loader:
            inputs = inputs.to(device)
            batch_predictions = model(inputs)
            predictions.append(batch_predictions.cpu().numpy())

    return np.concatenate(predictions)
```

You can then get the full set of predictions like this:

```
# Make predictions
train_predictions = predict(model, train_loader)
val_predictions = predict(model, val_loader)
```

Similarly, if you want to plot them, you could extend on this a little to pass in a loader and get back arrays of the predictions and the targets as well as an analytic, such as the MAE:

```python
def evaluate_predictions(model, loader):
    """Generate predictions and calculate metrics"""
    model.eval()
    device = torch.device("cuda" if torch.cuda.is_available() else "cpu")

    all_predictions = []
    all_targets = []

    with torch.no_grad():
        for inputs, targets in loader:
            inputs = inputs.to(device)
            outputs = model(inputs)
            all_predictions.extend(outputs.cpu().numpy())
            all_targets.extend(targets.cpu().numpy())

    predictions = np.array(all_predictions)
    targets = np.array(all_targets)


    # Calculate metrics
    mae = mean_absolute_error(targets, predictions)

    return predictions, targets, mae
```

And then you'd call this to get back the multiple responses like this:

```python
# Generate predictions
val_predictions, val_targets, val_mae
            = evaluate_predictions(model, val_loader)
```

This is now nice and easy to plot:

```python
def plot_predictions(val_pred, val_true):
    """Plot the predictions against actual values"""
    plt.figure(figsize=(15, 6))
    # Plot validation data
    offset = len(val_true)
    plt.plot(range(offset, offset + len(val_true)),
                val_true, 'b-', label='Validation Actual')
    plt.plot(range(offset, offset + len(val_pred)),
                val_pred, 'r-', label='Validation Predicted')
    plt.title('Time Series Prediction vs Actual')
    plt.xlabel('Time Step')
    plt.ylabel('Value')
    plt.legend()
    plt.grid(True)
    plt.show()
```

A plot of the results against the series is in Figure 11-4.

The MAE in this case is 5.33, which is slightly worse than for the previous prediction. This could be because we haven't tuned the convolutional layer appropriately, or it could be that convolutions simply don't help. This is the type of experimentation you'll need to do with your data.

Do note that this data has a random element in it, so values will change across sessions. If you're using code from Chapter 10 and then running this code separately, you will, of course, have random fluctuations affecting your data and thus your MAE.

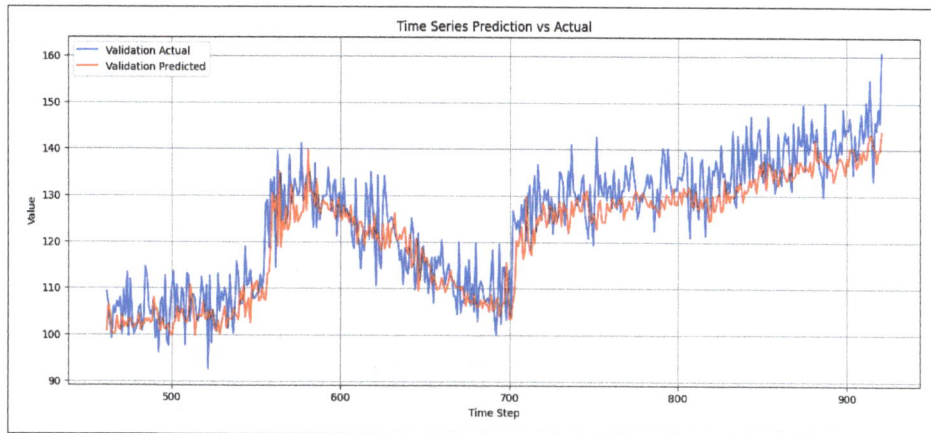

Figure 11-4. Convolutional neural network with time sequence data prediction versus actual

But when using convolutions, questions always come up. Why choose the parameters that we chose? Why 128 filters? Why size 3 × 1? The good news is that you can experiment with these things easily to explore different results.

Experimenting with the Conv1D Hyperparameters

In the previous section, you saw a 1D convolution that was hardcoded with parameters for things like filter number, kernel size, number of strides, etc. When you were training the neural network with it, it appeared that the MAE went up slightly, so you got no benefit from using the Conv1D. This may not always be the case, depending on your data, but it could be because of suboptimal hyperparameters. So, in this section, you'll see how you can do a neural architecture search to find the best results.

One of the nice things about how verbose PyTorch is, in particular for defining the neural network and the forward pass, is that it becomes pretty straightforward to change up the parameters that you use. The idea with a *neural architecture search* is to come up with sets of different parameters to try and then explore the impact that they have on the results by training for a short time and finding those that give the best results.

So, for example, here, we used a single `Conv1D` layer. But what if there were more? Similarly, we hardcoded a number of channels and a kernel size, and we also hardcoded the size of the dense layers and the LR for the optimizer. But what if, instead of hardcoding them, we created a set of options like this?

```python
# Define the search space
num_conv_layers_options = [1, 2]  # Reduced for initial testing
conv_channels_options = [
    [32],
    [64],
    [32, 16],
    [64, 32],
]
kernel_sizes = [3, 5]
dense_sizes_options = [
    [16],
    [32, 16],
    [64, 32],
]
learning_rates = [0.001, 0.0001]
```

With 4 options for the `conv` layers, 2 for the kernel sizes, 3 for the dense dimensions, and 2 for the LR, we have $4 \times 2 \times 3 \times 2$ options total, which is 48 combinations. This is called the *search space*.

Note that in this case, you might think it would be 96 because there are 2 layers options and 4 channels options. But in the code to define the search space, which you'll see in a moment, I only allowed `conv` channel options that match the number of layers, so there will just be 4 options in total for the `conv` layers.

These options will be loaded into a configurations array, with name-value pairs set up for the parameters, like this:

```python
# Generate valid configurations
configurations = []
for num_conv_layers in num_conv_layers_options:
    for channels in conv_channels_options:
        # Only use channel configs that match layer count
        if len(channels) == num_conv_layers:
            for kernel_size in kernel_sizes:
                for dense_sizes in dense_sizes_options:
                    for lr in learning_rates:
                        configurations.append({
                            'num_conv_layers': num_conv_layers,
                            'conv_channels': channels,
                            'kernel_size': kernel_size,
                            'dense_sizes': dense_sizes,
                            'learning_rate': lr
                        })
```

So now, we can loop through these configurations and set up our CNN1D model by using them like this. Note the use of the config[] array:

```
for idx, config in enumerate(configurations):
    print(f"\nTrying configuration {idx + 1}/{len(configurations)}:")
    print(config)

    try:
        model = CNN1D(
            input_size=input_size,
            num_conv_layers=config['num_conv_layers'],
            conv_channels=config['conv_channels'],
            kernel_size=config['kernel_size'],
            dense_sizes=config['dense_sizes']
        ).to(device)

        criterion = nn.MSELoss()
        optimizer = torch.optim.Adam(model.parameters(),
                                     lr=config['learning_rate'])
```

It would be really time-consuming to train each of the 48 combinations for the full set of epochs (say, 100), so we've introduced the idea of *early stopping*. First, let's train the model with the parameters we loaded from the configuration and a new parameter: early stopping patience:

```
trained_model, val_loss = train_model(
    model, train_loader, val_loader, criterion, optimizer,
    epochs=100, device=device, early_stopping_patience=10
)
```

Then, within the training loop, we can implement an early stopping like this:

```
# Early stopping check
if val_loss < best_val_loss:
    best_val_loss = val_loss
    best_model = deepcopy(model)
    patience_counter = 0
else:
    patience_counter += 1

if patience_counter >= early_stopping_patience:
    print(f'Early stopping triggered after {epoch} epochs')
    break
```

This keeps track of the loss for the best model and compares the current model with the best one. If the current model "loses" more times than our patience parameter (in this case, 10), then we'll throw it out and move to the next one. If it "wins," then we'll keep the current model as the best one.

Starting from this code, you can try to experiment with the hyperparameters for the number of filters, the size of the kernel, and the size of the stride, keeping the other parameters static.

After some experimentation, I discovered that 2 convolutional layers, with 64 and 32 filters (respectively), a kernel size of 5, two dense layers of 64 and 32, and an LR of .0001 gave the best MAE on the validation set, giving me a final result of 4.4439 MAE.

After training with this, the model had improved accuracy compared with both the naive CNN created earlier *and* the original DNN, giving the results shown in Figure 11-5.

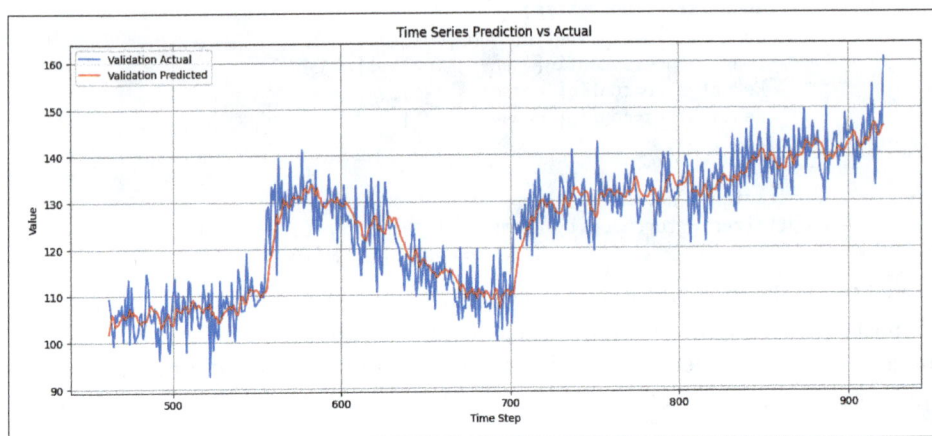

Figure 11-5. Optimized CNN time series predictions versus actual

Further experimentation with the CNN hyperparameters may improve this further.

Beyond convolutions, the techniques we explored in the chapters on NLP with RNNs, including LSTMs, may be powerful when working with sequence data. By their very nature, RNNs are designed for maintaining context, so previous values can have an effect on later ones. You'll explore using RNNs for sequence modeling next. But first, let's move on from a synthetic dataset and start looking at real data. In this case, we'll consider weather data.

Using NASA Weather Data

One great resource for time series weather data is the NASA Goddard Institute for Space Studies (GISS) Surface Temperature Analysis (*https://oreil.ly/6IixP*). If you follow the Station Data link (*https://oreil.ly/F9Hmw*), on the right side of the page, you can pick a weather station to get data from. For example, I chose the Seattle Tacoma (SeaTac) airport and was taken to the page shown in Figure 11-6.

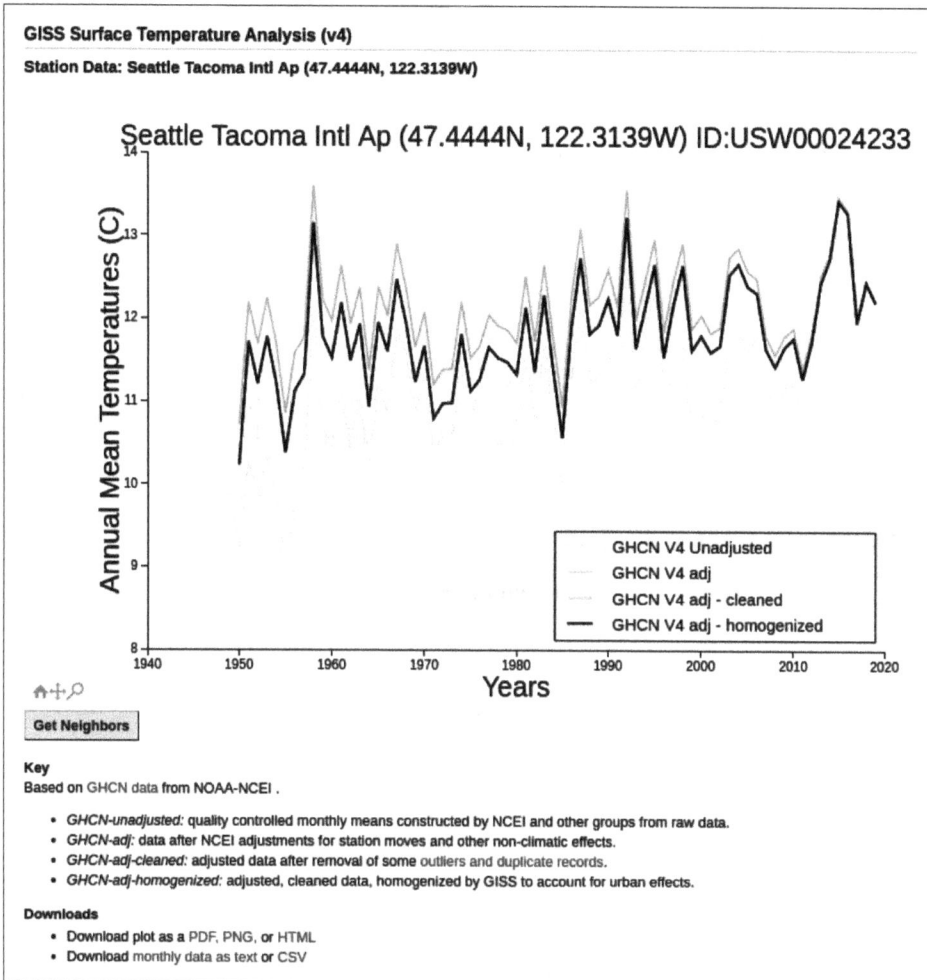

Figure 11-6. Surface temperature data from GISS

You can also see a link to download monthly data as CSV at the bottom of this page. If you select this link, a file called *station.csv* will be downloaded to your device, and if you open it, you'll see that it's a grid of data with a year in each row and a month in each column (see Figure 11-7).

	A	B	C	D	E	F	G	H	I	J	K	L	M
1	YEAR	JAN	FEB	MAR	APR	MAY	JUN	JUL	AUG	SEP	OCT	NOV	DEC
2	1950	-2.54	5.85	6.99	9.37	12.31	17.04	18.99	19.05	15.74	10.95	7.84	8.5
3	1951	4.14	6.19	5.49	11.15	13.73	17.57	19.38	17.86	16.43	11.85	8.38	3.87
4	1952	3.86	6.21	7.13	10.55	13.63	15.01	18.86	18.63	16.51	13.77	6.43	6.76
5	1953	8.25	5.95	7.49	9.63	12.85	14.45	18	18.49	16.77	13.17	9.56	7.12
6	1954	3.67	7.08	6.24	8.94	13.57	14.91	17.02	17.3	16.24	11.73	10.83	6.26
7	1955	5.38	4.86	5.37	8.39	11.78	15.83	17	17.49	15.3	11.62	5.23	4.96
8	1956	5.3	3.45	6.32	11.2	15.2	15.25	19.54	18.61	15.81	10.77	6.98	5.64

Figure 11-7. Exploring the data

As this is CSV data, it's pretty easy to process in Python, but as with any dataset, do note the format. When reading CSV, you tend to read it line by line, and often, each line has one data point that you're interested in. In this case, there are at least 12 data points of interest per line, so you'll have to consider this when reading the data.

Reading GISS Data in Python

The code to read the GISS data is shown here:

```python
def get_data():
    data_file = "station.csv"
    f = open(data_file)
    data = f.read()
    f.close()
    lines = data.split('\n')
    header = lines[0].split(',')
    lines = lines[1:]
    temperatures=[]
    for line in lines:
        if line:
            linedata = line.split(',')
            linedata = linedata[1:13]
            for item in linedata:
                if item:
                    temperatures.append(float(item))

    series = np.asarray(temperatures)
    time = np.arange(len(temperatures), dtype="float32")
    return time, series
```

This will open the file at the indicated path (yours will, of course, differ) and read in the entire file as a set of lines, where the line split is the new line character (\n). It will then loop through each line, ignoring the first line, and split them on the comma character into a new array called linedata. The items from 1 through 13 in this array will indicate the values for the months January through February as strings, and these values will then be converted into floats and added to the array called temperatures. Once it's completed, it will be turned into a NumPy array called series, and another

NumPy array called `time` will be created that's the same size as `series`. As it is created using `np.arange`, the first element will be 1, the second will be 2, etc. Thus, this function will return `time` in steps from 1 to the number of data points and will return `series` as the data for that time.

I have noticed that often, there will be "unfilled" data in some of the columns, and these are represented by the value 999.9. This will, of course, skew any predictive results you want to create. But fortunately, 999.9 values are usually at the *end* of the dataset, so they can easily be cropped. Here's a helper function to normalize the series while cropping out the 999.9 values:

```python
import numpy as np

def normalize_series(data, missing_value=999.9):
    # Convert to numpy array if not already
    data = np.array(data, dtype=np.float64)

    # Create mask for valid values (not NaN and not missing_value)
    valid_mask = (data != missing_value) & (~np.isnan(data))

    # Keep only valid values
    clean_data = data[valid_mask]

    # Normalize using only valid values
    mean = np.mean(clean_data)
    std = np.std(clean_data)
    normalized = (clean_data - mean) / std

    return normalized

time, series = get_data()
series_normalized = normalize_series(series)
```

You can now load this into a `torch.tensor` and turn it into a set of sliding windows with a target value as before. We discussed the helper function in Chapter 10:

```python
series_tensor = torch.tensor(series_normalized, dtype=torch.float32)
window_size = 48
features, targets = create_sliding_windows_with_target(series_tensor,
                    window_size=window_size, shift=1)
```

Once we have that, we can turn it into a `TensorDataset` and split it into subsets for training and validation:

```python
split_location = 800
# Create the full dataset
full_dataset = TensorDataset(features, targets)

# Calculate split indices
# Note: Since we're using windows, we need to account for the overlap
train_size = 800 - window_size + 1  # Adjust for window overlap
```

```
total_windows = len(full_dataset)
train_indices = list(range(train_size))
val_indices = list(range(train_size, total_windows))

# Create training and validation datasets using Subset
train_dataset = Subset(full_dataset, train_indices)
val_dataset = Subset(full_dataset, val_indices)
```

Now that we have the splits as datasets, we can create loaders for them that the neural network will use:

```
batch_size = 32
train_loader = DataLoader(train_dataset, batch_size=batch_size,
                          shuffle=True)
val_loader = DataLoader(val_dataset, batch_size=batch_size, shuffle=False)
```

And now, we're ready to train with this data. We can inspect the split by charting the data, and we can see this with the training/validation split in Figure 11-8.

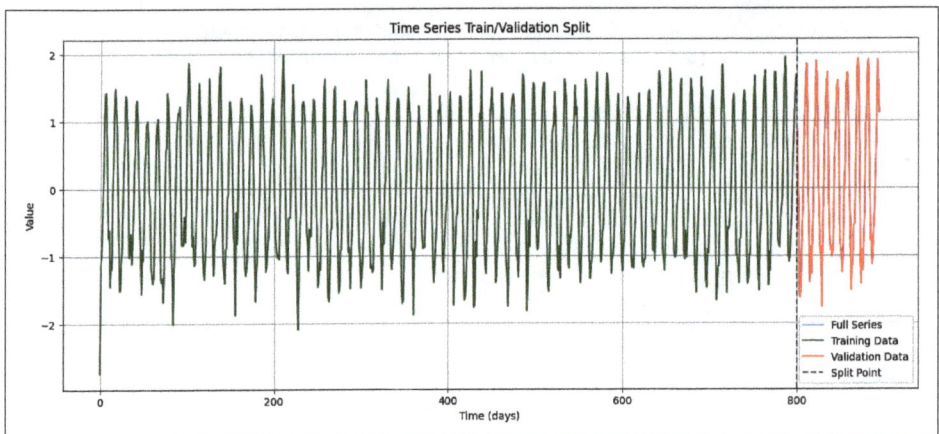

Figure 11-8. The time series train/validation split

In the next section, we'll explore creating a simple RNN-based neural network to see if we can predict the next values in the sequence.

Using RNNs for Sequence Modeling

Now that you have the data from the NASA CSV in a windowed dataset, it's relatively easy to create a model to train a predictor for it. (However, it's a bit more difficult to train a *good* one!) Let's start with a simple, naive model using RNNs. Here's the code:

```
class SimpleRNNModel(nn.Module):
    def __init__(self, input_size=1, hidden_size=100,
                       output_size=1, dropout_rate=0.3):
        super(SimpleRNNModel, self).__init__()
```

```python
        self.rnn1 = nn.RNN(input_size=input_size,
                           hidden_size=hidden_size,
                           batch_first=True,
                           dropout=dropout_rate)  # Add dropout to RNN

        self.rnn2 = nn.RNN(input_size=hidden_size,
                           hidden_size=hidden_size,
                           batch_first=True,
                           dropout=dropout_rate)  # Add dropout to RNN

        self.dropout = nn.Dropout(dropout_rate)  # Additional dropout layer
        self.linear = nn.Linear(hidden_size, output_size)

    def forward(self, x):
        out1, _ = self.rnn1(x)
        out2, _ = self.rnn2(out1)
        last_out = out2[:, -1, :]
        last_out = self.dropout(last_out)  # Add dropout before final layer
        output = self.linear(last_out)
        return output
```

In this case, as you can see, we use a basic RNN. RNNs are a class of neural networks that are powerful for exploring sequence models, and you first saw them in Chapter 7, when you were looking at NLP. I won't go into detail on how they work here, but if you're interested and you skipped that chapter, take a look back at it now. Notably, an RNN has an internal loop that iterates over the time steps of a sequence while maintaining an internal state of the time steps it has seen so far.

While training, you can use a loss function and optimizer like this:

```python
criterion = nn.MSELoss()
optimizer = torch.optim.Adam(model.parameters(), lr=learning_rate)
```

The full code is available in this book's GitHub repository (*https://oreil.ly/ pytorch_ch11*). Even one hundred epochs of training is enough to get an idea of how the model can predict values. Figure 11-9 shows the results.

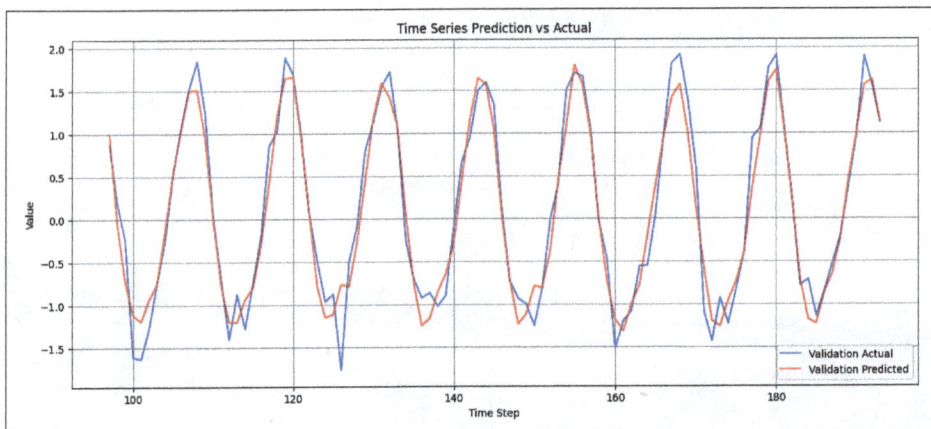

Figure 11-9. Results of the SimpleRNN time series prediction versus actual

As you can see, the results were pretty good. It may be a little off in the peaks and when the pattern changes unexpectedly (like at time steps 160–170), but on the whole, it's not bad. Now, let's see what happens if we train it for 1,500 epochs (see Figure 11-10).

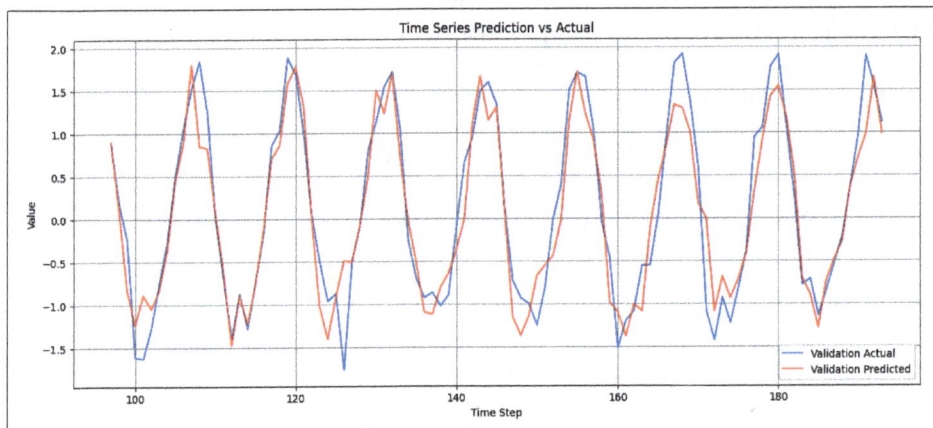

Figure 11-10. Time series prediction versus actual for RNN trained over 1,500 epochs

There's not much of a difference, except that some of the peaks are smoothed out. If you look at the history of loss on both the validation set and the training set, it looks like Figure 11-11.

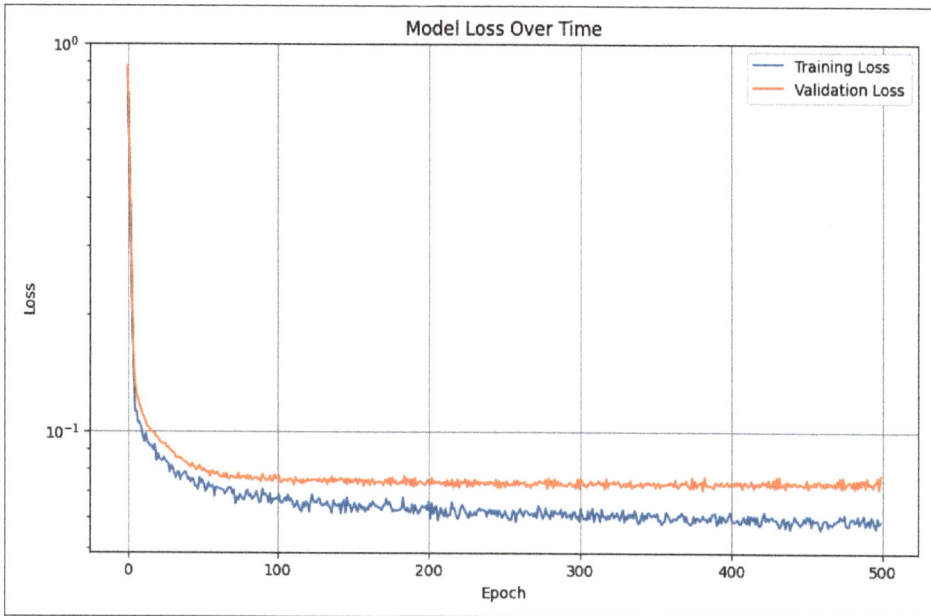

Figure 11-11. Training and validation model loss over time for the SimpleRNN

As you can see, there's a healthy match between the training loss and the validation loss, but as the epochs increase, the model begins to overfit on the training set. Perhaps a better number of epochs would be around five hundred.

One reason for this could be the fact that the data, being monthly weather data, is highly seasonal. Another is that there is a very large training set and a relatively small validation set.

Next, we'll explore using a larger climate dataset.

Exploring a Larger Dataset

The KNMI Climate Explorer (*https://oreil.ly/J8CP0*) allows you to explore granular climate data from many locations around the world. I downloaded a dataset (*https://oreil.ly/Ci9DI*) consisting of daily temperature readings from the center of England from the years 1772 to 2020. This data is structured differently from the GISS data, with the date as a string, followed by a number of spaces, followed by the reading. Go back to Chapter 4 to check the details on handling and managing large datasets.

I've prepared the data, stripping the headers and removing the extraneous spaces, so that there's only one space between the date and the reading. That way, it's easy to read with code like this:

```python
import numpy as np
def get_data():
    data_file = "tdaily_cet.dat.txt"
    f = open(data_file)
    data = f.read()
    f.close()
    lines = data.split('\n')
    temperatures=[]
    for line in lines:
        if line:
            linedata = line.split(' ')
            temperatures.append(float(linedata[1]))

    series = np.asarray(temperatures)
    time = np.arange(len(temperatures), dtype="float32")
    return time, series
```

This dataset has 91,502 data points in it, so before training your model, be sure to split it appropriately. I used a split time of 80,000, leaving 10,663 records for validation:

```python
split_location = 80000

features = features.unsqueeze(1)
# Create the full dataset
full_dataset = TensorDataset(features, targets)

# Calculate split indices
# Note: Since we're using windows, we need to account for the overlap
train_size = split_location - window_size + 1  # Adjust for window overlap
total_windows = len(full_dataset)
train_indices = list(range(train_size))
val_indices = list(range(train_size, total_windows))

# Create training and validation datasets using Subset
train_dataset = Subset(full_dataset, train_indices)
val_dataset = Subset(full_dataset, val_indices)
```

Everything else can remain the same. As you can see in Figure 11-12, after training for one hundred epochs, the plot of the predictions against the validation set looks pretty good.

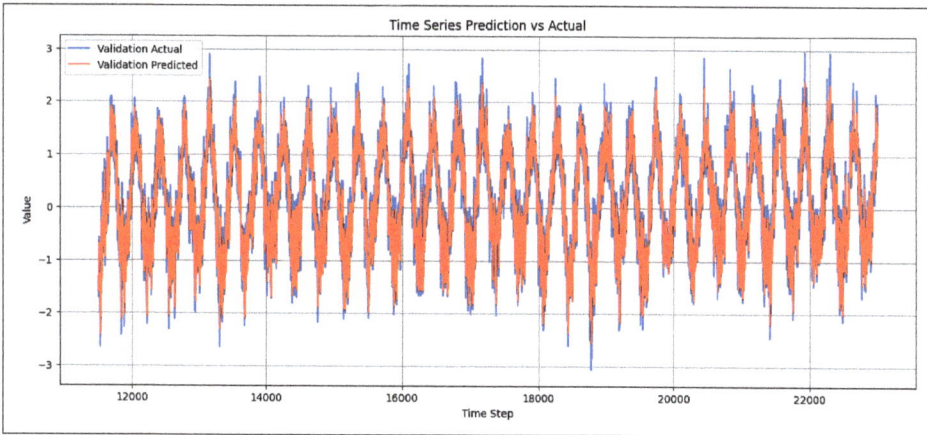

Figure 11-12. Plot of predictions against real data

There's a lot of data here, so let's zoom in to the last hundred days' worth (see Figure 11-13).

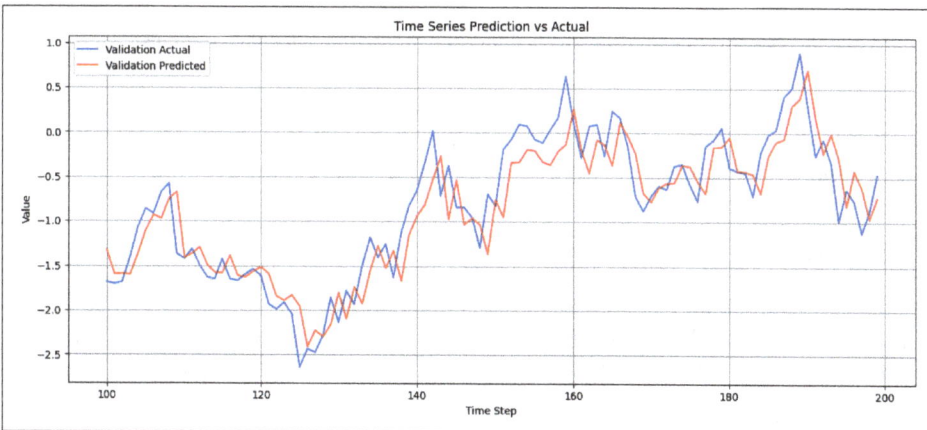

Figure 11-13. Results of time series prediction versus actual for one hundred days' worth of data

While the chart generally follows the curve of the data and is getting the trends roughly correct, it is pretty far off, particularly at the extreme ends, so there's room for improvement.

It's also important to remember that we normalized the data, so while our loss and MAE may look low, that's because they are based on the loss and MAE of normalized values that have a much lower variance than the real ones. As Figure 11-14 shows, a tiny amount of loss might lead you into having a false sense of security.

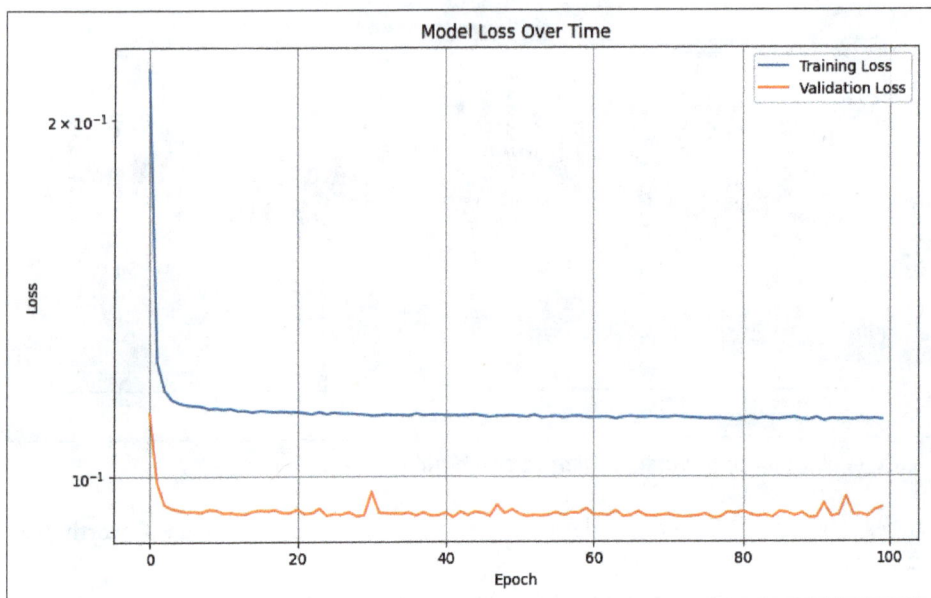

Figure 11-14. Training and validation model loss over time for large dataset

To denormalize the data, you can do the inverse of normalization: first, multiply by the standard deviation, and then add back the mean. At that point, if you wish, you can calculate the real MAE for the prediction set as you've done previously.

Using Other Recurrent Methods

In addition to the RNN, PyTorch has other recurrent layer types, such as gated recurrent units (GRUs) and long short-term memory layers (LSTMs), which we discussed in Chapter 7. It is relatively simple to just drop in these RNN types if you want to experiment.

So, for example, if you consider the simple, naive RNN that you created earlier, replacing it with a GRU becomes as easy as using nn.GRU:

```python
class SimpleRNNModel(nn.Module):
    def __init__(self, input_size=1, hidden_size=100,
                 output_size=1, dropout_rate=0.3):
        super(SimpleRNNModel, self).__init__()

        self.rnn1 = nn.GRU(input_size=input_size,
                           hidden_size=hidden_size,
                           batch_first=True,
                           dropout=dropout_rate)

        self.rnn2 = nn.GRU(input_size=hidden_size,
                           hidden_size=hidden_size,
```

```
                    batch_first=True,
                    dropout=dropout_rate)

        self.dropout = nn.Dropout(dropout_rate)
        self.linear = nn.Linear(hidden_size, output_size)
```

With an LSTM, it's similar:

```
# LSTM Optional Architecture
import torch.nn as nn

class SimpleLSTMModel(nn.Module):
    def __init__(self, input_size=1, hidden_size=100,
                       output_size=1, dropout_rate=0.3):
        super(SimpleLSTMModel, self).__init__()

        self.lstm1 = nn.LSTM(input_size=input_size,
                        hidden_size=hidden_size,
                        batch_first=True,
                        dropout=dropout_rate)  # Add dropout to LSTM

        self.lstm2 = nn.LSTM(input_size=hidden_size,
                        hidden_size=hidden_size,
                        batch_first=True,
                        dropout=dropout_rate)  # Add dropout to LSTM

        # Add more layers before final output
        self.fc1 = nn.Linear(hidden_size, hidden_size)
        self.relu = nn.ReLU()
        self.linear = nn.Linear(hidden_size, output_size)

    def forward(self, x):
        out1, _ = self.lstm1(x)  # LSTM returns (output, (h_n, c_n))
        out2, _ = self.lstm2(out1) # We ignore both hidden and cell states with _
        last_out = out2[:, -1, :]
        output = self.linear(last_out)
        return output
```

It's worth experimenting with these layer types as well as with different hyperparameters, loss functions, and optimizers. There's no one-size-fits-all solution, so what works best for you in any given situation will depend on your data and your requirements for prediction with that data.

Using Dropout

If you encounter overfitting in your models, where the MAE or loss for the training data is much better than with the validation data, you can use dropout. As discussed in earlier chapters, with dropout, neighboring neurons are randomly dropped out (ignored) during training to avoid a proximity bias. When you're using RNNs, there's also a *recurrent dropout* parameter that you can use.

What's the difference? Recall that when using RNNs, you typically have an input value and the neuron calculates an output value and a value that gets passed to the next time step. Dropout will randomly drop out the input values, and recurrent dropout will randomly drop out the recurrent values that get passed to the next step.

For example, consider the basic RNN architecture shown in Figure 11-15.

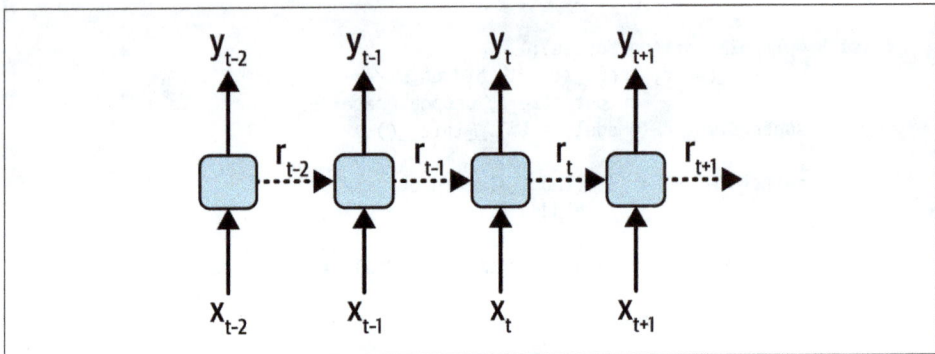

Figure 11-15. A recurrent neural network

Here, you can see the inputs into the layers at different time steps (x). The current time is t, and the steps shown are $t - 2$ through $t + 1$. The relevant outputs at the same time steps (y) are also shown, and the recurrent values passed between time steps are indicated by the dotted lines and labeled as r.

Using *dropout* will randomly drop out the x inputs, while using *recurrent dropout* will randomly drop out the r recurrent values.

You can learn more about how recurrent dropout works from a deeper mathematical perspective in the paper "A Theoretically Grounded Application of Dropout in Recurrent Neural Networks" by Yarin Gal and Zoubin Ghahramani (*https://oreil.ly/MqqRR*). One other thing to consider when using recurrent dropout is discussed by Gal in his research around uncertainty in deep learning (*https://oreil.ly/3v8IB*), in which he demonstrates that the same pattern of dropout units should be applied at every time step and that a similar constant dropout mask should also be applied at every time step.

To add dropout and recurrent dropout, you use the relevant parameters on your layers. For example, adding them to the basic GRU from earlier was as simple as using a parameter in the recurrent layers and adding another layer between the RNNs and the linears:

```
class SimpleRNNModel(nn.Module):
    def __init__(self, input_size=1, hidden_size=100,
                 output_size=1, dropout_rate=0.1):
        super(SimpleRNNModel, self).__init__()

        self.rnn1 = nn.GRU(input_size=input_size,
                           hidden_size=hidden_size,
                           batch_first=True,
                           dropout=dropout_rate)

        self.rnn2 = nn.GRU(input_size=hidden_size,
                           hidden_size=hidden_size,
                           batch_first=True,
                           dropout=dropout_rate)

        self.dropout = nn.Dropout(dropout_rate)
        self.linear = nn.Linear(hidden_size, output_size)
```

Each parameter takes a value between 0 and 1 that indicates the proportion of values to drop out. For example, a value of 0.1 will drop out 10% of the requisite values.

Training a model with dropout like this shows a much steeper learning curve, which is still trending downward at 100 epochs. The validation is quite flat, indicating that a larger validation set may be necessary. It's also quite noisy, and you'll often see noise like this in the loss when using dropout. It's an indication that you may want to tweak the amount of dropout as well as the parameters of the loss function and optimizer, such as the LR. You can see this in Figure 11-16.

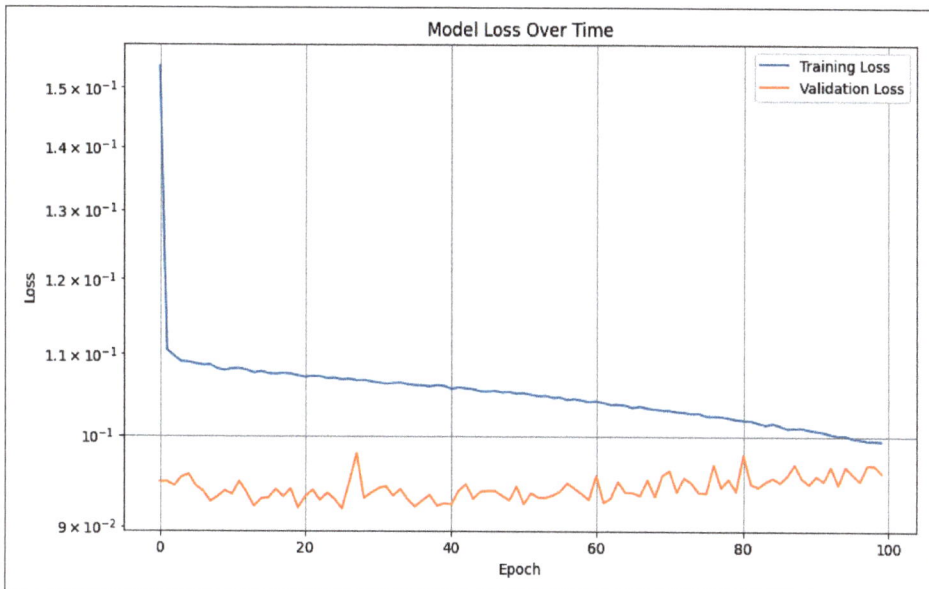

Figure 11-16. Training and validation loss over time using a GRU with dropout

As you've seen in this chapter, predicting time sequence data using neural networks is a difficult proposition, but tweaking their hyperparameters can be a powerful way to improve your model and its subsequent predictions.

Using Bidirectional RNNs

Another technique to consider when classifying sequences is to use bidirectional training. This may seem counterintuitive at first, as you might wonder how future values could impact past ones. But recall that time series values can contain seasonality, where values repeat over time, and when using a neural network to make predictions, all we're doing is sophisticated pattern matching. Given that data repeats, a signal for how data can repeat might be found in future values—and when using bidirectional training, we can train the network to try to spot patterns going from time t to time $t + x$, as well as going from time $t + x$ to time t.

Fortunately, coding this is simple. For example, consider the GRU from the previous section. To make this bidirectional, you simply add a bidirectional parameter. This will effectively train twice on each step—once with the sequence data in the original order and once with it in reverse order. The results are then merged before proceeding to the next step.

Here's an example:

```python
class BidirectionalGRUModel(nn.Module):
    def __init__(self, input_size=1, hidden_size=100,
                 output_size=1, dropout_rate=0.1):
        super(BidirectionalGRUModel, self).__init__()

        self.gru1 = nn.GRU(input_size=input_size,
                           hidden_size=hidden_size,
                           batch_first=True,
                           dropout=dropout_rate,
                           bidirectional=True)

        self.gru2 = nn.GRU(input_size=hidden_size * 2,
                           hidden_size=hidden_size,
                           batch_first=True,
                           dropout=dropout_rate,
                           bidirectional=True)

        # Additional layers
        self.fc1 = nn.Linear(hidden_size * 2, hidden_size)
        self.relu = nn.ReLU()
        self.dropout = nn.Dropout(dropout_rate)
        self.linear = nn.Linear(hidden_size, output_size)

    def forward(self, x):
        out1, _ = self.gru1(x)
        out2, _ = self.gru2(out1)
```

```
last_out = out2[:, -1, :]

# Additional processing
x = self.fc1(last_out)
x = self.relu(x)
x = self.dropout(x)
output = self.linear(x)
return output
```

A plot of the results of training with a bidirectional GRU with dropout on the time series is shown in Figure 11-17. While the MAE has improved slightly, the bigger impact is that the predicted curve has lost the "lag" compared with the single direction version.

Additionally, tweaking the training parameters—particularly window_size, to get multiple seasons—can have a pretty big impact.

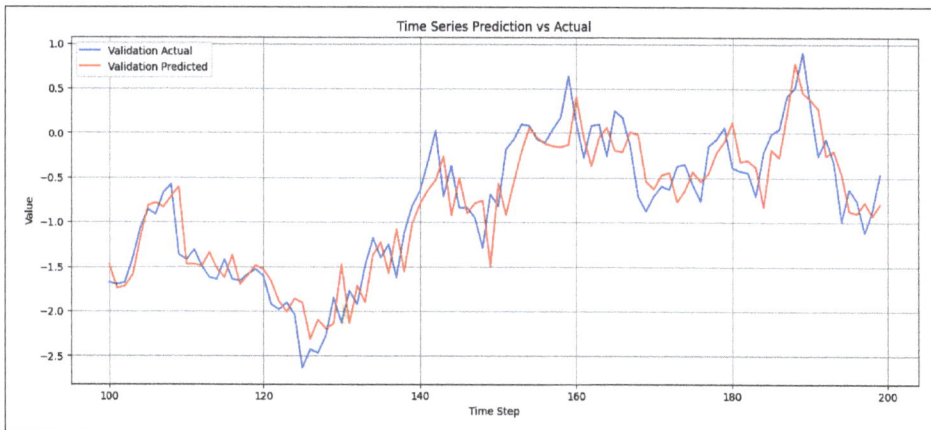

Figure 11-17. Time series prediction training with a bidirectional GRU

As you can see, you can experiment with different network architectures and different hyperparameters to improve your overall predictions. The ideal choices are very much dependent on the data, so the skills you've learned in this chapter will help you with your specific datasets!

Summary

In this chapter, you explored different network types for building models to predict time series data. You built on the simple DNN from Chapter 10, adding convolutions, and you experimented with recurrent network types such as simple RNNs, GRUs, and LSTMs. You also learned how to tweak hyperparameters and the network architecture to improve your model's accuracy, and you practiced working with some real-world datasets, including one massive dataset with hundreds of years' worth of temperature readings.

Now, you're ready to get started building networks for a variety of datasets, and you have a good understanding of what you need to know to optimize them!

## CHAPTER 12

Concepts of Inference

In the previous chapters of this book, you focused on *training* models using PyTorch and on how to create models that manage images (aka Computer Vision), text content (aka NLP), and sequence modelling. For the rest of this book, you'll cover a lot of content around *using trained models* to make predictions from new data (aka *inference*) and in particular using large generative models for text-to-text and text-to-image generative AI.

But before you jump into that, it's important for you to understand the underlying data transfer technology. We've touched on it a little in the training chapters, but as you go deeper into ML—in either training or inference—it's important for you to be able to understand the underlying concepts of tensors.

Ultimately, no matter what data type you have, you'll convert it into tensors to pass it *into* the model. Similarly, no matter the data type in which you want to present answers from the model to your users, you'll get them back as tensors as well!

In many cases, you'll have helper functions, such as the *transformers* that you'll see in Chapter 15 (which covers LLMs) and the *diffusers* that you'll see in Chapter 19 (which handles image generation). And while you won't be touching tensors with them, you'll still be using them under the hood.

Tensors

A *tensor* is an array that can have any number of dimensions. Tensors are typically used to represent numerical data for deep-learning algorithms; they're containers that can hold numbers in multiple dimensions.

Tensors can be simple scalar values (in zero dimensions), vectors (in one dimension), matrices (in two dimensions), and beyond (in three dimensions or more). In PyTorch, they're the fundamental data structure for all computation.

Tensors are also the source of the name *TensorFlow* for the alternative deep-learning framework from Google.

Here are some examples of tensors in PyTorch, which were created using torch. tensor:

```python
import torch

# Scalar (0D tensor)
scalar = torch.tensor(42)  # Single number

# Vector (1D tensor)
vector = torch.tensor([1, 2, 3, 4])  # Array of numbers

# Matrix (2D tensor)
matrix = torch.tensor([[1, 2, 3],
                       [4, 5, 6]])  # 2x3 grid of numbers

# 3D tensor
cube = torch.tensor([[[1, 2], [3, 4]],
                     [[5, 6], [7, 8]]])  # 2x2x2 cube of numbers
```

What makes tensors so useful for ML is this flexibility to hold different value types. Numbers can be 0D tensors, the embedding vectors representing text can be 1D, and images can be 3D, with dimensions for height, width, and pixel value. Plus, all of these can have an additional dimension added for batches. So, for example, a single image can be a 3D matrix, but 100 images instead of 100 3D matrices could be a single 4D tensor, with the fourth dimension being the index of the image!

When you're using torch.tensor, keep in mind that a lot of work and investment has been put into optimizing them to run on GPUs, which makes them extremely efficient for deep learning computation.

Image Data

Images are typically stored in formats like JPEG or PNG, which are optimized for *human* viewing as well as storage efficiency. Each dot (or pixel) in the image is usually composed of a number of values, with each value being the intensity of a color channel. Typically, an image will have 24 bits of data, with 8 bits assigned each to red, green, and blue channels. If you see 32 bits, then the additional 8 are for an alpha, or transparency, channel.

So, for example, a green pixel might have values 0 on the red, blue, and alpha channels and 255 on the green channel. If it's semi-transparent, it might still have a value of 0 on red and blue, 128 on alpha, and 255 on green.

An image file is typically *compressed*, which means that mathematical transforms have been applied to it to avoid wasted data and make the image smaller. An image can also contain metadata, file headers, and more. Then, once the image is loaded into memory, it is usually uncompressed to the 32-bits-per-pixel previously described.

ML models typically use values between –1 and 1, and not 0 to 255, as the native values of the image are stored. If we want to learn the details of an image, it's good for us to standardize the values by focusing on the meaningful *variations* between the pixel intensities as opposed to just their values. So, one method is to standardize by figuring out how far the pixel value is from the mean and standard deviations. This gives a broader spectrum of values that can lead to smoother loss curves and more effective learning.

In PyTorch, you achieve this with code like this:

```python
import torch
from torchvision import transforms
from PIL import Image

def prepare_image(image_path):
    # Load the image using PIL
    raw_image = Image.open(image_path)

    # Define the transformations
    preprocess = transforms.Compose([
        transforms.Resize(256),
        transforms.CenterCrop(224),
        transforms.ToTensor(),
        transforms.Normalize(
            mean=[0.485, 0.456, 0.406],
            std=[0.229, 0.224, 0.225]
        )
    ])

    # Apply transformations
    input_tensor = preprocess(raw_image)

    # Add batch dimension
    input_batch = input_tensor.unsqueeze(0)
    return input_batch
```

This code uses a very popular Python library called Pillow, or just PIL.

In this case, the `PIL Image.open` will read the image, decompress it into pixels, and then apply a set of transforms to those pixels. The transforms normalize the channels into ML-friendly values, as described earlier, and then convert them into tensors.

The `input_tensor` is then a 3D matrix/tensor, but if we want to have a number of them in a single batch, we can *unsqueeze* this tensor to add a new dimension, making it a 4D tensor.

You can see a single image in Figure 12-1 as a 3D tensor, with one dimension for each color depth.

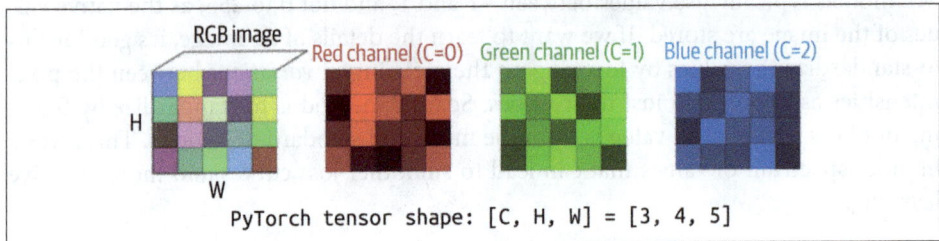

Figure 12-1. Tensor representing a full-color image

And then, if there's a batch of images, you can see it as a fourth dimension in Figure 12-2.

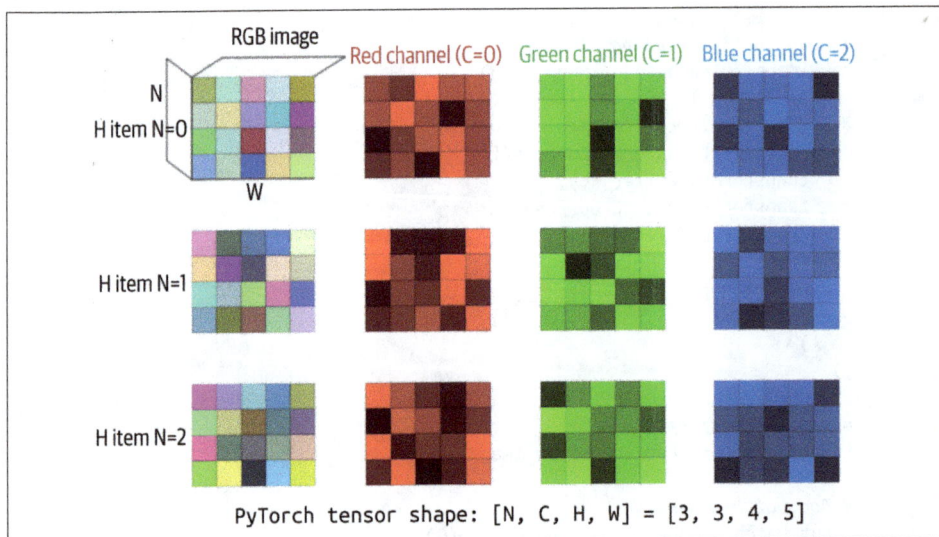

Figure 12-2. Tensor representing a batch of colored images

Text Data

Typically, text is stored in a string, like "The cat sat on the mat," but training a model or passing text like this to a pretrained model is unfeasible. Models, as you saw in earlier chapters, are trained on numeric data—and the best way to do that is to either *tokenize* the text, by turning words or subwords into numbers, or to *calculate embeddings* for the text, by turning them into vectors. And if you choose to calculate embeddings for the text, you can also encode sentiment about the text into the direction of the vector (see Chapter 6).

So, as a simple example, let's consider the following sentences:

```
texts = [
    "I love my dog",
    "The manatee became a doctor"
]
```

You can *tokenize* this text into a series of numbers by using a tokenizer. You can create your own tokenized series of numbers from the corpus, as you did in Chapter 5, or you can use an existing one, like this:

```
import torch
from transformers import BertTokenizer, BertModel

def text_to_embeddings(texts):
    # Load pretrained BERT tokenizer and model
    tokenizer = BertTokenizer.from_pretrained('bert-base-uncased')
    model = BertModel.from_pretrained('bert-base-uncased')
    model.eval()  # Set to evaluation mode

    # Tokenize the input texts
    encoded = tokenizer(
        texts,
        padding=True,        # Pad shorter sequences to match longest
        truncation=True,     # Truncate sequences that are too long
        return_tensors='pt'  # Return PyTorch tensors
    )
```

You'll learn a lot more about transformers, including how to use them for BERT, starting in Chapter 13.

The important line here is the last one, in which we ask the tokenizer to return tensors in PyTorch format. You can see the result of this in the output of the encodings, like this:

```
Encodings:
tensor([[ 101, 1045, 2293, 2026, 3899,  102,    0,    0],
        [ 101, 1996, 24951, 17389, 2150, 1037, 3460, 102]])
```

If you explore this closely, you'll see that the two sentences have been turned into a series of numbers. The first sentence, which has four words, has six numbers, and the second, which has six words, has eight numbers. Each has the number 101 at the front and 102 at the end, which gives us the two extra tokens. These are special tokens the encoder has added to indicate the start and end of the sentence.

A string can be represented as a 1D vector, but we have multiple strings here, so we can add a dimension to 1D to get 2D, and the second dimension gives us each string. So, value 0 in the second dimension is the first string, value 1 is the second string, etc.

The BERT model can also create embeddings from the sentence by getting an embedding for each word in the sentence and summing them all up to get an overall set of values. In the base model version of BERT, each embedding vector is a 1D matrix with 768 values. The embedding for a sentence is the same. Multiple sentences, just like the previous encodings, will have a second dimension.

Here's the code:

```
# Generate embeddings
with torch.no_grad():  # No need to calculate gradients
    outputs = model(**encoded)
    embeddings = outputs.last_hidden_state
```

Each embedding has 768 values, so I'm not going to print them all out, but for the embedding that represents the first sentence, you can see values like this:

```
First word embedding (first 5 values):
      tensor([ 0.0401,  0.3046,  0.0669, -0.1975, -0.0103])
```

Don't worry if you don't fully understand this yet—we'll be going into it in more detail starting in Chapter 13. The important point here is that the idea of a *tensor* is that the very flexible matrix, which can have any number of dimensions, gives you a consistent input into a model. You don't need to train models on different data types —they'll always be tensor in, tensor out.

Tensors Out of a Model

As noted earlier, the power of tensors is in their consistency—regardless of what type of data you pass *into* a model, when they're tensors, you can be consistent in your coding interface. The same applies for tensors *out* of a model.

So, for example, consider a dataset like ImageNet that contains 15 million images in over 21,000 classes. When you design a model to recognize images in this dataset, you'll need over 21,000 output neurons, each of which gives you a percentage likelihood that the image is of the representative class. So, for example, if neuron 0 represents "goldfish," the value coming out of it when you do inference will be the probability that the image contains a goldfish!

So, instead of outputting the classification, the model will expose the values of *each* of its output neurons. These values are often called *logits*.

This list of values is a 1D vector of values—so of course, a tensor is the appropriate data type.

Here's a simulated example, with a set of representative outputs from multiple images passed into the model and the list of class names:

```python
# ImageNet class labels (simplified - just a few examples)
class_names = [
    'tench', 'goldfish', 'great white shark', 'tiger shark', 'hammerhead shark',
    'electric ray', 'stingray', 'rooster', 'hen', 'ostrich', 'brambling',
    'goldfinch', 'house finch', 'junco', 'indigo bunting', 'robin', 'bulbul',
    'jay', 'magpie', 'chickadee'
]

# Simulate model output for demonstration
# This would normally come from model(input_tensor)
example_output = torch.tensor([
    [ 1.2,  4.5, -0.8,  2.1,  0.3,  # First image predictions
     -1.5,  0.9,  3.2, -0.4,  1.1,
      0.5, -0.2,  1.8,  0.7, -1.0,
      2.8,  1.6, -0.6,  0.4,  1.3],
    [-0.5,  5.2,  0.3,  1.4, -0.8,  # Second image predictions
      0.9,  1.2,  2.8,  0.6,  1.5,
     -1.1,  0.4,  2.1,  0.2, -0.7,
      1.9,  0.8, -0.3,  1.6,  0.5]
])
```

Note that because the output is *tensors*, we can use many of the functions built into PyTorch that are optimized for tensors to work with them.

When dealing with the output values, we want to find the best ones and maybe limit them to a range. Softmax is perfect for that, when it converts the raw output into probabilities—and TopK is used to pick the best *k* values. Here's an example in which we can use Softmax and TopK functions that manage tensors, regardless of their dimensionality:

```python
def interpret_output(output_tensor, top_k=5):
    # Apply softmax to convert logits to probabilities
    probabilities = torch.nn.functional.softmax(output_tensor, dim=1)

    # Get top k probabilities and class indices
    top_probs, top_indices = torch.topk(probabilities, k=top_k)

    # Convert to numpy for easier handling
    top_probs = top_probs.numpy()
    top_indices = top_indices.numpy()

    return top_probs, top_indices
```

The values can then be converted to NumPy so that other code—like printing out the values—will work more easily.

We can see the output of this here, where Softmax and TopK were used to interpret the data and then print it out:

```
Image 1 Predictions:
-----------------------
Raw logits (first 5): [1.2000000476837158, 4.5, -0.800000011920929,
                        2.0999999046325684, 0.30000001192092896]

Top 5 Predictions:
1. goldfish: 52.3%
2. rooster: 14.2%
3. robin: 9.5%
4. tiger shark: 4.7%
5. house finch: 3.5%
```

The full code for this can be found in the GitHub repository (*https://github.com/lmoro ney/PyTorch-Book-FIles*) for this book.

Summary

In this chapter, you took a brief look at tensors and the underlying idea behind them —that they are a flexible data structure that can be used to represent the best way to put data *into* an ML model, regardless of what it represents, and even batch it. It also provides a consistent way to manage outputs from a model—in which values are typically emitted via neurons that are arranged in the output layer as a list. Thus, by being able to handle tensors, you can build a foundation for how data flows in *and* out of a model.

With that, we're now going to switch gears from model training to inference, in particular with generative AI, starting with getting models from registries and hubs.

## CHAPTER 13

Hosting PyTorch Models for Serving

In the earlier chapters of this book, we looked at many scenarios for training ML models, including those for computer vision, NLP, and sequence modeling. But that was just the first step—a model is of little use without a way for other people to use its power! In this chapter, we'll take a brief tour of some of the more popular tools that allow you to give them a way to do that.

You should note that taking a trained PyTorch model to a production-ready service will involve a lot more than just deploying it, and that the machine learning operations (MLOps) discipline is designed with that in mind. When you get into the world of serving these models, you'll need to understand new challenges, such as handling real-time requests, managing computational resources, ensuring reliability, and maintaining performance under varying loads.

Ultimately, MLOps is about bridging the gap between data science and software engineering. That's beyond the scope of this chapter, but there are some great books about it out there from O'Reilly, including *Implementing MLOps in the Enterprise* by Yaron Haviv and Noah Gift and *LLMOps* by Abi Aryan.

This chapter will introduce two popular approaches to serving PyTorch models in production environments.

We'll begin with TorchServe, the official serving solution from PyTorch, which provides a robust framework that's designed specifically for serving deep-learning models at scale. TorchServe offers out-of-the-box solutions for standard serving requirements like model versioning, A/B testing, and metrics collection. It's also an excellent choice for teams who are looking for a production-ready solution with minimal setup.

Then, we'll explore how to build serving solutions using the popular Flask framework, which is for developers who need more flexibility or have more straightforward serving requirements. Flask's simplicity and extensive ecosystem also make it an excellent choice for smaller-scale deployments and proof-of-concept services.

As you work through the chapter, you'll take a hands-on approach in which you'll take some of the models you created in earlier chapters, let us walk you through how to deploy them, and then call their hosting servers for inference.

Introducing TorchServe

TorchServe is PyTorch's default serving framework that's designed for performance and flexibility. You can find it at *pytorch.org/serve*.

TorchServe's goal was originally to be a reference implementation on how to properly serve models with a modular extensible architecture, but it has grown beyond that into a fully performant professional-grade framework that is likely more than enough for any serving needs.

It's also built on a modular architecture with the aim to handle the complexity of serving models at scale. To this end, it's built from the following key components:

The model server
> The PyTorch model server is the central component that handles the lifecycle of models and handles all inference requests. It provides the endpoints for model management and inference, supporting REST and gRPC calls.

Model workers
> These are independent processes that load models and perform inference on them. Each one is isolated, so in a multimodel serving environment, they are designed to continue operating should issues in one model arise.

Frontend handlers
> These are custom Python classes that handle preprocessing, inference, and post-processing for specific model types. As you'll see in a moment, when we get hands-on, frontend handlers are complementary to the training code for the model, and it's good practice to create separate handler classes.

The model store
> Model serving in PyTorch uses "model archives," or MAR files, for the servable object. Once you've trained your model, you'll convert it into this format. The model store is where these are kept.

You can see the architecture diagram for a TorchServe system in Figure 13-1.

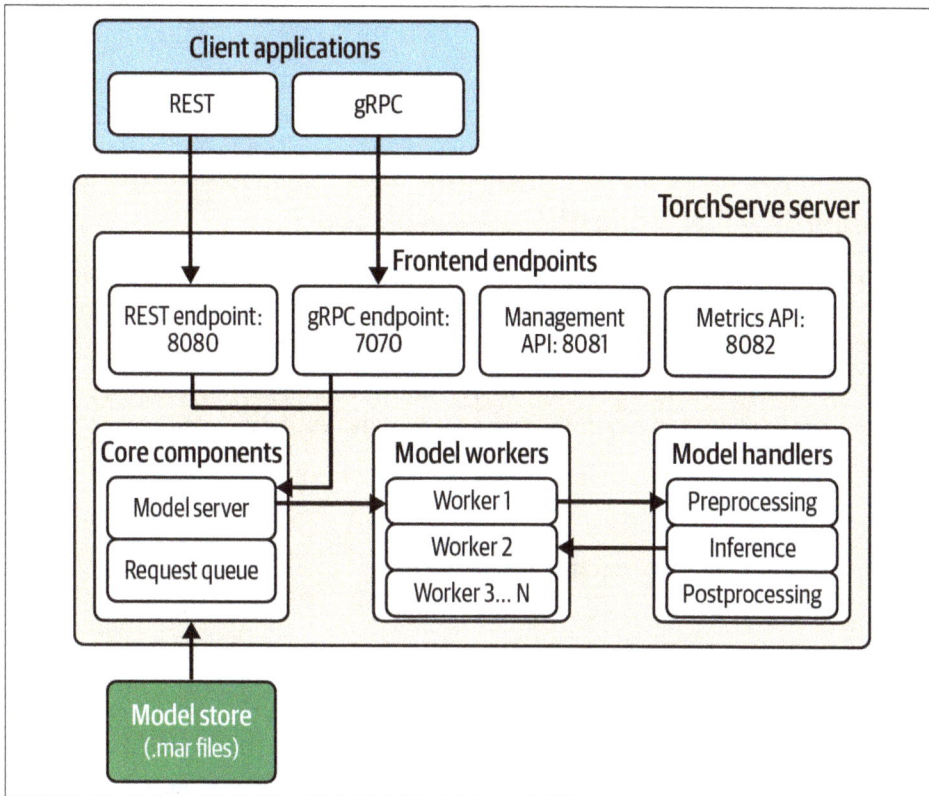

Figure 13-1. The TorchServe server infrastructure

For inference, the client application will call the server infrastructure over REST (via default port 8080) or gRPC (via default port 7070). It also provides management endpoints at 8081 and 8082, respectively.

The endpoint will then call the core model server, which in turn will spawn the appropriate number of model workers. The workers will then interface with the model handlers to do preprocessing, inference, and postprocessing. The model itself will be in the model store, and should it not be in memory, the model server will use the request queue to load it.

In the next section, we'll explore setting up a TorchServe server and using it to provide inference for the first model you created in Chapter 1.

Setting Up TorchServe

I find it's easiest to learn something if I go through it step-by-step with a simple but representative scenario. So, for TorchServe, we'll install the environment first and work from there.

Preparing Your Environment

I strongly recommend using a virtual environment, and I'll be stepping through this chapter using venv, which is a freely available one from the Python community that you can find in the Python documentation (*https://oreil.ly/FNsPt*).

Even if you've used virtual environments before, I'd recommend starting with a clean one to ensure you are going to install and use the right set of dependencies.

You can create a virtual environment like this:

```
python3 -m venv chapter13env
```

And once you've done that, you can start it with this:

```
source chapter13env/bin/activate
```

Then, you'll be ready to install torchserve. To get started, I recommend installing torchserve, the model archiver, and the workflow archiver like this:

```
pip install torchserve torch-model-archiver torch-workflow-archiver
```

You aren't limited to these, and you'll often need to install other dependencies. One of the difficulties of working with TorchServe is that errors may be buried in log files, so it's hard to figure out which dependencies you'll also need. At a minimum, I've found that starting from a clean system, I also needed to install a JDK after version 11 and PyYAML. Your mileage may vary.

Once you've done this, you can change to the directory where you want to work, and within that, you can create a subdirectory called `model_store`:

```
mkdir model_store
```

With that in place, the first thing you'll need to do is set up the configuration file for your PyTorch server. It will be a file called *config.properties*.

Setting Up Your config.properties File

There's a lot going on in this file, and you can learn more about it on the PyTorch site (*https://oreil.ly/czQh5*). But the important settings are the inference and management addresses, which are set to 0.0.0.0 and 8080/8081, respectively (similar to what's shown in Figure 13-1). Also, it's important to set the directory of the model store to be the one you just created. I've also set it to log debug messages to help catch dependency issues:

```
# config.properties
inference_address=http://0.0.0.0:8080
management_address=http://0.0.0.0:8081
metrics_address=http://0.0.0.0:8082
number_of_netty_threads=32
job_queue_size=1000
model_store=model_store
default_response_timeout=120
default_workers_per_model=1
log_level=DEBUG
```

Make sure these settings are in a file with the name *config.properties*.

Defining Your Model

In Chapter 1, you created a model that learned the relationship between two sets of numbers, when the equation describing this relationship was $y = 2x - 1$. When you're getting the model ready for serving, it's best practice to have a separate file for the model definition and the model training. The model definition file will then be used in another Python file called the *handler*, which you'll see momentarily. So, to that end, you should create a model definition file, like this:

```python
import torch
import torch.nn as nn

class SimpleLinearModel(nn.Module):
    def __init__(self):
        super().__init__()
        self.linear = nn.Linear(1, 1)

    def forward(self, x):
        return self.linear(x)
```

Then, save this into a file called *linear.py* so that you can then load and train it with code like the following. Also note the `import` for `SimpleLinearModel` that is bolded:

```python
import torch
import torch.nn as nn
import torch.optim as optim
from linear import SimpleLinearModel

def train_model():
    model = SimpleLinearModel()
    optimizer = optim.SGD(model.parameters(), lr=0.01)
    criterion = nn.MSELoss()

    xs = torch.tensor([[-1.0], [0.0], [1.0], [2.0], [3.0], [4.0]],
                      dtype=torch.float32)
    ys = torch.tensor([[-3.0], [-1.0], [1.0], [3.0], [5.0], [7.0]],
                      dtype=torch.float32)

    for _ in range(500):
```

```
            optimizer.zero_grad()
            outputs = model(xs)
            loss = criterion(outputs, ys)
            loss.backward()
            optimizer.step()

        return model
    # Save model
    model = train_model()
    torch.save(model.state_dict(), "model.pth")
```

In particular, note the last line, where, after training the model, it's saved as a *model.pth* file, along with this piece of code: `mode.state_dic()`, which saves out the *state dictionary* (aka the current state of the trained model). I find that this way of saving out a model works best with TorchServe.

Then, you run this code to get the file. You'll use this later to create a *.mar* file that goes in the model store. We'll look at that shortly, but first, we'll need a handler file. You'll explore that next.

Creating the Handler File

Once you've trained and saved the model, you'll need to create the handler file. It's the job of the handler to do the heavy lifting of serving inference—it loads your model, handles data preprocessing, does the inference, and handles any postprocessing to turn the inferred values back into data your users may want.

This file should inherit from the `base_handler` torch class like this:

```
from ts.torch_handler.base_handler import BaseHandler
```

Once you've done this, you'll need to create a model handler class that overrides this base class and implements a number of methods.

Let's start with the class declaration and initialization. It's pretty straightforward: just reporting on class initialization:

```
class ModelHandler(BaseHandler):
    def __init__(self):
        super().__init__()
        self.initialized = False
        logger.info("ModelHandler initialized")
```

Note that an `initialized` property is set to `False` in this code. That may seem odd, but the idea here is that this code will initialize the *class* but it won't be ready to use until the `initialize` custom function is called. That function will then load the model and get it ready for inference, and at that point, we will set `self.initialized` to be `True`. Note that the model is initialized as a `SimpleLinearModel`, so you'll need to import it in the same way as you did the training code.

Here's the code:

```python
def initialize(self, ctx):
    self.manifest = ctx.manifest
    properties = ctx.system_properties
    model_dir = properties.get("model_dir")

    # Load model
    serialized_file = "model.pth"
    model_pt_path = os.path.join(model_dir, serialized_file)
    self.device = torch.device("cuda:" + str(properties.get("gpu_id"))
                    if torch.cuda.is_available() else "cpu")

    # Initialize model
    self.model = SimpleLinearModel()
    state_dict = torch.load(model_pt_path, weights_only=True)
    self.model.load_state_dict(state_dict)
    self.model.to(self.device)
    self.model.eval()

    self.initialized = True
    return self
```

It begins by reading `ctx.manifest`—which is information in the *MAR* file (such as the model directory) that lets you use the model. We'll see how to create that file shortly.

The rest of the code is pretty straightforward: we create an instance of the model (called `SimpleLinearModel`, as shown in the previous code) and load its weights from where they were saved in (*model.pth*). We can then push the model to the device, where we will do inference, such as the CPU or CUDA, if it's available. Then, we'll put the model into evaluation mode for inference. I have found that TorchServe works best if you save the model with its state dictionary, so be sure to load that back, as shown.

At that point, we set `self.initialized` to `True`, and we're good to go for inference!

The next method we need to override is the *preprocess method*, which will take the input data and turn it into the format the model needs. There are many ways in which you could post the data to the backend server, and it's in the preprocess function that you'd handle them. You could, for example, just take basic parameters, or you could allow your user to post a JSON file. It's up to you, and this flexibility in the Torch-Serve architecture opens up those possibilities.

For the sake of simplicity, I'm just going to take a basic parameter in this code:

```python
def preprocess(self, data):
    # Get the value directly without decoding
    value = float(data[0].get("body"))
    tensor = torch.tensor([value], dtype=torch.float32).view(1, 1)
    return tensor.to(self.device)
```

As you can see, the main purpose of this is not only to *get* the parameters but also to *reformat* them into the format the model needs. So, it gets the parameter from the request call and turns it into a single-dimension tensor, which it will then return. This tensor will be used in the inference method.

The next method to override is the inference method, and here's where we will pass the data into the model and get its response back. Here's the code:

```python
def inference(self, data):
    """Run inference on the preprocessed data"""
    with torch.no_grad():
        results = self.model(data)
    return results
```

This receives the preprocessed data from the previous step, so we can just get the output by passing in this data. We run it with `torch.no_grad()` because we aren't interested in backpropagation, just a straight inference.

Finally, there's postprocessing. The end user isn't expecting a tensor back, but they are expecting more human-readable data, so we do the reverse of the preprocess step and cast the result back to a float by using NumPy:

```python
def postprocess(self, inference_output):
    """Return inference result"""
    return inference_output.tolist()
```

These steps are, as you can see, very customized to this model, and I've deliberately made them bare-bones so you can see the flow—but the general architecture remains the same for other models, regardless of the complexity. The goal is to give you a standard way to approach the problem, and code that extends the `BaseHandler` like this makes it easy for you to take advantage of all the unseen aspects of the server infrastructure—not least, passing the data around!

Creating the Model Archive

Starting with a trained model, you can create the archive file for it by calling `torch-model-archiver`, which is a command-line tool that's provided as part of TorchServe.

There are a few things you need to note and be careful of here. Here's the command, and I'll discuss the parameters afterward:

```
torch-model-archiver
    --model-name simple_linear
    --version 1.0
    --serialized-file model.pth
    --handler model_handler.py
    --model-file models/linear.py
    --export-path model_store
    --force
    --extra-files models/linear.py,models/__init__.py
```

Before running it, make sure you have a *model_store* (or similar) directory that you will store the archived model in. Don't mix it up with your source code! This can simply be a subdirectory, and you'll specify that directory in the `export-path` parameter.

The `model-name` parameter will specify the name of the model in the model store. You can call it whatever you want—it doesn't have to be the class name. In this case, I called it `simple-linear`.

The `version` parameter can be whatever you want, and you'll use it for tracking. As you train new versions of your model for new and different scenarios or bug fixes, you'll likely want to keep track of which version does what. You can let the server know about that here.

When you trained the model, you saved the weights and the state dictionary out to a file. You specify this file with the `serialized-file` parameter. In this case, it's `model.pth`.

The handler file is specified with the `handler` parameter.

A common problem I have encountered occurs when the model training file contains the model definition *and* the handler also contains it. TorchServe gets confused as to where to get the model definition, and that's why I separated it out in this case. But if you need to do that, it's important to specify *where* the model definition is, and you can do that with the `model-file` parameter. Note that following this methodology, you should also point to the location of the model file by using the `extra-files` parameter. To make the model file importable to both the model training and model handler as a package, I put the file in a directory called *models* and put an empty file called _ _init_ _.py in there. To make sure that the model archiver uses these, *both* of these files are specified in the `extra-files` parameter.

The `force` parameter just overwrites any existing model archive in the *model_store* directory. When you're learning, this parameter saves you from having to delete the model archive manually when trying different things. However, in a production system, you should use it carefully!

Once this line runs correctly, your *model_store* directory should contain a *simple_linear.mar* file.

Starting the Server

Once you've created your archive, you can start TorchServe and have it load that model. Here's an example command:

```
torchserve
  --start
  --model-store model_store
  --ts-config config/config.properties
  --disable-token-auth
  --models simple_linear=model_store/simple_linear.mar
```

The `start` parameter instructs TorchServe to start. You can also use `--stop` in the same way to stop it from executing (and filling your terminal with a wall of text).

The `model-store` parameter should also point to the directory where you store the *.mar* file that you created earlier.

The `ts-config` parameter should point to the configuration properties file you created earlier.

When you're learning and testing, you can use `--disable-token-auth` so that commands you send to the server to test your models don't need authentication. However, in proper production systems, you probably wouldn't want to use it!

The `models` parameter is a list of models that you want the server to make available to your users. In this case, there'll just be one, and it's the `simple_linear` model we defined. If you give the path to this model as the value, you'll see that it's the location of the *mar* file in the `model_store`.

If all goes well, you should then see TorchServe start in your terminal and give you a wall of status text, a little like in Figure 13-2.

Note that if this text is constantly scrolling, it's likely that there was an error in starting up TorchServe. From experience, I would say that there are dependencies that TorchServe needs (like PyYAML) that haven't been installed. If that's the case, then the configuration file was set to debug, and you can inspect the *models_log.log* file in the *logs* directory to see what's going on.

You may also see errors like the one in Figure 13-2, where it can't find the `nvgpu` module. This module is used by TorchServe to do GPU-based inference with an Nvidia GPU. Because I was running on a Mac in this case, you can safely ignore the error, and all inference will just run on the CPU, as per the code in the handler.

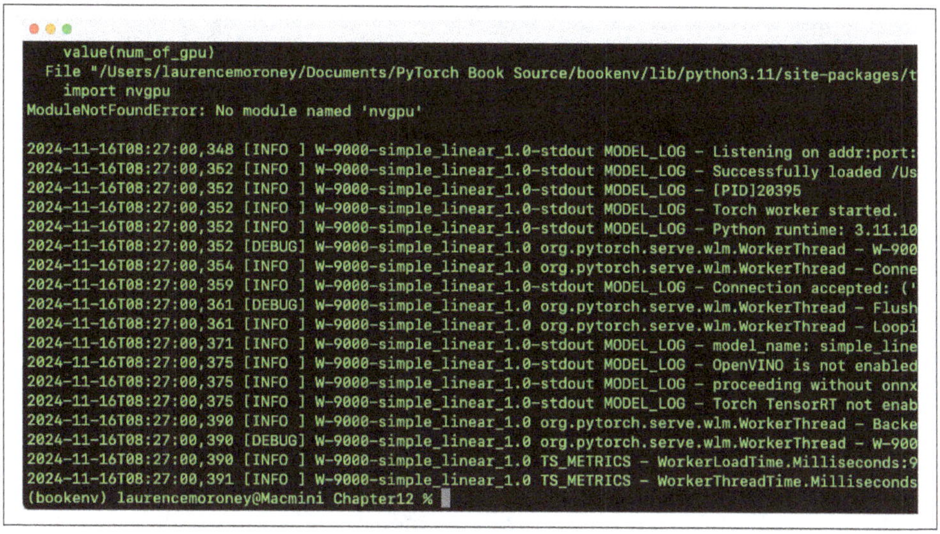

```
    value(num_of_gpu)
    File "/Users/laurencemoroney/Documents/PyTorch Book Source/bookenv/lib/python3.11/site-packages/t
      import nvgpu
ModuleNotFoundError: No module named 'nvgpu'

2024-11-16T08:27:00,348 [INFO ] W-9000-simple_linear_1.0-stdout MODEL_LOG - Listening on addr:port:
2024-11-16T08:27:00,352 [INFO ] W-9000-simple_linear_1.0-stdout MODEL_LOG - Successfully loaded /Us
2024-11-16T08:27:00,352 [INFO ] W-9000-simple_linear_1.0-stdout MODEL_LOG - [PID]20395
2024-11-16T08:27:00,352 [INFO ] W-9000-simple_linear_1.0-stdout MODEL_LOG - Torch worker started.
2024-11-16T08:27:00,352 [INFO ] W-9000-simple_linear_1.0-stdout MODEL_LOG - Python runtime: 3.11.10
2024-11-16T08:27:00,352 [DEBUG] W-9000-simple_linear_1.0 org.pytorch.serve.wlm.WorkerThread - W-900
2024-11-16T08:27:00,354 [INFO ] W-9000-simple_linear_1.0 org.pytorch.serve.wlm.WorkerThread - Conne
2024-11-16T08:27:00,359 [INFO ] W-9000-simple_linear_1.0-stdout MODEL_LOG - Connection accepted: ('
2024-11-16T08:27:00,361 [DEBUG] W-9000-simple_linear_1.0 org.pytorch.serve.wlm.WorkerThread - Flush
2024-11-16T08:27:00,361 [INFO ] W-9000-simple_linear_1.0 org.pytorch.serve.wlm.WorkerThread - Loopi
2024-11-16T08:27:00,371 [INFO ] W-9000-simple_linear_1.0-stdout MODEL_LOG - model_name: simple_line
2024-11-16T08:27:00,375 [INFO ] W-9000-simple_linear_1.0-stdout MODEL_LOG - OpenVINO is not enabled
2024-11-16T08:27:00,375 [INFO ] W-9000-simple_linear_1.0-stdout MODEL_LOG - proceeding without onnx
2024-11-16T08:27:00,375 [INFO ] W-9000-simple_linear_1.0-stdout MODEL_LOG - Torch TensorRT not enab
2024-11-16T08:27:00,390 [INFO ] W-9000-simple_linear_1.0 org.pytorch.serve.wlm.WorkerThread - Backe
2024-11-16T08:27:00,390 [DEBUG] W-9000-simple_linear_1.0 org.pytorch.serve.wlm.WorkerThread - W-900
2024-11-16T08:27:00,390 [INFO ] W-9000-simple_linear_1.0 TS_METRICS - WorkerLoadTime.Milliseconds:9
2024-11-16T08:27:00,391 [INFO ] W-9000-simple_linear_1.0 TS_METRICS - WorkerThreadTime.Milliseconds
(bookenv) laurencemoroney@Macmini Chapter12 %
```

Figure 13-2. Starting TorchServe

Testing Inference

Once the server is successfully up and running, you can test it by using `curl` from another terminal.

So, to get an inference from the model, you can `curl` like this:

```
curl -X POST http://127.0.0.1:8080/predictions/simple_linear -H
                      "Content-Type: text/plain" -d "5.0"
```

Notice that it is an HTTP POST to the predictions endpoint. We specify the `simple_linear` model name as defined with the *.mar* file earlier, and we can then add the header (with the `-H` parameter) as plain text containing the data `5.0`.

As you may recall, the model learned the linear relationship $Y = 2x - 1$, so in this case, x will be 5 and the result will be a number close to 9.

The return should look like this:

```
[
  8.997674942016602
]
```

Your value may vary, based on how your model was trained, but it should be a value very close to 9.

You can also use the management endpoint to inspect the models that the server is hosting, like this:

```
curl http://localhost:8081/models
```

The response will contain the name and location of the *.mar* file for each model on the server:

```json
{
    "models": [
        {
            "modelName": "simple_linear",
            "modelUrl": "model_store/simple_linear.mar"
        }
    ]
}
```

Note that earlier, when you did inference on the model, you used the `predictions` endpoint followed by a model name, which in that case was `simple_linear`. This key should map to a model name in this models collection, or you would have gotten an error.

Finally, if you want to explore the details of a specific model, you can call the management URL (via port 8081, as earlier) with the model's endpoint and the name of the model you want to know more about:

```
curl http://localhost:8081/models/simple_linear
```

The server will return some detailed specs on the model, along with information you can use to help debug any issues. Here's an example:

```json
[
    {
        "modelName": "simple_linear",
        "modelVersion": "1.0",
        "modelUrl": "model_store/simple_linear.mar",
        "runtime": "python",
        "minWorkers": 1,
        "maxWorkers": 1,
        "batchSize": 1,
        "maxBatchDelay": 100,
        "responseTimeout": 120,
        "startupTimeout": 120,
        "maxRetryTimeoutInSec": 300,
        "clientTimeoutInMills": 0,
        "parallelType": "",
        "parallelLevel": 0,
        "deviceType": "gpu",
        "continuousBatching": false,
        "useJobTicket": false,
        "useVenv": false,
        "stateful": false,
        "sequenceMaxIdleMSec": 0,
        "sequenceTimeoutMSec": 0,
        "maxNumSequence": 0,
        "maxSequenceJobQueueSize": 0,
        "loadedAtStartup": true,
```

```
      "workers": [
        {
          "id": "9000",
          "startTime": "2024-11-16T08:26:59.394Z",
          "status": "READY",
          "memoryUsage": 0,
          "pid": 20395,
          "gpu": true,
          "gpuUsage": "failed to obtained gpu usage"
        }
      ],
      "jobQueueStatus": {
        "remainingCapacity": 1000,
        "pendingRequests": 0
      }
    }
  }
]
```

Notice, for example, that the `deviceType` is expecting `gpu`. However, since I don't have `nvgpu` on this system (see the earlier note about when you ran the server), it can't load from the GPU, and the workers reported on that. It's OK for that to be the case in my dev box, which doesn't have the Nvidia GPU—but should you be running this on a server that *does* have the Nvidia GPU, that message is something you'd want to follow up on, and it's likely an issue in your handler file.

Going Further

The foregoing was a bare-bones example to help you understand the nuts and bolts of how TorchServe works. As you use more sophisticated models that take in more complex data, the basic pattern you followed here should follow suit. In particular, the breakdown of preprocessing, inference, and postprocessing in the handler file is an enormous help! Additionally, the PyTorch ecosystem has add-ons for common scenarios to help you avoid having to write preprocessing code to begin with! For example, if you are interested in image classification and are worried about taking an image and turning it into tensors so that you can do inference on that image, the `ImageClassifier` class builds on the base handler to do this for you, and you can have image classification without needing to write a preprocessor. To see more of this in action, take a look at the open source examples at the PyTorch repository. In particular, you can go to this GitHub page (*https://oreil.ly/YA4v4*) for an example of how to create a handler for MNIST images.

You'll find many more useful examples in that repo, but I would still recommend going through the steps to get a bare-bones example like the one we showed here up and running first. There are a lot of steps and a lot of concepts, and it's easy to get lost in the maze.

Serving with Flask

While TorchServe is extremely powerful, a great alternative that's super easy to use is Flask. Flask is a lightweight and flexible web framework for Python that enables efficient development of web applications and APIs.

You can use Flask to build everything from minimal single-endpoint services to complex web applications, starting with just a few lines of code. It perfectly complements PyTorch by giving you the ability to host models for inference, as we'll explore in this section.

As a microframework, Flask provides the essential components for web development —routing, request handling, and templating—while allowing you to select additional functionality as needed. It is highly extensible, with stuff like a backend database, authentication, etc., and there's also a vibrant ecosystem of extensions available to use off-the-shelf. Because of all of this, Flask has become a standard tool in web development, powering applications across industries at all scales.

In this chapter, we'll just explore Flask from the hosting model's perspective, but I'd encourage you to dig deeper into the framework if you're interested in serving Python code—not just PyTorch!

Now, let's take a look at the same example that we used for TorchServe. This will help you see how simple Flask makes serving applications!

Creating an Environment for Flask

First, to use Flask, you'll need to install it and any dependencies. If you've been using the same environment as earlier in this chapter, you can simply update it with this:

```
pip install flask
```

Then, just ensure that you have a model definition and training file—exactly the same as those from earlier in this chapter—and that you have trained a model and saved it with its state dictionary as a file called *model.pth*.

With those in hand, all you'll need is a single Python file that I'm going to call *app.py*, which will be the primary server application that Flask will use. We'll explore the code in that file next.

Creating a Flask Server in Python

To create a server with Flask, you implement an app that creates a new Flask instance by using the following code:

```
app = Flask(__name__)
```

Then, on the app object, you can specify routes, such as predict for doing prediction on models. To serve, you implement the code for the endpoints at these routes. Here's an example of a full Flask server for our simple app:

```python
from flask import Flask, request, jsonify
import torch
from model_def import SimpleLinearModel

app = Flask(__name__)

# Load the trained model
model = SimpleLinearModel()
model.load_state_dict(torch.load("model.pth"))
model.eval()

@app.route("/predict", methods=["POST"])
def predict():
    value = float(request.form.get('value', 0))
    input_tensor = torch.tensor([[value]], dtype=torch.float32)
    with torch.no_grad():
        prediction = model(input_tensor)

    return jsonify({
        "input": value,
        "prediction": prediction.item()
    })

if __name__ == "__main__":
    app.run(port=5001)
```

Flask documentation and samples tend to use port 5000. If you're using a Mac as your development box, you might have issues with this as it conflicts with the port used by Airplay. To that end, I've used 5001 in this sample.

In this case, we declare the SimpleLinearModel and load it along with its state dictionary. Then, we put it into eval() mode to get it ready for inference.

Then, it becomes as simple as creating a route that we call predict and then implementing the inference within it. As you can see, we handle getting the value from the HTTP POST, converting it into a tensor, and getting the prediction back when we pass that tensor to the model.

To make the return a little friendlier, I used jsonify to turn it into a name-value pair. As you can see, that's much simpler than using TorchServe, but for that simplicity, you give up power. There's no set of base classes to handle preprocessing, post-processing, etc., and if you want to scale or implement multiple worker threads, you'll have to do it yourself.

I think this is a really useful and powerful server mechanism for smaller-scale environments, as well as for learning how to serve. For large-scale production environments, it's a lot of extra work, but it can definitely handle the load.

For inference, you can `curl` a POST to the model like this:

```
curl -X POST -d "value=5" http://localhost:5001/predict
```

And the response will be the JSON payload:

```
{"input":5.0,"prediction":8.993191719055176}
```

In addition to TorchServe and Flask, there are many other serving options, such as ONNX and FastAPI.

Summary

In this chapter, we explored two popular approaches to serving PyTorch models in production environments. You started with TorchServe, which is PyTorch's official serving solution that offers a robust framework with built-in support for model versioning, A/B testing, and metrics collection. It's also designed to be highly scalable at runtime, with a worker-thread architecture that's configurable based on the needs of your app. And while TorchServe requires more setup and understanding of its components like model workers and frontend handlers, I think it's worth investing the time in rolling up your sleeves and understanding all the different components and how they work together. To that end, you explored step-by-step how to take the simple linear model example from Chapter 1, save it, archive it, build a handler for it, and launch the server with the model details.

Then, you explored how to use Flask as a lightweight alternative that's extremely quick and simple to get up and running. You saw how its minimalist approach makes it ideal for smaller-scale deployments or proof-of-concept services. It's not limited to those, but as you move to production scale, you'll likely need to implement more code. Of course, that's not necessarily a disadvantage, as it gives you more granular control over your serving environment.

Both approaches have their place in the ML serving ecosystem. TorchServe shines in enterprise environments requiring comprehensive model management, while Flask's simplicity makes it perfect for smaller projects or learning environments with a smooth glide path toward scalable production. Of course, you aren't limited to just these two, and new frameworks are coming online all the time—in particular, one called FastAPI, which is rapidly growing in popularity. Which one you should choose ultimately depends on your specific needs around scaling, monitoring, and deployment complexity.

Next, in Chapter 14, you're going to look at third-party models that have been pretrained for you and various registries and hubs that you can load them from.

Using Third-Party Models and Hubs

The success of the open source PyTorch framework has led to the growth of supplementary ecosystems. In this chapter, we'll look at the various options of pretrained models and the associated tools and resources used to download, instantiate, and use them for inference.

While the PyTorch framework provides the foundation for deep learning, the community has created numerous repositories and hubs that store models that are ready to use and extend, making it easier for you to use and extend existing work rather than starting from scratch. I like to call this "standing on the shoulders of giants."

Since the advent of generative AI, these hubs have exploded in popularity, and many scenarios of generative ML models within workflows have grown out of this. As a result, when it comes to using pretrained models, there are many options. You might use them directly for inference, taking advantage of those trained on massive datasets that would be impractical to replicate. Or you might use these models as starting points for fine-tuning, adapting them to specific domains or tasks while retaining their learned features. This can take the form of low-rank adaption (LoRA), as we'll discuss in Chapter 20, or *transfer learning*, in which knowledge from one task is applied to another. Transfer learning or other fine-tuning has become a standard practice, especially when working with limited data or computational resources.

The advantages of using pretrained models extend beyond saving computational resources and time. These models often represent state-of-the-art architectures, and they've been trained on diverse, high-quality datasets that you may not have direct access to.

Additionally, the providers generally release the model with extensive documentation, performance benchmarks, and community support, giving you a long head start. Given the importance of responsible AI, these models often come with model cards that help you understand any research and work done so you can navigate any potential responsibility issues.

There is no "One Hub to Rule Them All," so it's useful to understand each of the major ones and how you can make the most of them. To that end, we'll look at some of the more popular ones in this chapter.

Hugging Face has become the de facto standard for transformer models, while PyTorch Hub offers officially supported implementations. Platforms like Kaggle provide competition-winning models, and GitHub-based TorchHub enables direct access to research implementations.

I think it's important for you to understand these resources and how to use them effectively. As the field of deep learning continues to advance, these hubs play an increasingly crucial role in widening access to state-of-the-art models and enabling rapid development of AI applications. And as the role of AI developer matures and grows, I'm personally seeing huge growth in the careers of software developers who don't train models from scratch and instead use or fine-tune existing ones. To that end, I hope this chapter helps you grow!

The Hugging Face Hub

In recent years, particularly with the rise of generative AI, the Hugging Face Hub has emerged as a leading platform for discovering and using pretrained ML models, particularly for NLP. Much of its usefulness (and a significant driver of its success) is the open source availability of two things: a transformers library (which makes using pretrained language models very easy to use) and a diffusers library (which does the same for text-to-image generative models like stable diffusion).

As a result, what started as a repository for transformer-based models has evolved into a comprehensive ecosystem supporting computer vision, audio processing, and reinforcement learning models. It has grown into a one-stop shop combining version control for models, documentation, and model cards—and because of the PyTorch-friendly libraries like transformers and diffusers, using these models with your Python and PyTorch skills is relatively easy.

Collaboration has also been one of the keys to the Hub's success. You can download, use, and fine-tune models with just a few lines of code, and many developers and organizations have shared their models or fine-tunes with the community. There were over 900,000 publicly available models at the time of writing, so there's plenty to choose from!

Using Hugging Face Hub

Before rolling up your sleeves to code with Hugging Face Hub, you should get an account and use it to get an API token.

Getting a Hugging Face token

This section will walk you through the *HuggingFace.co* (*http://huggingface.co*) user interface as it existed at the time of writing. It may have changed by the time you're reading this, but the principles are still the same. Hopefully, they'll still apply!

Start by visiting *Huggingface.co* (*http://huggingface.co*), and if you don't already have an account, you can use the Sign Up button at the top right to create one (see Figure 14-1).

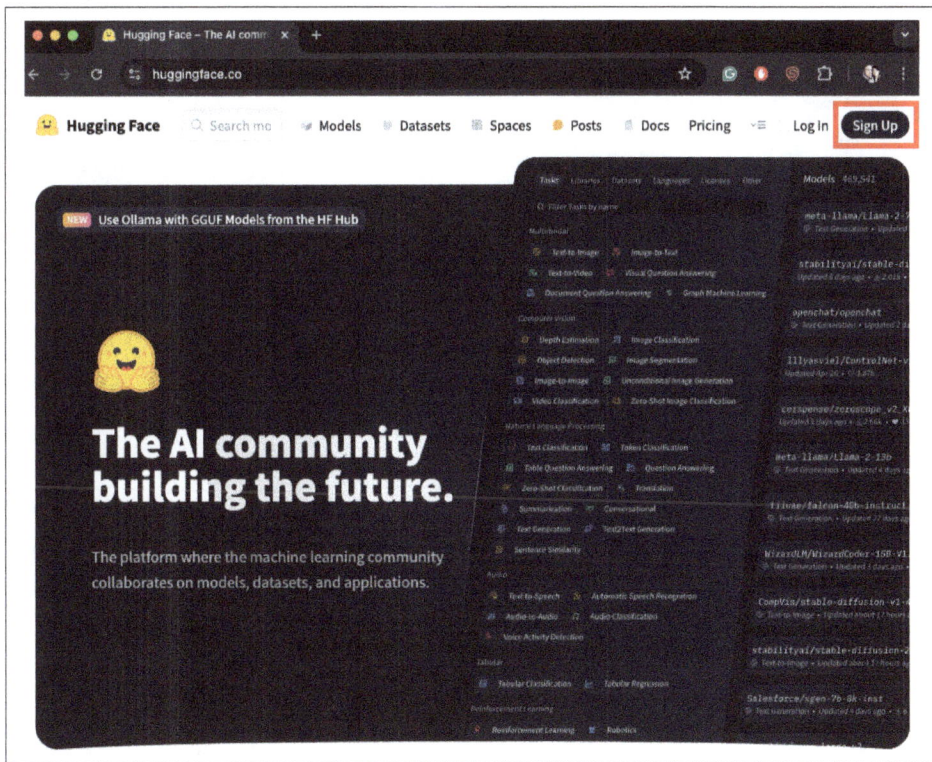

Figure 14-1. Signing up for Hugging Face

Once you've signed up and gotten an account, you can sign in, and in the top-right-hand corner of the page, you'll see your avatar icon. Select this and a drop-down menu will appear. On this menu, you'll see an option to Access Tokens, and you can select it to view your access tokens (see Figure 14-2).

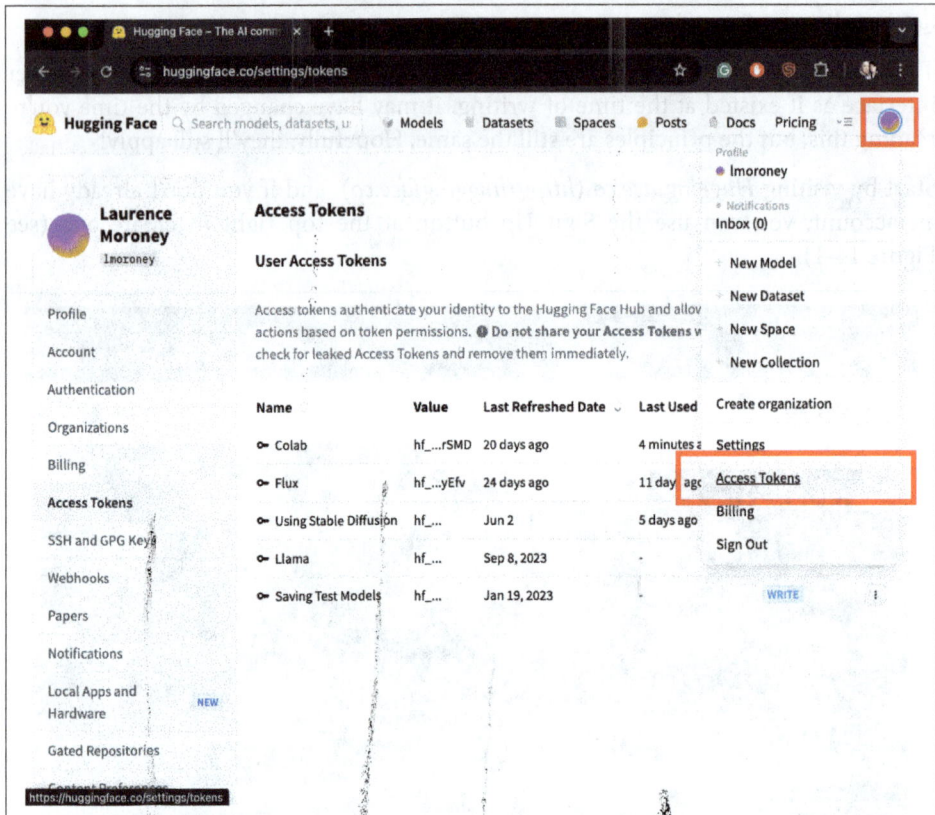

Figure 14-2. Access tokens

On this page, you'll see a Create New Token button, which will take you to a screen where you can specify your token details. Select the Read tab and give the token a name. For example, in Figure 14-3, you can see where I created a new Read token called PyTorch Book.

You'll also see a pop-up asking you to save your access token (see Figure 14-4). Note that it tells you that you will not be able to see the token again after you close this dialog modal, so be sure to hit the Copy button to have the token ready for the next steps.

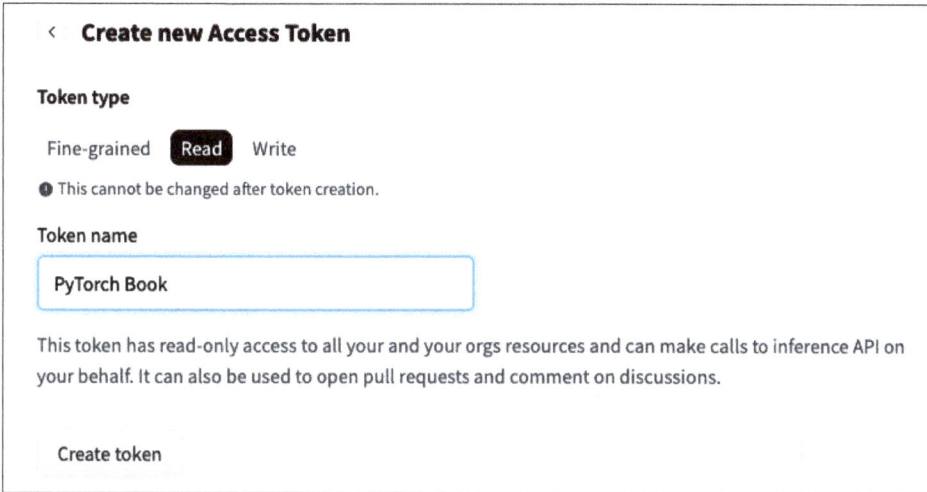

Figure 14-3. Creating an access token

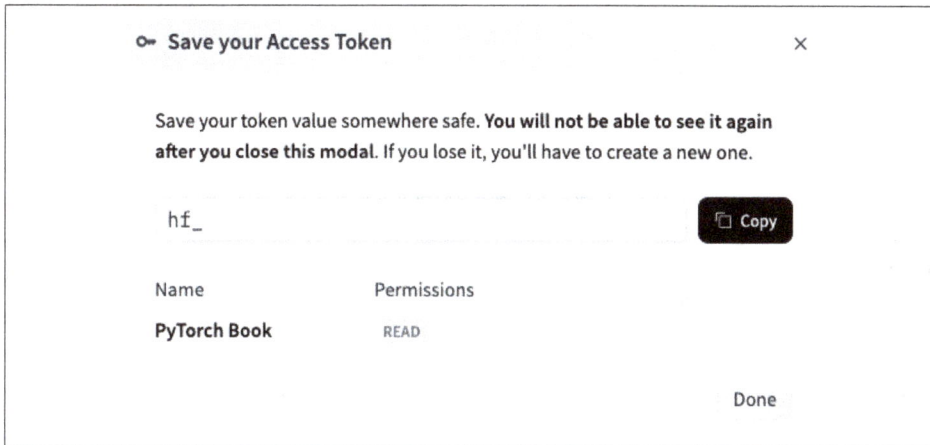

Figure 14-4. Saving your access token

If you *do* forget the token, you'll have to Invalidate and Refresh it on the token list screen. To do this, you select the three dots to the right of the token and then select Invalidate and Refresh from the drop-down menu (see Figure 14-5).

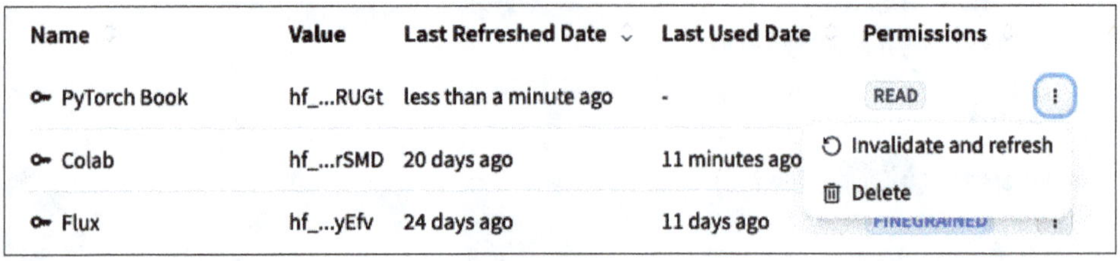

Name	Value	Last Refreshed Date	Last Used Date	Permissions
PyTorch Book	hf_...RUGt	less than a minute ago	-	READ
Colab	hf_...rSMD	20 days ago	11 minutes ago	Invalidate and refresh
Flux	hf_...yEfv	24 days ago	11 days ago	Delete

Figure 14-5. Invalidating and refreshing a token

Then, go back to the dialog from Figure 14-4 with a new token value. Copy it if you want to use it.

Now that you have a token, let's explore how to configure Colab to use it.

Getting permission to use models

Many models on Hugging Face will require additional permission to use them. In those cases, you should always check the model page and apply for permission on the link provided. Your permission to use the model will be tracked using the Hugging Face token. If you do *not* have permission, you'll see an error like this:

```
GatedRepoError: 401 Client Error.
(Request ID: [...])
Cannot access gated repo for url [...]
Access to model [...] is restricted.
You must have access to it and be authenticated to access it.
Please log in.
```

When this happens, the easiest thing to do is use the model name to find its landing page on the Hugging Face Hub and follow the steps to get permission to use it from there.

Configuring Colab for a Hugging Face token

If you want to use models from Hugging Face in Google Colab, then you need to configure a Colab secret in which code executing in Colab will read the token value, send it to Hugging Face on your behalf, and grant you access to the object.

It's pretty easy to do. First, in Colab, you select the key icon on the left of the screen (see Figure 14-6).

You should see a list of secrets that looks like the one in Figure 14-7. Don't worry if you don't have any API keys there yet. At the bottom of the list is a button that says Add new secret, and you'll select that.

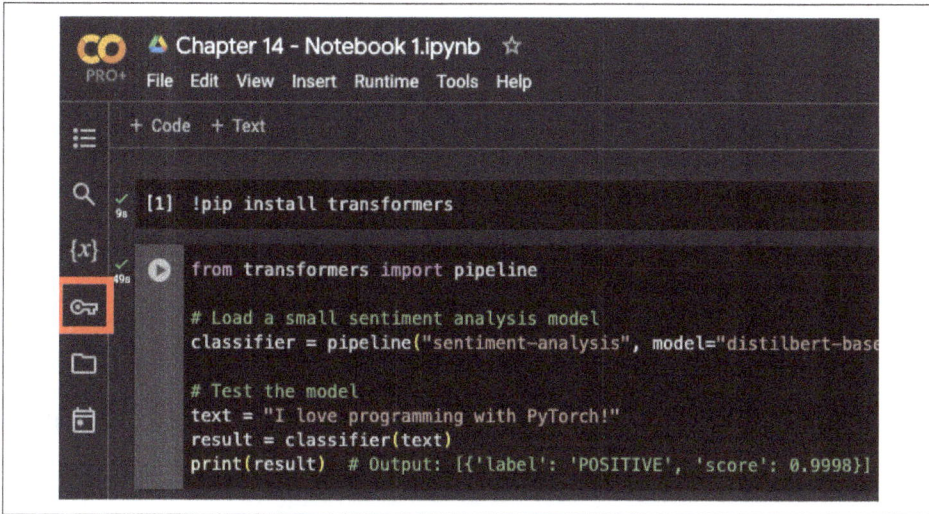

Figure 14-6. Selecting the Colab secrets

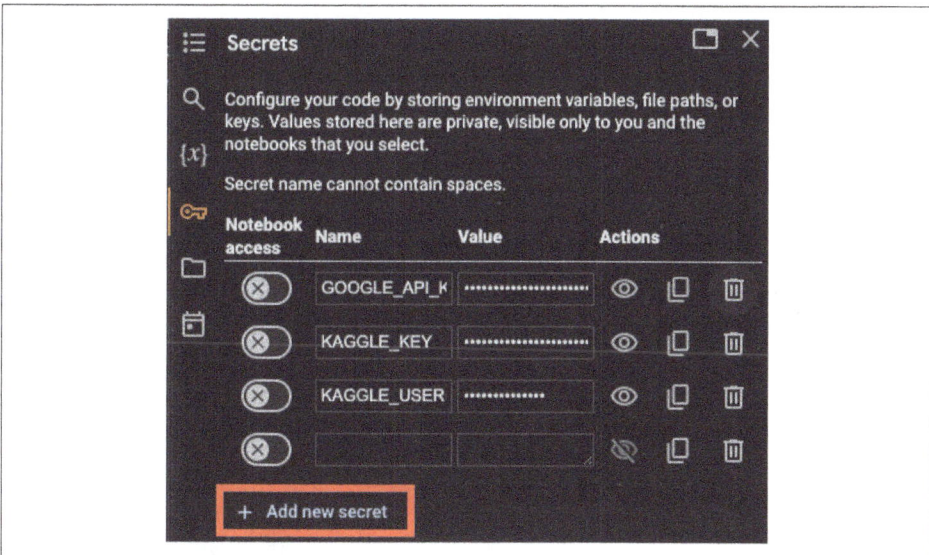

Figure 14-7. List of Colab secrets

Use the name "HF_TOKEN" in the Name field, and paste the value of the key into the Value field. Then, flip the switch to give Notebook Access to the secret (see Figure 14-8).

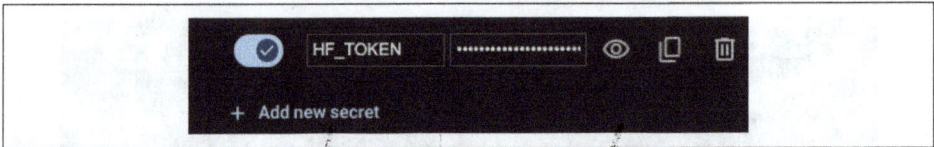

Figure 14-8. Configuring the HF_TOKEN in Colab

Your code in Colab will now use this token to access Hugging Face.

Using the Hugging Face token in code

If you just want to use the token directly in your code, whether in Colab or not, you'll have to log in to the Hugging Face Hub in your code and pass the key to it. It's pretty straightforward, with the Hugging Face Hub libraries providing the required support.

To start, just import the login class like this:

```
from huggingface_hub import login
```

You can then pass the token to the login class and initialize by using it in your Python session like this:

```
login(token="YOUR_TOKEN_HERE")
```

The Hugging Face classes will then use the token for the remainder of your session.

Using a Model From Hugging Face Hub

Once you have your token set up, getting and using a model is very simple. For this walk-through, we'll explore using a language model for text classification and sentiment analysis. This will require you to use the transformers library, so be sure to have it installed with this:

```
pip install transformers
```

The `transformers` API offers a pipeline class that lets you download and use a model based on its name in the Hugging Face repository. This one was fine-tuned using the SST sentiment analysis dataset from Stanford:

```
# Load a small sentiment analysis model
classifier = pipeline("sentiment-analysis",
            model="distilbert-base-uncased-finetuned-sst-2-english")
```

Pipeline offers much more than just downloading. It encapsulates what's needed to perform common tasks with models. The first parameter, which in this case is `sentiment analysis`, describes the overall pipeline task that you'll do. Transformers offer a variety of task types, including this, text classification, text generation, and a whole lot more.

When using the `pipeline` class, a number of key steps take place under the hood. These include the following:

Tokenization
In this step, the text is converted into tokens (as we discussed in Chapter 4).

Input processing
In this step, special tokens are added and the text is converted into tensors.

The model forward pass
In this step, the tokenized input is passed through the model's layers to get a result.

Output processing
In this step, the output is decoded from tensors back to the desired labels.

You can see this workflow in Figure 14-9.

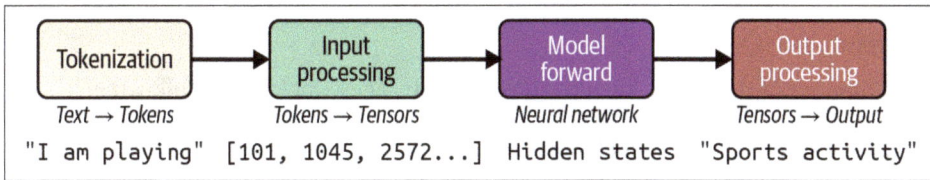

Figure 14-9. NLP pipeline flow

Then, when you want to use it, the burden of coding is removed from you as the developer and you just use the model like this:

```
# Test the model
text = "I love programming with PyTorch!"
result = classifier(text)
print(result)  # Output: [{'label': 'POSITIVE', 'score': 0.9998}]
```

All of the steps required for text classification and sentiment analysis are encapsulated and abstracted away from you. It makes your code much simpler!

Similarly, if you want to use the diffusers library, it comes with a number of pipelines that are often associated with a model type. So, for example, if you want to use the popular Stable Diffusion model for text to image—in which you give a prompt and the model will draw an image based on that prompt—you can do so very easily.

Let's explore this with an example.

First, from the diffusers library, you can import the pipeline that supports Stable Diffusion like this:

```
from diffusers import StableDiffusionPipeline
```

With this, you can specify the name of the model in the Hugging Face repository and use it to initialize the pipeline:

```python
import torch
from diffusers import StableDiffusionPipeline

model_id = "CompVis/stable-diffusion-v1-4"
device = "cuda"


pipe = StableDiffusionPipeline.from_pretrained(model_id,
                                        torch_dtype=torch.float16)
pipe = pipe.to(device)
```

Similar to the preceding text example, the pipeline encapsulates and abstracts a number of steps away from you. This means you can write relatively simple code like this:

```python
prompt = "a cute colorful cartoon cat"
image = pipe(prompt).images[0]

image.save("cat.png")
```

But a number of steps have been handled for you. These include the following:

1. In the text encoding step, Stable Diffusion uses a technology called CLIP to take the text prompts and turn them into embeddings that the model can understand.

2. An initial image is then constructed from random noise.

3. The embeddings are then fed into the model, which uses a process of denoising to create pixels and features that match the embeddings.

4. The final output of the model is then converted from tensors into an RGB image.

The overall process of creating an image from text is beyond the scope of this chapter, but it's well explained in this video from Google Research (*https://oreil.ly/zNjUT*).

The important thing to note here is that because the image-generation process begins by creating random noise, any images you create with the preceding code will be different. So, don't be alarmed if you're not getting the same picture consistently! There are ways of guiding this noise by using a seed, which we'll discuss in later chapters.

PyTorch Hub

One of the primary reasons for PyTorch's success—particularly in the research community—is the foresight of the developers in creating a hub where people could share their models. While this functionality has been massively superseded by Hugging Face Hub, as described earlier in this chapter, it's still worth looking at because many new and innovative models (or updates to existing ones) like YOLO are often shared on the Hub.

YOLO is "You Only Look Once,"' a popular and efficient object detection model.

As with Hugging Face Hub, the primary benefit of PyTorch Hub is that it gives you access either to models that you may not have the compute resources to train yourself or to the required data used to train them. At its core, PyTorch Hub functions as a centralized repository where researchers and developers can publish, share, and access models that have been trained on diverse datasets across various domains.

In this section, we'll explore PyTorch Hub and the APIs that you'll use to access models within it. Unfortunately, the APIs aren't as consistent as they could be, and it can sometimes be a little bit of a struggle to understand everything. But hopefully, this chapter will help!

We'll start with the PyTorch Vision libraries, which are composed of image classifiers, object detectors, and other computer vision models.

Using PyTorch Vision Models

Before you begin, you'll need to ensure that you have torchvision installed. Use this:

```
pip install torchvision
```

When you have installed it, you'll see the install version. This is really important when using Hub, in particular when you want to list the models to see what's available. You can also see the versions of these on GitHub (*https://oreil.ly/KIiFD*).

So, to list the models that are available, you'll use code like this:

```
models = torch.hub.list('pytorch/vision:v0.20.1')
for model in models:
        print(model)
```

Note the version number (which you can get from the GitHub page we just mentioned).

At the time of writing, there were close to a hundred models on this list. Do note that your version of the tag should match your version of torchvision, so if you are having problems, you can use this code to see your current version:

```
print(torchvision.__version__)
```

You can also choose a model from those available and load it into memory like this:

```
# Load ResNet-50 from PyTorch Hub
model = torch.hub.load('pytorch/vision:v0.20.1', 'resnet50',
                       pretrained=True)

# Set the model to evaluation mode
model.eval()
```

The model will be downloaded, cached, and then placed into evaluation mode.

Next up, you'll need to get your data ready for inference, and that requires you to have some domain knowledge of the model. So, for example, in the code we just cited, we used resnet50 as the model. This model (ResNet) is a very popular one for image classification that uses CNNs. A great place to go to learn more about this is the PyTorch Hub site (*https://pytorch.org/hub*).

From here, you can dig into model details—such as the size of the desired input, the labels that it can classify, etc. Then, with this information in hand, you can write inference code for the model. Here's an example:

```
# Load and preprocess the image
image_path = "example.jpg"  # Replace with your image path
image = Image.open(image_path).convert('RGB')

preprocess = transforms.Compose([
    transforms.Resize(256),
    transforms.CenterCrop(224),
    transforms.ToTensor(),
    transforms.Normalize(mean=[0.485, 0.456, 0.406],
                         std=[0.229, 0.224, 0.225]),
])
input_tensor = preprocess(image)
input_batch = input_tensor.unsqueeze(0)
```

You saw similar code in Chapter 3 and Chapter 4, where you explored building your own image classifier. This code resizes the images to 256 × 256 and then crops a 224 × 224 image from the center of that. Images are typically 32 bit RGB, in which each pixel is represented by 8 bits of alpha, 8 bits of red, 8 bits of green, and 8 bits of blue. However, for image classification, the neural network usually expects normalized values (i.e., between 0 and 1), so the transform to normalize the image performs this.

When you're using models in PyTorch, even though you know the desired dimensions (in this case, 224 × 224), you'll also need to batch the images for inference, even if you're just doing a single image. The input_tensor.unsqueeze(0) adds this extra dimension to the input tensor to handle this.

Next up, you'll do the actual inference, which just means you'll load the model onto the appropriate device—which is cuda if you have a GPU and the CPU otherwise. You'll then pass the input batch to the model to get an output:

```
# Perform inference
device = torch.device("cuda" if torch.cuda.is_available() else "cpu")
model.to(device)
input_batch = input_batch.to(device)

with torch.no_grad():
    output = model(input_batch)

_, predicted_idx = torch.max(output, 1)
```

The `predicted_idx` is the output for the class that has the highest probability of matching the input image. You'll see something numeric, like `tensor([153])`, in the output here. Recall that models' output layers will be neurons that correspond to the index of the required label. In the case of ResNet, the number 153 is a Maltese dog.

To decode this, you can use code like the following. You can find the URL of the labels file by digging into the model page on PyTorch Hub:

```
# Get class labels and map to prediction
url = "https://raw.githubusercontent.com/anishathalye/imagenet-simple-labels/
                           master/imagenet-simple-labels.json"
class_labels = json.load(urllib.request.urlopen(url))
predicted_label = class_labels[predicted_idx]
print("Predicted Label:", predicted_label)
```

This code will then print out the label for `predicted_idx`. So, in the case of `tensor([153])`, this will output the label for a Maltese dog.

Natural Language Processing

The PyTorch Hub for NLP ultimately directs to two different repositories: those implemented using Hugging Face Transformers (*https://oreil.ly/j1w3-*) as outlined earlier in this chapter and those from Facebook's fairseq research team (*https://oreil.ly/mmQeY*).

If you use the fairseq models, you may encounter a lot of sharp edges. Therefore, I thoroughly recommend setting up an environment with Python 3.11 (no later than that).

Within that, you can set up fairseq2 like this:

```
pip install fairseq
```

You'll likely also need other dependencies like hydra-core, OmegaConf, and requests.

Once you have the full system set up, you can use fairseq models like this:

```
import torch

en2de = torch.hub.load(
    'pytorch/fairseq','transformer.wmt19.en-de.single_model')

en2de.translate('Hello Pytorch', beam=5)
# 'Hallo Pytorch'
```

Do note that the environment will be very picky about which versions of PyTorch, pip, and many other libraries you can use. It can make for a very brittle experience, and unless you really want to use the models from the fairseq repository, I'd recommend just going with the Hugging Face transformer versions.

Other Models

PyTorch Hub also has repos for a variety of other model types, including audio, reinforcement learning, generative AI, and more. I've found the best way to explore them is to browse at the PyTorch Hub (*https://pytorch.org/hub*) and use the links on the landing pages to navigate to the requisite GitHub.

Summary

This chapter explores the ecosystem of pretrained models and model repositories for PyTorch, focusing on two major platforms: Hugging Face Hub and PyTorch Hub. While PyTorch Hub was the granddaddy that started the ball rolling, Hugging Face Hub has rapidly taken over as the go-to resource for pretrained models.

We took a look at how to use the transformers and diffusers libraries from Hugging Face, which encapsulate model loading and instantiation. With these, you have the keys to over 900,000 publicly available models. As a bonus, many of these have comprehensive documentation and model cards to get you up and running quickly and responsibly. You also got hands-on with using them, including setting up your account and getting authentication from Hugging Face using tokens.

Hugging Face APIs offer pipelines that encapsulate many of the common tasks of using models, such as tokenization and sequencing for NLP under the hood, making your coding surface much easier. We explored these with a text sentiment analysis scenario as well as another for image classification.

While PyTorch Hub has a lot less in it and accessing the models can be brittle in comparison, it's worth looking at because it's still well used in the research community. We looked at how to access PyTorch Vision models, prepare data for inference, and handle model outputs. The Hub also includes practical examples of using pretrained models like ResNet50 for image classification.

Ultimately, you should consider the advantages of using pretrained models, which have been built by expert researchers who have used expensive hardware and high-quality datasets that you may not otherwise have access to. To that end, you may find that using and fine-tuning existing models rather than training from scratch might be better for your scenario. We're going to explore that over the next few chapters, starting with Chapter 15, where we will go deeper into using LLMs with Hugging Face Transformers.

Transformers and transformers

With the paper "Attention Is All You Need" by Ashish Vaswani et al. (*https://oreil.ly/R7og7*) in 2017, the field of AI was changed forever. While the abstract of the paper indicates something lightweight and simple—an evolution of the architecture of convolutions and recurrence (see Chapters 4 through 9 of this book)—the impact of the work was, if you'll forgive the pun, transformative. It utterly revolutionized AI, beginning with NLP. Despite the authors' claim of the simplicity of the approach, implementing it in code is and was inherently complex. At its core was a new approach to ML architecture: *Transformers* (which we capitalize to indicate that we're referring to them as a concept).

In this chapter we'll explore the ideas behind Transformers at a high level, demonstrating the three main architectures: encoder, decoder and encoder-decoder. Please note that we will just be exploring at a very high level, giving an overview of how these architectures work. To go deep into these would require several books, not just a single chapter!

We'll then explore *transformers*, which we lowercase to indicate that they are the APIs and libraries from Hugging Face that are designed to make using Transformer-based models easy to use. Before transformers, you had to read the papers and figure out how to implement the details for yourself for the most part. So, the Hugging Face transformers library has widened access to models created using the Transformer architecture and has become the de facto standard for using the many models that have been created using the transformer-based architecture.

> Just to clarify, for the rest of this chapter, I'll refer to the architecture, models, and concepts as Transformers (with a capital *T*) and the Hugging Face libraries as transformers (with a lowercase *t*) to prevent confusion.

Understanding Transformers

Since the publication of the original paper mentioned in the introduction to this chapter, the field of Transformers has evolved and grown, but its underlying basis has remained pretty much the same. In this section, we'll explore this.

When working with LLMs anywhere (not just with Hugging Face), you'll hear of the terms *encoder*, *decoder*, and *encoder-decoder*. Therefore, I think it's a good idea for you to get a high-level understanding of them. Each of these architectures represents a different approach to text management—be it processing, classification, or generation. They have specific strengths for particular scenarios, and to optimize for your scenario, it's good to understand them so you can choose the appropriate ones.

Encoder Architectures

Encoder-only architectures (e.g., BERT, RoBERTa) generally excel at *understanding* text because of how rigorous they are in processing it. They're bidirectional in nature, being able to "see" the entire input sequence at once. With that nature of understanding, they're particularly effective for tasks that require a deep understanding and comprehension of the text and its semantics. So, they're particularly suited for tasks such as classification, named-entity recognition, and extraction of meaning for things like question answering. Their strength is transforming text into rich, contextual representations, but they're not designed to *generate* new text.

You can see the encoder-based architecture in Figure 15-1.

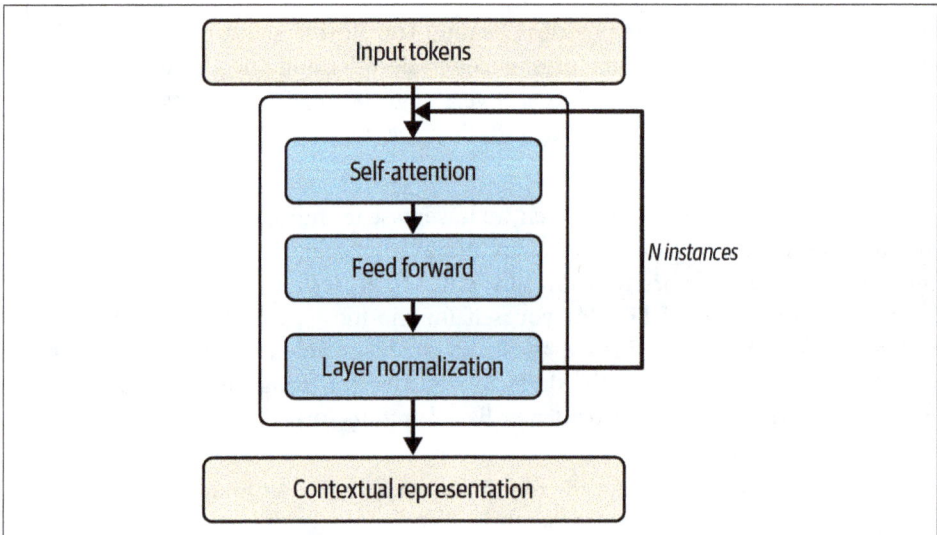

Figure 15-1. Encoder-based architecture

Let's explore this architecture in a little more detail. It begins with the tokenized inputs, which are then passed to the self-attention layer.

The self-attention layer

Self-attention is the core mechanism that allows tokens to "pay attention" to other tokens in the input sequence. So, for example, consider the sentence "I went to high school in Ireland, so I had to study how to speak Gaelic." The last word in this sentence is *Gaelic*, and it's effectively triggered by the word *Ireland* earlier in the sentence. If a model pays attention to the entire sentence, it can predict the word *Gaelic* to be the next word. On the other hand, if the model didn't pay attention to the entire sentence, then it might interpret from the sentence something more appropriate to "how to speak," such as *politely* or another adjective.

However, the self-attention mechanism—by considering the entire sentence—can understand context like that more granularly. It works by having each token in the sentence get three vectors associated with it. These are the query (Q) vector (aka "What am I looking for that's relevant to this token?"), the key (K) vector (aka "What tokens might reference me?"), and the value (V) vector (aka "What type of information do I carry?"). The representations in these vectors are learned over time, in much the same process as we saw in earlier chapters of this book. An attention score is then calculated as a function of these, and using Softmax, the embeddings for the words will be updated with the attention details, bending word embeddings closer to one another when there are similarities learned between them.

Do note that self-attention is generally bidirectional, so the order of the words doesn't matter.

The self-attention mechanism usually also has the context of *heads*, which are effectively multiple, parallel instances of the three vectors (Q, K, and V) that we saw previously, which can learn different representations and effectively specialize in different aspects of the input. A high-level representation of these heads is shown in Figure 15-2.

Thus, each head has its own set of learned weights for Q, K, and V vectors. The processing and learning for these vectors is done in parallel, with their results concatenated, and their final output projection is then a combination of information from each of the heads. As models have grown larger over time, one of the factors for this growth is the number of heads. For example, BERT-base has 12 heads and BERT-large has 16 heads. Architectures that also use a decoder, such as GPT, have grown similarly. GPT-2 had 12 attention heads, whereas GPT-3 grew to 96!

Returning to Figure 15-1, the self-attention instance (in which Figure 15-2 can be encapsulated into the self-attention box from Figure 15-1) then outputs to a *feedforward network* (FFN), which is often more accurately referred to as a *position-wise feedforward network*.

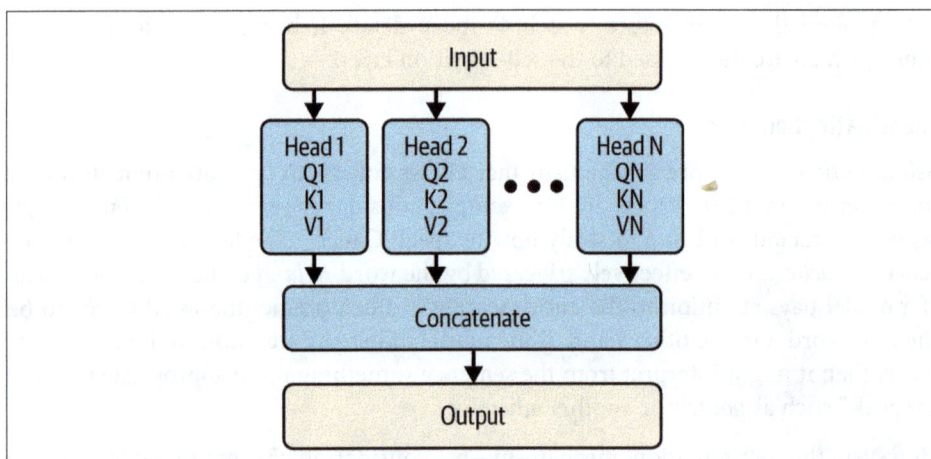

Figure 15-2. Multihead self-attention

The feedforward network layer

The FFN layer is vital in supporting the model's capacity to learn complex patterns in the text. It does this by introducing nonlinearity into the model.

Why is that important? First of all, let's understand the difference between linearity and nonlinearity. A *linear equation* is one for which the value is relatively easy to predict. For example, consider an equation that determines a house price. A linear version of this might be the cost of the land plus a particular dollar amount per square foot, and every house would follow the same formula. But as we know, house prices are far more complex than this—they don't (unfortunately) follow a simple linear equation.

Understanding sentiment can be the same. So, for example, if you were to assign coordinates on a graph (like we did when explaining embeddings in Chapter 7) to the words *good* and *not*, where *good* might be +1 and *not* might be −1, then a linear relationship between these for *not good* would give us 0, which is neutral, whereas *not good* is clearly negative. So, we need equations that are more nuanced (i.e., nonlinear) when capturing sentiment and effectively understanding our text.

That's the job of the FFN. It achieves this nonlinearity by expanding the dimensions of its input vector, applying a transformation to that, and using ReLU to "remove" the negative values (and thus remove the linearity) before restoring the vector back to its original dimensions. You can see this in Figure 15-3.

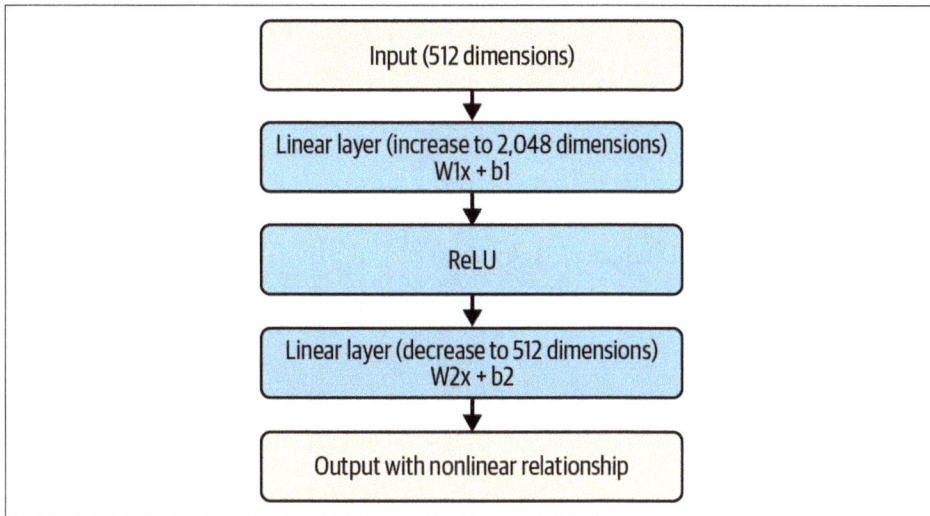

Figure 15-3. A feedforward network

The underlying math and logic behind how it works are a little beyond the scope of this book, but let's explore them with a simple example. Consider this code, which simulates what's happening in Figure 15-3:

```python
import torch
import torch.nn as nn

# Simplified example with small dimensions
d_model = 2  # Input/output dimension
d_ff = 4     # Hidden dimension

# Create some sample input
x = torch.tensor([[-1.0, 2.0]])

# First linear layer (2 → 4)
W1 = torch.tensor([
    [1.0, -1.0],
    [-1.0, 1.0],
    [0.5, 0.5],
    [-0.5, -0.5]
])
b1 = torch.tensor([0.0, 0.0, 0.0, 0.0])
layer1_out = torch.matmul(x, W1.t()) + b1
print("After first linear layer:", layer1_out)

# Apply ReLU
relu_out = torch.relu(layer1_out)
print("After ReLU:", relu_out)
# Notice how negative values became 0; this is the nonlinear operation!
```

```
# Second linear layer (4 → 2)
W2 = torch.tensor([
    [1.0, -1.0, 0.5, -0.5],
    [-1.0, 1.0, 0.5, -0.5]
])
b2 = torch.tensor([0.0, 0.0])
final_out = torch.matmul(relu_out, W2.t()) + b2
print("Final output:", final_out)
```

We start with a simple 2D tensor: `[-1.0, 2.0]`.

The first linear layer has the following weights and biases:

```
W1 = torch.tensor([
    [1.0, -1.0],
    [-1.0, 1.0],
    [0.5, 0.5],
    [-0.5, -0.5]
])
b1 = torch.tensor([0.0, 0.0, 0.0, 0.0])
```

When we pass our 2D tensor through this layer to get `layer1_out`, the matrix multiplication gives us a 4D output that looks like this:

```
After first layer: [-3.0, 3.0, 0.5, -0.5]
```

There are two negative values in this layer (–3 and –0.5), so when we pass it through the ReLU, they are set to zero and our matrix becomes this:

```
Output: [0.0, 3.0, 0.5, 0.0]
```

This process is called *bending*. By taking these values out, we're now introducing nonlinearity into the equation. The relationships between the values have become much more complex, so a process that attempts to learn the parameters of those relationships will have to deal with that complexity, and if it succeeds in doing so, it will avoid the linearity trap.

Next, we'll convert back to a 2D tensor by going through another layer, with weights and biases like this:

```
W2 = torch.tensor([
    [1.0, -1.0, 0.5, -0.5],
    [-1.0, 1.0, 0.5, -0.5]
])
b2 = torch.tensor([0.0, 0.0])
```

The output of the layer will then look like this:

```
Final output: [-2.75, 3.25]
```

So, the effect of the FFN is to take in a vector and output a vector of the same dimension, with linearity removed.

We can explore this with our simple code. Consider what happens when we take our input and apply simple, linear changes to it. So, if we take [−1.0, 2.0] and double it to [−2.0, 4.0], the nonlinearity introduced by the FFN will mean that the output won't be a simple doubling. And similarly, if we negate it, the output again won't be a simple negation:

```
Input: [-1.0, 2.0] → Output: [-2.75, 3.25]
Input: [-2.0, 4.0] → Output: [-5.5, 6.5]    # Not a simple doubling!
Input: [1.0, -2.0] → Output: [2.75, -3.25]  # Not a simple negation!
```

Over time, the parameters that are learned for the weights and biases should maintain the relationships between the tokens and allow the network to learn more nuanced, nonlinear equations that define the overall relationships between them.

Layer normalization

Referring back to Figure 15-1, the next step in the process is called *layer normalization*. At this point, the goal is to stabilize the data flowing through the neural network by removing outliers and high variance. Layer normalization does this by calculating the mean and variance of the input features, which it then normalizes and scales/shifts before outputting (see Figure 15-4).

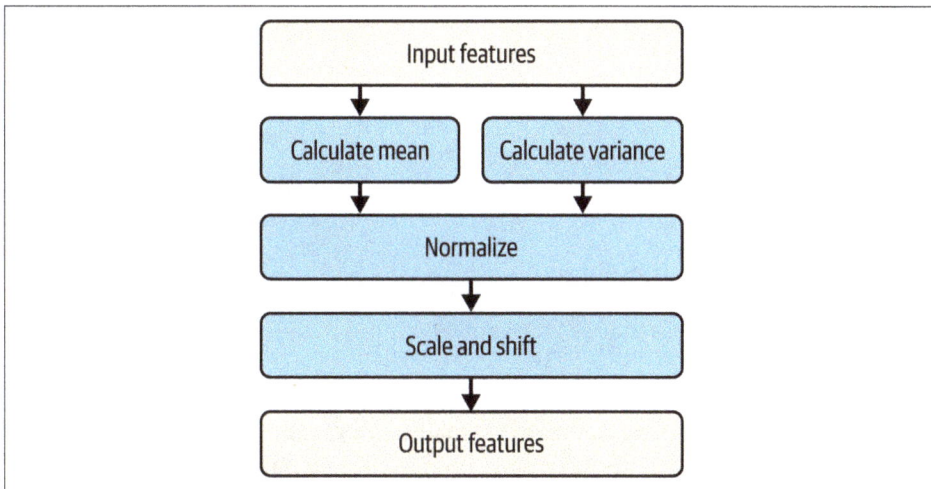

Figure 15-4. Layer normalization

From a statistical perspective, the idea of removing the outliers from using mean and variance and then normalizing them is quite straightforward. I won't go into details on the statistics here, but that's generally the goal of doing these types of calculations.

The *Scale and Shift* box then becomes a mystery. Why would you want to do this? Well, if you dig a little bit into the logic, the idea here is that the process of *normalization* will drive the mean of a set of values to 0 and the standard deviation to 1. The process itself can destroy distinctiveness in our input features by making them too alike. So, if there's a process that we can use to return some level of variance to them with parameters that are learned, we can clean up the data without destroying it—meaning we won't throw the baby out with the bathwater!

Therefore, multiplying our outputs by values with offsets can change this. These values are typically called *gamma* and *beta* values, and they act a little like weights and biases. It's probably easiest to show this in code.

So, consider this example, in which we'll take an input feature containing some values and then normalize them. We'll see that the normalized values will have a mean of 0 and a standard deviation of 1:

```python
import torch

# Create sample feature values
features = torch.tensor([5.0, 1.0, 0.1])
print("\nOriginal features:", features)
print("Original mean:", features.mean().item())
print("Original std:", features.std().item())

# Standard normalization
mean = features.mean()
std = features.std()
normalized = (features - mean) / std
print("\nJust normalized:", normalized)
print("Normalized mean:", normalized.mean().item())
print("Normalized std:", normalized.std().item())

# With learnable scale and shift
gamma = torch.tensor([2.0, 0.5, 1.0])  # Learned parameters
beta = torch.tensor([1.0, 0.0, -1.0])  # Learned parameters
scaled_shifted = gamma * normalized + beta
print("\nAfter scale and shift:", scaled_shifted)
print("Final mean:", scaled_shifted.mean().item())
print("Final std:", scaled_shifted.std().item())
```

But when we move the values through the scale and shift by using gamma and beta values, we get a new set of parameters that maintain a closer relationship to the originals but with massive variance (aka noise) removed. The output of this code should look like this:

```
Original features: tensor([5.0000, 1.0000, 0.1000])
Original mean: 2.0333333015441895
Original std: 2.6083199977874756

Just normalized: tensor([ 1.1374, -0.3962, -0.7412])
Normalized mean: 0.0
Normalized std: 1.0

After scale and shift: tensor([ 3.2748, -0.1981, -1.7412])
```

```
Final mean: 0.44515666365623474
Final std: 2.5691161155700684
```

Like the FFN, this is effectively destroying and then reconstructing the features in a clever way. In this case, it's designed to do it to remove variance, which has the effect of amplifying or dampening features as needed by shifting the overall distribution to better ranges for activation functions.

I like to think of this as what you do with your TV to get a better image—sometimes, you adjust the contrast or the brightness. By finding the optimal values of these settings, you can see the important details of a particular image better. Think of the contrast as the scale and the brightness as the shift. If a network can learn these for your input features, it will improve its ability to understand them!

Repeated encoder layers

Referring back to Figure 15-1, you'll see that the self-attention, feedforward, and layer normalization layers can be repeated N times. Typically, smaller models will have 12 instances and larger ones will have 24. The deeper the model, the more computational resources are required. More layers provide more capacity to learn complex patterns, but of course, this means longer training time, more GPU memory overhead, and potentially greater risks of overfitting. Additionally, larger models can impact the corresponding data requirements, leading to a possibly negative knock-on effect for complexity.

For the most part, there's a trade-off between the depth of the model (the number of layers) and the width (the size of each layer, including the number of heads). In some cases, models can reuse the same layer multiple times to reduce the overall parameter count—and ALBERT is an example of this.

The Decoder Architecture

While the encoder architecture specializes in understanding text by having attention across the entire context of the input sequence, the decoder architecture serves as the generative powerhouse. It is designed to produce sequential outputs one element at a time. While the encoder processes all inputs simultaneously (or as many of them as it can, based on the parallelization of the system), the decoder operates autoregressively. It generates each output token while considering both the encoded input representations *and* the previously generated outputs. The goal is to maintain coherence and contextual relevance through the process.

You can see a diagram of the encoder architecture in Figure 15-5.

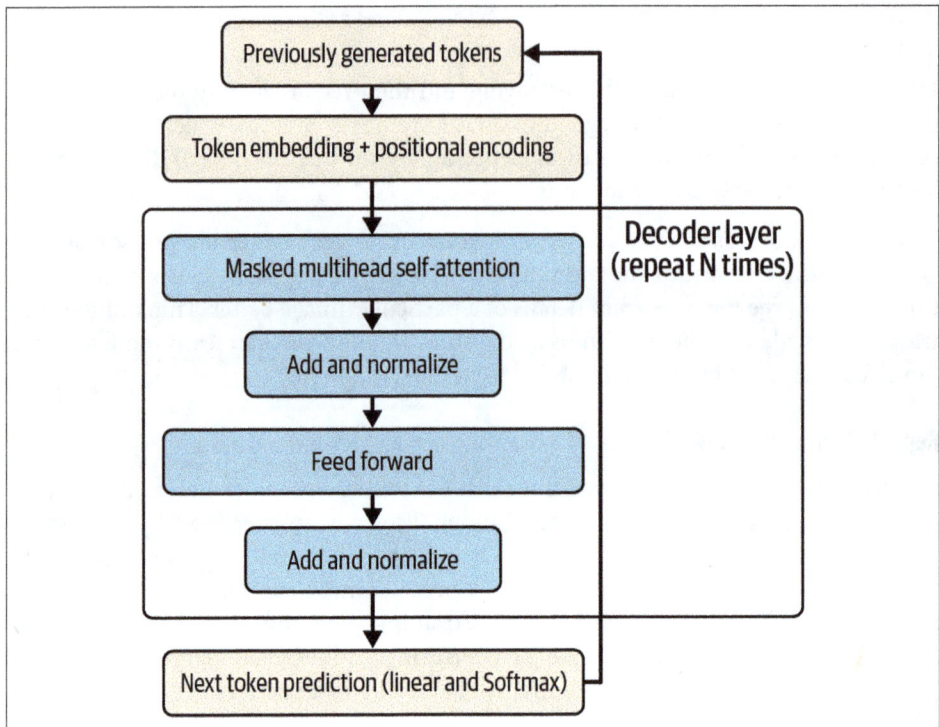

Figure 15-5. The decoder architecture

Let's explain this from the top down. The first box is the previously generated tokens. In a pure decoder architecture, this is the set of tokens that have already been generated or provided. When they're provided, they're typically called the prompt. So, say you provide the following tokens:

["If", "you", "are", "happy", **"and"**, "you", "know", "it"]

after one is run through the decoder, the token for "clap" will be generated. You will now have these:

["If", "you", "are", "happy", **"and"**, "you", "know", "it", "clap"]

These tokens will flow into the box for token embedding + positional encoding.

Understanding token and positional encoding

This transforms each token into a vector representation called an *embedding* (as explained in Chapter 5), which clusters words of similar semantic meaning in a similar vector space. Remember that these embeddings are learned over time.

So, for example, if we think about the words *it* and *clap* at the end of the aforementioned token list, they may have token embeddings that look like this:

```
"it" -> [0.2, -0.5, 0.7] (position 7)
"clap" -> [-0.3, 0.4, 0.1] (position 8)
```

I have simplified the embedding to just three dimensions for readability.

The next step is to perform the *positional* encoding, which is a huge innovation in Transformers. In addition to an encoding in a vector space for the semantics and meaning of the word, an innovative method using sine and cosine waves is performed to encode the word's position and the impact of this position on neighbors.

So, for example, given that we have 3D vectors for our encodings, we'll create a 3D vector using sine waves for the odd-numbered indices and cosine waves for the even-numbered ones, like this:

```
Position 7 -> [sin(7), cos(7), sin(7)] = [.122, .992, .122]
Position 8 -> [sin(8), cos(8), sin(8)] = [.139, .990, .139]
```

We then add these together to get this:

```
"It" -> [0.2+.122, -0.5+.992, 0.7+.122] = [0.322, 0.492, 0.822]
"Clap" -> [-0.3+.139, 0.4+.990, 0.1+.139] = [-0.161, 1.390, 0.239]
```

While this may seem arbitrary, the positional encodings actually come from a specific formula. These are shown here:

```
# For Even-numbered dimensions
PE(pos, d) = sin(pos / 10000^(d/d_model))

# For Odd-numbered dimensions
PE(pos, d) = cos(pos / 10000^(d-1/d_model))
```

If you plot these values as a table, they'll look like Table 15-1.

Table 15-1. Positional encodings

Position	Dimension 0	Dimension 1	Dimension 2	Dimension 3	Dimension 4	Dimension 5	Dimension 6	Dimension 7
0	0.000	1.000	0.000	1.000	0.000	1.000	0.000	1.000
1	0.841	0.540	0.100	0.995	0.010	1.000	0.001	1.000
2	0.909	−0.416	0.199	0.980	0.020	1.000	0.002	1.000
3	0.141	−0.990	0.296	0.955	0.030	1.000	0.003	1.000

The ultimate goal here is to have a relationship between each position on the list and every other position that, in some dimensions, has tokens that are typically far apart being a little closer and those that are typically closer being a little further apart!

So, imagine you have an input sequence of four tokens in positions 0 through 3, as charted in the table. The token in position 3 is as far away as possible from the token in position 0, so they are at extreme ends of the sequence.

With a positional encoding like this, we are given the possibility that in some dimensions, they are closer together. You can see in the first column that the values for dimension 0 places position 3 closer to position 0 than either of the others, whereas in the column for dimension 2, they are further apart. By using these positional encodings, we're opening up the possibility that words can be clustered, even if they're far apart in a sentence.

Think back to the sentence "I went to high school in Ireland, so I had to study how to speak Gaelic." In that case, the final token "Gaelic" was most accurate because it described a language in "Ireland," which was earlier in the sentence. Without positional encoding, this would have been missed!

Also, the positional encodings are *added* to the token embeddings, so they provide a sort of pressure to keep together the words that might be semantically related in different parts of the sentence, but they don't completely override the embeddings.

This is then fed into the multihead masked attention. We'll look at that next.

Understanding multihead masked attention

Earlier in this chapter, in the section on attention, we saw how the Q, K, and V vectors for each token are learned and used to update the embeddings of the word with attention to the other words. The idea behind *masked attention* updates this to ignore words that we shouldn't be paying attention to. In other words (sic), the goal is that we should only pay attention to *previous* positions in the sequence.

So, imagine you have a sequence of eight words and you want to predict the ninth. When you're processing the third word in the sentence, it should only pay attention to the second and first word but nothing after them. You can achieve this with a triangular matrix like this:

```
1 0 0
1 1 0
1 1 1
```

So, imagine this for a set of words like *the*, *cat*, and *sat* (see Table 15-2).

Table 15-2. Simple masked attention

	the	cat	sat
the	1	0	0
cat	1	1	0
sat	1	1	1

When processing *the*, using this method means we can only pay attention to *The* itself. When processing *cat*, we can pay attention to both *the* and *cat*. When processing sat, we can pay attention to *the*, *cat*, and *sat*.

So, recalling that the K, Q, and V vectors will amend the embeddings for the word in a way that bends them closer together for instances where the words may not have close syntactical meaning but are impacting one another through attention (like *Ireland* and *Gaelic* in the earlier example), the goal of the masked attention layer will only do this bending for words that we're allowed to pay attention to.

When this is performed multiple times, in parallel, across multiple heads, and aggregated together, we get an attention adjustment of the embeddings that's very similar to the one we did with the encoder—except that the masking prevents any amendment of the token from words *after* it in the sequence, particularly those that are generated.

Adding and normalizing

Next, we take the attention output from the masked attention layer and add it to the original input. This is called making a *residual connection*.

So, for example, the process might look like this:

```
Original input (word "cat" embedding):
[0.5, -0.3, 0.7, 0.1]  # Contains basic info about "cat"

Attention output (learned changes):
[0.2, 0.1, -0.1, 0.3]  # Contains contextual updates based on other words

After adding (final result):
[0.7, -0.2, 0.6, 0.4]  # Original meaning of "cat" PLUS contextual information
```

As we think about this, we'll see that we don't *replace* the original information with the attention mechanism but instead we *add* the new learned embeddings from the attention mechanism, thus preserving the original information. Over time, this will have the effect of helping the network learn. If some attention updates aren't useful, then the network will just make them close to zero through the backpropagation of gradients.

This is beyond the scope of this book, and it's generally found in the papers behind the creation of these models. But ultimately, the goal here is to get rid of a problem called the *vanishing gradient problem*—in which, if the original input was *not* maintained, then the gradients of the attention layer can get smaller and smaller with successive layers, thus limiting the number of layers you can use. But if you always add the gradients to the original input, then there will be a floor—such as the [0.5, -0.3, 0.7, 0.1] for the cat gradient previously mentioned—so the small changes from the attention gradients won't push these values close to zero and cause the overall gradients to vanish.

This is then pushed through a layer normalization, as described in the encoder chapter, to remove outliers while keeping the knowledge of the sequences intact.

The feedforward layer

The feedforward layer operates in exactly the same way as those layers used in encoders (see earlier in this chapter), with the goal of reducing any linear dependencies in the token sequence. The output from this is again added to the original data and then normalized, the logic being that the process of removing the outliers with the FFN should also prevent the gradients from vanishing and thus preserve important information. The normalization also keeps the values in a stable range, as repeatedly adding as we're doing here might push some values far above 1.0 or below –1.0, and normalized values in these ranges tend to be better for matrix calculation.

We can repeat this process of masked attention -> add and normalize -> feedforward -> add and normalize multiple times before we get to the next layer, where we'll use the learned values to predict the next token.

The linear and Softmax layers

The linear and Softmax layers are responsible for turning the decoder's representations into probabilities for the next token.

The linear layer will learn representations for each of the words in the dictionary with the transposed size of the decoder's representations. This is a bit of a mouthful, so let's explore it with an example.

Say our decoder output, having flowed through all the layers, is a 4D representation, like in Table 15-3.

Table 15-3. Simulated decoder output

0.2	–0.5	0.8	–0.3

We now have a weights matrix for each word in our vocabulary that is learned during training, and that matrix might look like the one in Table 15-4.

Table 15-4. Weights matrix for words in our vocab

cat	dog	sat	mat	the
1.0	0.5	2.0	0.3	0.7
-0.3	0.8	1.5	0.4	0.2
2.0	0.3	2.1	0.5	0.8
0.4	0.6	0.9	0.2	0.5

Note that the decoder representation is 1×4 and that each matrix for each word is 4×1. That's the transposition, and multiplying them out is now easy.

So, for *cat*, our final score will be this:

$$(0.2 \times 1.0) + (-.5 \times -.3) + (0.8 \times 2.0) + (-0.3 \times 0.4) = 1.8$$

We can then get final scores for each word, as in Table 15-5:

Table 15-5. Final scores for each word

cat	dog	sat	mat	the
1.8	-0.2	1.1	0.2	0.5

Using the Softmax function, these are then turned into probabilities, as in Table 15-6.

Table 15-6. Probabilities from Softmax function

cat	dog	sat	mat	the
47.5%	6.4%	23.6%	9.6%	12.9%

And then we can take the highest-probability word as the next token, in a process called *greedy decoding*. Alternatively, we can take 1 from k possible top values, in a process called *top-k decoding*, in which we pick, for example, the top three probabilities and choose one value from there.

This is then fed back into the top box as the new token list so that the process can continue to predict the next token.

And that's pretty much how decoders work, at least from a high level. In the next section, we'll look at how they can be combined in the encoder-decoder architecture.

The Encoder-Decoder Architecture

The *encoder-decoder architecture*, also known as *sequence-to-sequence*, combines the two aforementioned architecture types. It does this to tackle tasks that require transformation between input and output sequences of varying lengths. It's proven to be very effective for machine translation in particular, but it also can be used in models for text summarization and question answering.

As you can see in Figure 15-6, it's very similar to the decoder architecture, for the most part. The difference is the addition of a cross-attention layer that takes in the output from the encoder and injects it into the middle of the workflow.

The encoder will process the entire input sequence, creating a rich contextual representation that captures the full meaning of the input. The decoder layer can then query this, combining it with its representations to allow the decoder to focus additionally on relevant parts of the input when generating each new token.

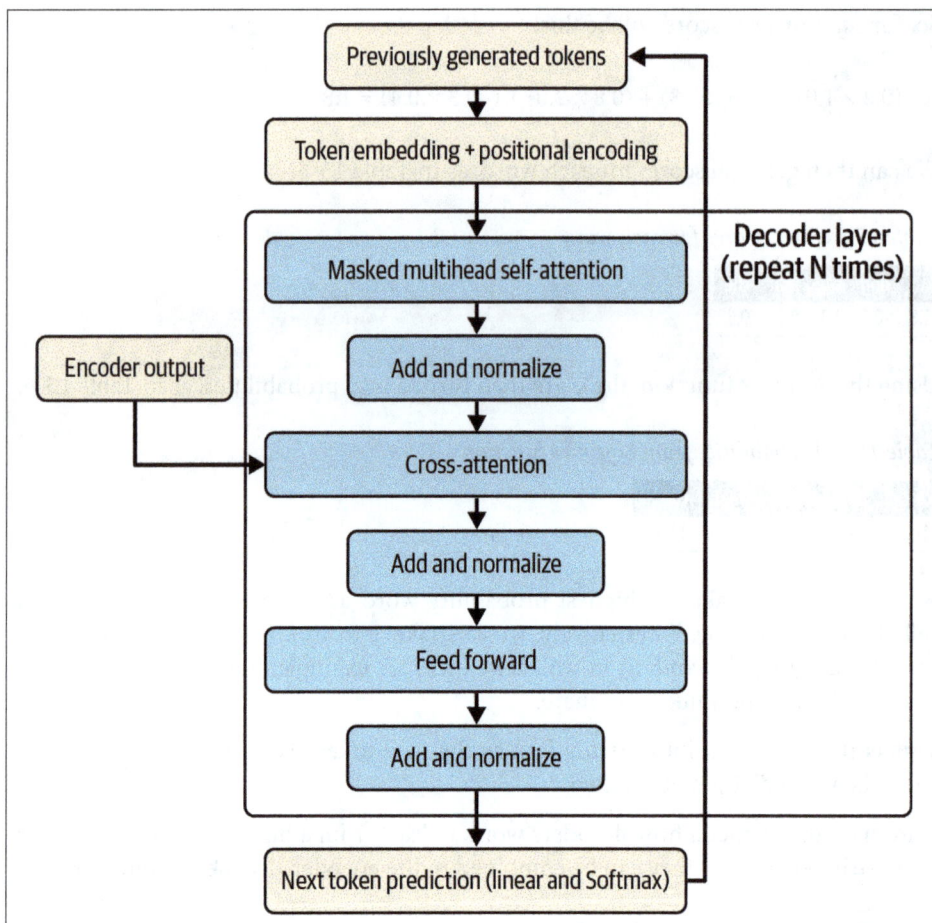

Figure 15-6. The encoder-decoder architecture

Now, you might wonder at this point why the encoder-decoder architecture needs the encoder's output, which it merges with cross-attention. Why can't it just unmask in its own self-attention block? The fundamental reason boils down to the fact that this is more powerful for the following reasons:

Separation of concerns and parameter focus

If the decoder self-attention block were to be unmasked, it would have to handle the tasks of both understanding the input *and* generating the output simultaneously. That could lead to issues with learning because there's a poor target. But if we separate them, each can focus on its own specialized role.

Quality

If we separate concerns, each role can build up a rich representation that's suitable for its task. In particular with the encoder, we have a well-known, battle-tested architecture for artificial understanding that we know works for that task.

The major innovation here is the *cross-attention block*. We can demonstrate the intuition behind this with the analogy of a human language translator. When a person translates a sentence from French to English, they don't just memorize the entire French corpus and then write English. Instead, while writing the English words, they actively look at different parts of the French sentence, focusing on the most relevant parts of the sentence for the word they are currently writing.

In French, the sentence "Le chat noir" translates to "The black cat," but the noun and adjective are reversed. A straight translation would be "The cat black." The human translator, when paying attention, would know this and would focus on other words in the French sentence. Cross-attention does the same thing. As the decoder generates each new word, it needs to refer to the source material to figure out what word to generate next. The words that have gone before in its output may not be enough.

You can see this mechanism in action in Figure 15-7.

Ultimately, an encoder creates a rich, contextual representation of the input because it artificially understands the source sentence. Cross-attention is a selective spotlight that highlights what the model believes to be the most relevant parts of this understanding for each word it generates, and this highlighting makes the model more effective at generating the correct words. You can see how this is very effective for machine translation, but it's not much of a stretch to understand how it might be useful for other tasks!

The mechanism for cross-attention works with the K, Q, and V vectors as before, but the innovation here is that the Q vector will code from the decoder, while the K and V vectors will come from the encoder.

This concludes a very high-level look at Transformers and how they work. There's lots more detail—in particular, about *how* they learn things like the Q, K, and V vectors—that's beyond the scope of this chapter, and I'd recommend reading the original paper to learn more.

Now, let's switch gears to the transformers (with a lowercase *t*) API from Hugging Face, which makes it really easy for you to use Transformer-based models in code.

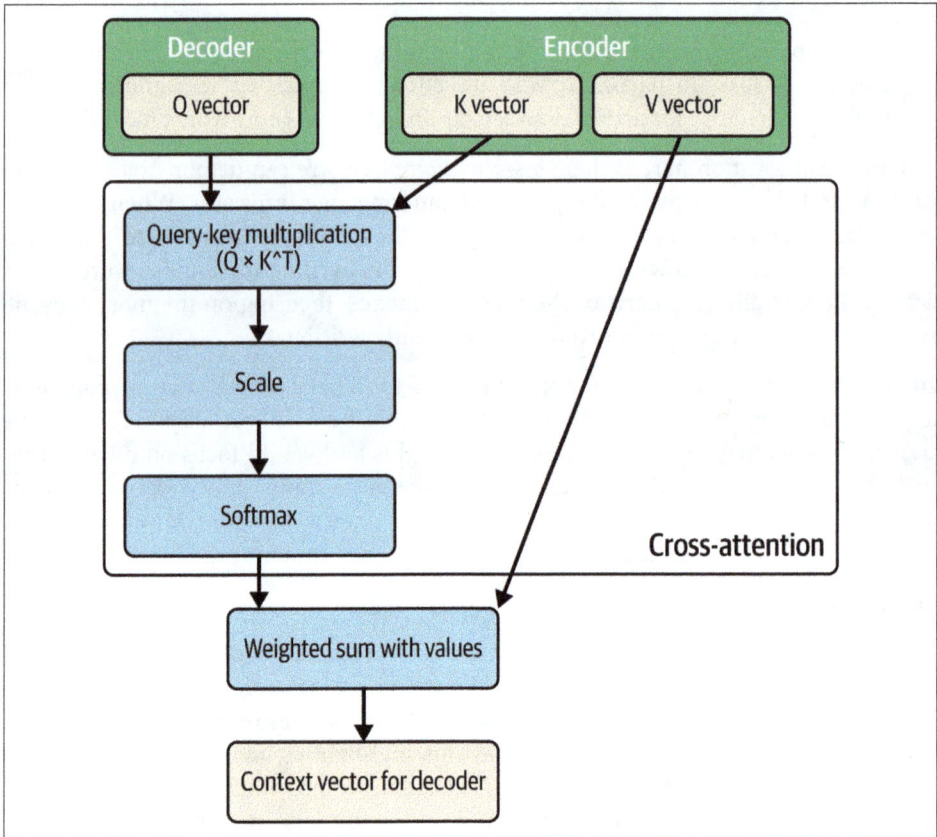

Figure 15-7. Cross-attention

The transformers API

At their core, transformers provide an API for working with pretrained models that use Python and PyTorch. The library's success stems from three key innovations: a simple interface for using pretrained models, an extensive collection of pretrained models (as we explored in Chapter 14), and a vibrant community that frequently contributes improvements and new models.

The library has evolved beyond its NLP roots to support multiple types of models, including those that support computer vision, audio processing, and multimodal tasks. Originally, the goal of the transformer-based architecture was to be effective in learning how a sequence of tokens is followed by another sequence of tokens. Then, innovative models built on this idea to allow concepts such as sound to be expressed as a sequence of tokens, and as a result, transformers could then learn sound patterns.

For developers, transformers offer multiple abstraction levels. The high-level pipeline API, which we explored in Chapter 14, enables immediate use of models for common tasks, while lower-level interfaces provide fine-grained control for custom implementations. Also, the library's modular design allows developers to mix and match components like tokenizers, models, and optimizers.

Perhaps most powerfully, transformers emphasize transparency and reproducibility. All model implementations are open source, well documented, and accompanied by model cards describing their capabilities, limitations, and ethical considerations. It's a wonderful learning process to crack open the transformers library on GitHub and explore the source code for common models like GPT and Gemma.

This commitment to openness has made the transformers API an invaluable tool for you, and it's something that's well worth investing your time to learn!

Getting Started with transformers

In Chapter 14, we explored how to access a model from the Hugging Face Hub, and we saw how easy it was to instantiate one and then use it with the various pipelines. We'll go a little deeper into that in this chapter, but let's begin by installing the requisite libraries:

```
pip install transformers
pip install datasets   # for working with HuggingFace datasets
pip install tokenizers   # for fast tokenization
```

Many of the models in the Hugging Face Hub, which are accessible via transformers, need you to have an authentication token on Hugging Face, along with permission to use that model. In Chapter 14, we showed you how to get an access token. After that, depending on the model you choose to use, you may need to ask permission on its landing page if you want access to it. You also learned how to use the transformers library in Google Colab, but there are, of course, many other ways to use transformers other than in Colab!

Once you have a token, you can use it in your Python code like this:

```
from huggingface_hub import login
login(token="your_token_here")
```

Or, if you prefer, you can set an environment variable:

```
export HUGGINGFACE_TOKEN="your_token_here"
```

With that, your development environment can now support development with transformers. Next, we'll look at some core concepts within the library.

Core Concepts

There are a number of important core concepts of using transformers that you can take advantage of. The simplicity of the transformers design hides the core concepts from you until you need them, but let's take a look at some of them here in a little more detail.

We'll start with the pipelines that you learned about in Chapter 14.

Pipelines

The `pipeline` class implements transformers' most useful abstraction, with a goal of making complex transformer operations accessible with minimal code. You can get the core default functionality of the model by using the appropriate pipeline with its defaults, and you can also override the defaults to create custom functionality.

Pipelines encapsulate all of ML processing—from data preprocessing to model inference to result formatting—in a single method.

To see the power of pipelines in action, let's look at an example:

```
from transformers import pipeline

# Basic sentiment analysis
classifier = pipeline("sentiment-analysis")
result = classifier("I love working with transformers!")
```

In this case, we didn't specify a model but rather a scenario (`sentiment-analysis`), and the pipeline configuration chose the default model for that scenario and did all of the initialization for us. As you saw in Chapters 5 and 6, when you're dealing with text, you need to tokenize and sequence it. You also need to know which tokenizer was used by a particular model so you can ensure that your text is encoded correctly and then converted to tensors in the correct, normalized format. You also need to parse and potentially detokenize the output, yet none of that code is present here. Instead, you simply say that you want sentiment analysis and then point the pipeline toward the text you want to classify.

We're not just limited to text sentiment analysis, of course. We can also do text classification, generation, summarization, and translation. Other scenarios include entity recognition and question answering. As more multimodal transformer-based models come online, there's also image classification and segmentation, object detection, image generation, and audio scenarios including speech recognition and generation.

While there are default models for a scenario, like we saw just now, there's also the ability to override the defaults and send custom parameters to a model.

So, for example, with text generation, we can use the default experience like this:

```
generator = pipeline("text-generation")
text = generator("The future of AI is")
print(text)
```

Or we can customize it further by passing parameters for things like the number of tokens to generate (`max_length`), the temperature (how creative it will be), and the number of tokens to evaluate when outputting a new one (`top_k`):

```
# Text generation with specific parameters
generator = pipeline(
    "text-generation",
    model="gpt2",
    max_length=50,
    temperature=0.7,
    top_k=50
)
text = generator("The future of AI is")
```

This gives you the flexibility you need to have fine control of any particular model, meaning it lets you set your desired parameters to override the default behavior. Importantly, the pipeline's abstraction handles the crucial steps of using the model.

To reiterate: when you're using a pipeline, you're getting more than just the model download. Depending on your scenario, you'll typically get the following steps:

Model loading

When you specify a model name, the pipeline API automatically downloads and caches the model and any associated tools, such as the tokenizer.

Preprocessing

This takes your raw inputs in their typical data format (strings, bitmaps, wave files, etc.) and turns them into model-compatible formats, even when multiple steps are needed. For example, it can tokenize a string and then turn the resulting tokens into embeddings or tensors.

Tokenization

As mentioned previously, this helps you with the specific tokenization strategy that a particular model needs. One tokenization scheme does not fit all!

Batching

This abstracts away the need for you to calculate optimal batch sizes. It will efficiently handle batching for you while respecting memory constraints.

Post-processing

Models output tensors of probabilities, and the pipeline will turn them into a human-readable format that you or your code can work with.

Now that we've had a quick look at pipelines and what they are (and you'll be using them a lot in this book), let's continue our tour of the core concepts and switch to tokenizers.

Tokenizers

We've spoken about tokenizers a lot, and we've even built our own in Chapter 5 and Chapter 6. But prebuilt tokenizers can be very powerful and useful tools as you create your own apps. Hugging Face transformers give you the `AutoTokenizer` class, and they make your life a lot easier when you're dealing with tokenizers by handling many of the complex scenarios for you. Even if you are creating your own models and training them from scratch, using an AutoTokenizer is probably a much smarter way of handling this task, rather than rolling your own.

To go a little deeper, the tokenizer is fundamental to how transformer models will process text. It's the first step in converting your raw text into a format that the model can work with. It's also often overlooked in discussions of building models, and that's a big mistake. The tokenization strategy is vital in the design of any system, and a badly designed one can negatively impact your overall model performance. Therefore, it's important to have a well-designed tokenizer for the task at hand.

At its core, the tokenizer's job is to break down text into smaller, numeric units called tokens. They can be words, parts of words (aka *subwords*), or even individual characters. To choose the right tokenization strategy, you'll need to make trade-offs among vocabulary size, the length of the sequences of tokens you will use in the model architecture, and your desire and requirements to handle words that are uncommon.

The transformers library supports multiple approaches, with subword tokenization being the most common one. It's a nice balance between character- and word-level tokenization that allows for less frequent words to still be in the corpus, because they're made of more common subwords. For example, the word *antidisestablishmentarianism* isn't a frequently used term, but it is made up of the letter combinations *anti, dis, est, ab, lish, ment, ari, an,* and *ism,* which are! It's also a fun word to use with AI models that interpret your speech, to see if it can complete the word before you do!

As you can see, this can give you terrific *vocabulary efficiency,* which is the ability to maintain a manageable vocabulary size while still being able to capture meaningful semantics. It can also handle *out-of-vocabulary* effectively, which means being effective and handling unseen words by breaking them down into known subwords.

Here's a really interesting example of this. In the very early days of transformers (pre-GPT), I worked on a project in which I created a transformer that I trained with the scripts of the TV show *Stargate,* and then I worked with the producers and actors on the show to do a table read of an AI-generated script. Given that the TV show is science fiction, with a lot of made-up words (aka technobabble), I used a subtoken tokenizer, following the logic that it could make up new words too. However, it ended up getting a little too creative! You can see the actors struggling with the new words in this video of the table read (*https://oreil.ly/A42Ko*).

So now, let's explore some common tokenizers in the ecosystem and see how they work. Note that tokenizers are associated with their model type, so when you explore models in the Hugging Face Hub (see Chapter 14), you'll be able to find their tokenizer. You must use the correct tokenizer with each model, or it won't be able to understand your input. Depending on the licenses associated with the tokenizers, you could also use them to train or tune your own models, instead of rolling your own as we did in Chapters 4 and 5.

The WordPiece tokenizer

The WordPiece tokenizer, which is associated with the BERT model, is a common tokenizer that's highly efficient at managing subwords. It starts with a basic vocabulary and then iteratively adds the most frequent combinations it sees. The subwords are denoted by ##. While generally created for English, it also works well for similar languages that have clear word boundaries denoted by spaces and other punctuation.

So now, let's consider an example sentence that contains complex words:

```
# Let's use a sentence with some interesting words to tokenize
text = "The ultramarathoner prequalified for the
        immunohistochemistry conference in neuroscience."
```

To load the tokenizer, you instantiate an instance of AutoTokenizer and call the `from_pretrained` method, passing the name of the tokenizer. For Bert's WordPiece, you can use `bert-base-uncased` as the tokenizer name:

```
from transformers import AutoTokenizer

# Load BERT tokenizer which uses WordPiece
tokenizer = AutoTokenizer.from_pretrained('bert-base-uncased')
```

Then, to tokenize the text, all you need to do is call the `tokenize` method and pass it your string. This will output the list of tokens:

```
# Tokenize the text
tokens = tokenizer.tokenize(text)
print("Tokens:", tokens)
```

This will then output the list of tokens from the sentence, which looks like this:

```
Tokens: ['the', 'ultra', '##mara', '##th', '##one', '##r', 'pre',
         '##qual', '##ified', 'for', 'the', 'im', '##mun', '##oh', '##isto',
         '##chemist', '##ry', 'conference', 'in', 'neuroscience', '.']
```

Long words that aren't commonly used (like *marathoner* and *immunohistochemistry*) are broken into subwords, whereas others (like *conference* and *neuroscience*) are kept as whole words. This is based on the training corpus used in BERT, and the decisions about which ones are common enough and which ones are not (for example, I would have expected *qualified* to be quite common) were made by the original researcher.

If you want to see the IDs for these tokens, you can get them with the encode method, like this:

```
# Get the token IDs
token_ids = tokenizer.encode(text)
print("\nToken IDs:", token_ids)
```

The list of IDs is shown here:

```
Token IDs: [101, 1996, 14357, 2108, 2339, 3840, 2837, 13462, 2005, 1996, 19763,
            2172, 3075, 7903, 5273, 13172, 1027, 2005, 23021, 1012, 102]
```

Note that the first and last tokens are 101 and 102, which are special tokens for the start and end of the sentence that the tokenizer inserted and which are expected by the model.

Now, say that you decode the list of token IDs back into a string, like this:

```
# Decode back to show special tokens
decoded = tokenizer.decode(token_ids)
print("\nDecoded with special tokens:", decoded)
```

Then, you'll see how the sentence has these special tokens inserted:

```
Decoded with special tokens: [CLS] the ultramarathoner prequalified for the
                             immunohistochemistry conference in
                             neuroscience. [SEP]
```

I would recommend that you continue to experiment with the tokenizer to understand how it manages text by turning it into tokens and special characters. Having this knowledge is often important when you're debugging model behavior or if you're doing some kind of fine-tuning to manage how your vocabulary will be used.

Byte-pair encoding

The GPT family uses a format called *byte-pair encoding* (BPE), which is a data compression algorithm as well as a tokenization one. It starts with the vocabulary of individual characters, which progressively learns common byte or character pairs to merge into new tokens.

The algorithm initially splits the training corpus into characters, assigning tokens to each. It then iteratively merges the most frequent pairs into new tokens, adding them to the vocabulary. The process continues for a predetermined number of merges. So, for example, over time, common patterns in words become their own tokens. This tends to veer toward the beginnings of the words with common prefixes (like *inter*) or the ends of words with common suffixes (like *er*) ending up having their own token. Instead of using the ## string to determine the beginning of a subword, BPE uses a special character (usually Ġ).

Here's the code you can use to tokenize the same sentence. You'll use the `gpt2` Auto-Tokenizer from OpenAI:

```python
from transformers import AutoTokenizer

# Load GPT-2 tokenizer which uses BPE
tokenizer = AutoTokenizer.from_pretrained('gpt2')

# Same sentence as before
text = "The ultramarathoner prequalified for the immunohistochemistry
        conference in neuroscience."

# Tokenize the text
tokens = tokenizer.tokenize(text)
print("Tokens:", tokens)
```

The output is shown here:

```
Tokens: ['The', 'Ġult', 'ram', 'ar', 'athon', 'er', 'Ġpre', 'qualified', 'Ġfor',
         'Ġthe', 'Ġimmun', 'oh', 'ist', 'ochemistry', 'Ġconference',
         'Ġin', 'Ġneuroscience', '.']
```

Over time, you'll see that the splits will be slightly different, which reflects the difference between the training sets. BERT was trained on Wikipedia and the Toronto BookCorpus, while GPT-2 was trained on web text.

SentencePiece

SentencePiece, which is used by the T5 model, is a unique tokenizer. It treats all input text as a raw sequence of Unicode characters, which gives it strong support for non-English languages. As part of this, it treats whitespaces like any other characters. That makes it effective for languages like Japanese and Chinese that don't always have clear word boundaries, and it also removes the need for language-specific preprocessing. In fact, while it was being built, this tokenizer learned its subword units directly from raw sentences in multiple languages.

Here's how to use it:

```python
from transformers import AutoTokenizer

# Load T5 tokenizer which uses SentencePiece
tokenizer = AutoTokenizer.from_pretrained('t5-base')

# Same sentence as before
text = "The ultramarathoner prequalified for the immunohistochemistry
        conference in neuroscience."
```

```
# Tokenize the text
tokens = tokenizer.tokenize(text)
print("Tokens:", tokens)
```

This produces the following set of tokens:

```
Tokens: ['_The', '_ultra', 'marathon', 'er', '_pre', 'qualified', '_for',
         '_the', '_immun', 'oh', 'ist', 'ochemistry', '_conference', '_in',
         '_neuroscience', '.']
```

As mentioned, where it's particularly powerful is with non-English languages. So, for example, consider this code:

```
# Let's also try a multilingual example with mixed scripts
text2 = "Tokyo 東京 is beautiful! Preprocessing in 2024 costs $123.45"
tokens2 = tokenizer.tokenize(text2)
print("\nMultilingual example tokens:", tokens2)
```

It will output the following:

```
Multilingual example tokens: ['_Tokyo', '_東', '京', '_is', '_beautiful', '!',
                              '_Pre', '-', 'processing', '_in', '_2024',
                              '_costs', '_$', '123', '.', '45']
```

Note how the Japanese characters for Tokyo were split into multiple tokens and the numbers were kept whole (i.e., 123 and 45).

Given that transformers were initially designed to improve machine translation, you can see that SentencePiece, which predates generative AI like GPT, was designed with internationalization in mind!

Summary

In this chapter, we looked at Transformers, the architecture that underpins modern LLMs, and transformers, the library from Hugging Face that makes Transformers easy to use.

We explored how the original Transformer architecture revolutionized AI through its use of attention mechanisms, in which context vectors for words were amended based on learned details of where the sequence might be paying appropriate attention to other parts of the sequence. We also looked at encoders that excel at artificial understanding of text, decoders that can intelligently generate text, and encoder-decoders that bring the best of both for sequence-to-sequence models. We also double-clicked into their architecture to understand how the mechanisms such as attention, feedforward, normalization, and many other parts of the architecture work.

We then looked into transformers, which form the library from Hugging Face that makes downloading and instantiation of Transformer-based models (including the entire inference pipeline) very easy. There's a whole lot more still in there, and hopefully, this gave you a good head start.

In the next chapter, you're going to go a little further and explore how to adapt LLM models to your specific needs, taking custom data and using it to fine-tune or prompt tune models to make them work for your specific use cases. Get ready to turn theory into practice!

## CHAPTER 16

Using LLMs with Custom Data

In Chapter 15, we looked at Transformers and how their encoder, decoder, and encoder-decoder architectures work. The results of their revolutionizing NLP can't be disputed! Then, we looked at transformers, which form the Python library from Hugging Face that's designed to make it easier to use Transformers.

Large Transformer-based models, which are trained on vast amounts of text, are very powerful, but they aren't always ideal for specific tasks or domains. In this chapter, we'll look at how you can use transformers and other APIs to adapt these models to your specific needs.

Fine-tuning allows you to customize pretrained models with your specific data. You could use this approach to create a chatbot, improve classification accuracy, or develop text generation for a more specific domain.

There are several techniques for doing this, including traditional fine-tuning and parameter-efficient tuning with methods like LoRA and parameter-efficient fine-tuning (PEFT). You can also get more out of your LLMs with retrieval-augmented generation (RAG), which we'll explore in Chapter 18.

In this chapter, we'll explore some hands-on examples, starting with traditional fine-tuning.

Fine-Tuning an LLM

Let's take a look, step by step, at how to fine-tune an LLM like BERT. We'll take the IMDb database and fine-tune the model on it to be better at detecting sentiment in movie reviews. There are a number of steps involved in doing this, so we'll look at each one in detail.

Setup and Dependencies

We'll start by setting up everything that we need to do fine-tuning with PyTorch. In addition to the basics, there are three new things that you'll need to include:

Datasets
> We covered datasets in Chapter 4. We're going to use these to load the IMDb dataset and the built-in splits for training and testing.

Evaluate
> This library provides metrics for measuring load performance.

transformers
> As we covered in Chapters 14 and 15, the transformers Hugging Face library is designed to make using LLMs much easier.

We'll use some classes from the Hugging Face transformers library for this chapter's fine-tuning exercise. These include the following:

AutoModelForSequenceClassification
> This class loads pretrained models for classification tasks and adds a classification head to the top of the base model. This classification head is then optimized for the specific classification scenario you are fine-tuning for, instead of being a generic model. If we specify the checkpoint name, it will automatically handle the model architecture for us. So, to use the BERT model with a linear classifier layer, we'll use `bert-base-uncased`.

AutoTokenizer
> This class automatically initializes the appropriate tokenizer. This converts text to the appropriate tokens and adds the appropriate special tokens, padding, truncation, etc.

TrainingArguments
> This class lets us configure the training settings and all the hyperparameters, as well as setting up things like the device to use.

Trainer
> This class manages the training loop on your behalf, handling batching, optimization, loss, backpropagation, and everything you need to retrain the model.

DataCollatorWithPadding
> The number of records in the dataset doesn't always line up with the batch size. This class therefore efficiently batches examples to the appropriate batch sizes while also handling details like attention masks and other model-specific inputs.

We can see this in code here:

```
# 1. Setup and Dependencies
import torch
from datasets import load_dataset
from transformers import (
    AutoModelForSequenceClassification,
    AutoTokenizer,
    TrainingArguments,
    Trainer,
    DataCollatorWithPadding
)
import evaluate
import numpy as np
```

Now that the dependencies are in place, we'll load the data.

Loading and Examining the Data

Next up, let's load our data by using the datasets API. We'll also explore the test and training dataset sizes. You can use the following code:

```
# 2. Load and Examine Data
dataset = load_dataset("imdb")  # Movie reviews for sentiment analysis
print(f"Train size: {len(dataset['train'])}")
print(f"Test size: {len(dataset['test'])}")
```

That will output the following:

```
Train size: 25000
Test size: 25000
```

The next step is to initialize the model and the tokenizer.

Initializing the Model and Tokenizer

We'll use the bert-base-uncased model in this example, so we need to initialize it by using AutoModelForSequenceClassification and getting its associated tokenizer:

```
# 3. Initialize Model and Tokenizer
model_name = "bert-base-uncased"
tokenizer = AutoTokenizer.from_pretrained(model_name)
model = AutoModelForSequenceClassification.from_pretrained(
    model_name,
    num_labels=2
)
device = torch.device("cuda" if torch.cuda.is_available() else "cpu")
model = model.to(device)
```

Note the AutoModelForSequenceClassification needs to be initialized with the number of labels that we want to classify for. This defines the new classification head with two labels. The IMDb database that we'll be using has two labels for positive and negative sentiment, so we'll retrain for that.

At this point, it's also a good idea to specify the device that the model will run on. Training with this model is computationally intensive, and if you're using Colab, you'll likely need a high-RAM GPU like an A100. Training with that will take a couple of minutes, but it can take many hours on a CPU!

Preprocessing the Data

Once we have the data, we want to preprocess it just to get what we need to train. The first step in this, of course, will be to tokenize the text, and the preprocess function here handles that, giving a sequence length of 512 characters with padding:

```
# 4. Preprocess Data
def preprocess_function(examples):
    result = tokenizer(
        examples["text"],
        truncation=True,
        max_length=512,
        padding=True
    )
    # Trainer expects a column called labels, so copy over from label
    result["labels"] = examples["label"]
    return result

tokenized_dataset = dataset.map(
    preprocess_function,
    batched=True,
    remove_columns=dataset["train"].column_names
)
```

One important note here is that the original data came with columns for text denoting the review and label being 0 or 1 for negative or positive sentiment. However, we don't *need* the text column to train the data, and the Hugging Face Trainer (which we will see in a moment) expects the column containing the label to be called labels (plural). Therefore, you'll see that we remove all of the columns in the original dataset, and the tokenized dataset will have the tokenized data and a column called labels instead of label, with the original values copied over.

Collating the Data

When we're dealing with passing sequenced, tokenized data into a model in batches, there can be differences in batch or sequence size that need processing. In our case, we shouldn't have to worry about the sequence size because we used a tokenizer that forces the length to be 512 (in the previous set). However, as part of the transformers library, the collator classes are still equipped to deal with it, and we'll be using them to ensure consistent batch sizing.

So ultimately, the role of the `DataCollatorWithPadding` class is to take multiple examples of different lengths, provide padding if and when necessary, convert the inputs into tensors, and create attention masks if necessary.

In our case, we're really only getting the conversion to tensors for input to the model, but it's still good practice to use `DataCollatorWithPadding` if we want to change anything in the tokenization process later.

Here's the code:

```
# 5. Create Data Collator
data_collator = DataCollatorWithPadding(tokenizer=tokenizer)
```

Defining Metrics

Now, let's define some metrics that we want to capture as we're training the model. We'll just do accuracy, where we compare the predicted value to the actual value. Here's some simple code to achieve that:

```
# 6. Define Metrics
metric = evaluate.load("accuracy")

def compute_metrics(eval_pred):
    predictions, labels = eval_pred
    predictions = np.argmax(predictions, axis=1)
    return metric.compute(predictions=predictions, references=labels)
```

It's using `evaluate.load` from Hugging Face's evaluate library, which provides a simple standardized interface that's specifically designed for tasks like this one. It can handle the heavy lifting for us, instead of requiring us to roll our own metrics, and for an evaluate task, we simply pass it the set of predictions and the set of labels and have it do the computation. The evaluate library is prebuilt to handle a number of metrics, including f1, BLEU, and many others.

Configuring Training

Next up, we can configure *how* the model will retrain by using the `TrainingArgu ments` object. This offers a large variety of hyperparameters you can set—including those for the learning rate, weight decay, etc., as used by the `optimizer` and `loss` function. It's designed to give you granular control over the learning process while abstracting away the complexity.

Here's the set that I used for fine-tuning with IMDb:

```
# 7. Configure Training
training_args = TrainingArguments(
    output_dir="./results",
    learning_rate=2e-5,
    per_device_train_batch_size=32,
```

```
    per_device_eval_batch_size=8,
    num_train_epochs=3,
    weight_decay=0.01,
    logging_dir='./logs',
    logging_steps=500,
    evaluation_strategy="epoch",
    save_strategy="epoch",
    load_best_model_at_end=True,
    metric_for_best_model="accuracy",
    push_to_hub=False,
    gradient_accumulation_steps=4,
    gradient_checkpointing=True,
    report_to="none",
    fp16=True
)
```

It's important to note and tweak hyperparameters for different results. In addition to the aforementioned ones for the optimizer, you'll want to consider the batch sizes. You can set different parameters for training or evaluation.

One very useful parameter—in particular for training sessions that are longer than the three epochs here—is `load_best_model_at_end`. Instead of always using the final checkpoint, it will keep track of the best checkpoint according to the specified metric (in this case, accuracy) and will load that one when it's done. And because I set the `evaluation` and `save` strategies to `epoch`, it will only do this at the end of an epoch.

Note also the `report_to` parameter: the training uses `weights and biases` as the backend for reporting by default. I set `report_to` to `none` to turn off this reporting. If you want to keep it, you'll need a Weights and Biases API key. You can get this very easily from the status window or by going to the Weights and Biases website (*https:// oreil.ly/yMX1A*). As you train, you'll be asked to paste in this API key. Be sure to do so before you walk away, particularly if you are paying for compute units on Colab.

There's a wealth of parameters to experiment with, and being able to parameterize easily like this also allows you easily to do a neural architecture search with tools like Ray Tune (*https://oreil.ly/fDAhG*).

Initializing the Trainer

As with the training parameters, transformers give you a trainer class that you can use alongside them to encapsulate a full training cycle.

You initialize it with the model, the training arguments, the data, the collator, and the metrics strategy that you've previously initialized. All the previous steps build up to this. Here's the code you'll need:

```
# 8. Initialize Trainer
trainer = Trainer(
    model=model,
```

```
    args=training_args,
    train_dataset=tokenized_dataset["train"],
    eval_dataset=tokenized_dataset["test"],
    tokenizer=tokenizer,
    data_collator=data_collator,
    compute_metrics=compute_metrics,
)
```

Training and Evaluation

With everything now set up, it becomes as simple as calling the `train()` method on
the trainer to do the training and the `evaluate()` method to do the evaluation.

Here's the code:

```
# 9. Train and Evaluate
train_results = trainer.train()
print(f"\nTraining results: {train_results}")

eval_results = trainer.evaluate()
print(f"\nEvaluation results: {eval_results}")
```

As an example, while you could train this model with the free tiers in Google Colab,
your experience in timing might vary. With the CPU alone, it can take many hours. I
trained this model with the T4 High Ram GPU, which costs 1.6 compute units per
hour. The entire training process was about 50 minutes, but I'll round that up to an
hour to include all the downloading and setup. At the time of writing, a pro Colab
subscription gets one hundred compute Units with the US$9.99 per month subscrip-
tion. You could also choose the A100 GPU, which is much faster (training took me
about 12 minutes with it) but also more expensive, at about 6.8 compute units per
hour.

After training, the results looked like this:

```
Training results:
TrainOutput(global_step=585,
            training_loss=0.18643947177463108,
            metrics={'train_runtime': 597.9931,
            'train_samples_per_second': 125.42,
            'train_steps_per_second': 0.978,
            'total_flos': 1.968912649469952e+16,
            'train_loss': 0.18643947177463108,
            'epoch': 2.9923273657289})
Evaluation results:
            {'eval_loss': 0.18489666283130646,
            'eval_accuracy': 0.93596,
            'eval_runtime': 63.8406,
            'eval_samples_per_second': 391.601,
            'eval_steps_per_second': 48.95,
            'epoch': 2.9923273657289}
```

We can see quite high accuracy on the evaluation dataset (about 94%) after only three epochs, which is a good sign—but of course, there may be overfitting going on that would require a separate evaluation. But after about 12 minutes of work fine-tuning an LLM, we're clearly moving in the right direction!

Saving and Testing the Model

Once we've trained the model, it's a good idea to save it out for future use, and the `trainer` object makes this easy:

```
# 10. Save Model
trainer.save_model("./final_model")
```

Once we've saved the model, we can start using it. To that end, let's create a helper function that takes in the input text, tokenizes it, and then turns those tokens into a set of input vectors of keys and values (k, v):

```
# 11. Example Usage
def predict_sentiment(text):
    inputs = tokenizer(
        text,
        truncation=True,
        padding=True,
        return_tensors="pt"
    )
    inputs = {k: v.to(device) for k, v in inputs.items()}
```

We can then use PyTorch in inference mode to get the outputs from those inputs and turn them into a set of predictions:

```
with torch.no_grad():
    outputs = model(**inputs)
    predictions = torch.nn.functional.softmax(outputs.logits, dim=-1)
```

The returned predictions will be a tensor with two dimensions. Neuron 0 is the probability that the prediction is negative, and neuron 1 is the probability that the prediction is positive. Therefore, we can look at the positive probability and return a sentiment and confidence with its values. We could also have done the same with the negative one; it's purely arbitrary:

```
positive_prob = predictions[0][1].item()
return {
    'sentiment': 'positive' if positive_prob > 0.5 else 'negative',
    'confidence': positive_prob if positive_prob > 0.5 else 1 - positive_prob
}
```

We can now test the prediction with code like this:

```
# Test prediction
test_text = "This movie was absolutely fantastic! The acting was superb."
result = predict_sentiment(test_text)
print(f"\nTest prediction for '{test_text}':")
print(f"Sentiment: {result['sentiment']}")
print(f"Confidence: {result['confidence']:.2%}")
```

And the output would look something like this:

```
Test prediction for 'This movie was absolutely fantastic!
                     The acting was superb.':
Sentiment: positive
Confidence: 99.16%
```

We can see that this statement is positive, with high confidence!

In this process, you can see how, step by step, you can fine-tune an existing LLM on new data to turn it into a classification engine! In many circumstances, this may be overkill (and training your own model instead of fine-tuning an LLM may be quicker and cheaper), but it's certainly worth evaluating this process. Sometimes, even untuned LLMs will work well for classification! In my experience, using the general artificial-understanding nature of LLMs will lead to the creation of far more effective classifiers with stronger results.

Prompt-Tuning an LLM

A lightweight alternative to fine-tuning is *prompt tuning*, in which you can adapt a model to specific tasks. With prompt tuning, you do this by prepending trainable *soft prompts* to each input instead of modifying the model weights. These soft prompts will then be optimized during training.

These soft prompts are like learned instructions that guide the model's behavior. Unlike discrete text prompts (such as `Classify the sentiment`), the idea of soft prompts is that they exist in the model's embedding space as continuous vectors. So, for example, when processing "This movie was great," the model would see "[V1][V2] …[V20]This movie was great." In this case, [V1][V2]...[V20] are vectors that will help steer the model toward the desired classification.

Ultimately, the advantage here is efficiency. So instead of fine-tuning a model, amending its weights for each task, and saving the entire model for reuse, you only need to save the soft prompt vectors. These are much smaller, and they can help you have a suite of fine-tunes that you can easily use to guide the model to a specific task without needing to manage multiple models.

Prompt tuning like this can actually match or exceed the performance of full fine-tuning, particularly with larger models, and it's significantly more efficient.

Now, let's explore how to prompt-tune the BART LLM with the IMDb dataset in direct comparison to the fine-tuning earlier in this chapter.

Preparing the Data

Let's start by preparing our data, loading it from the IMDb dataset, and setting up the virtual tokens. Here's the code you'll need:

```
# Data preparation
dataset = load_dataset("imdb")
tokenizer = AutoTokenizer.from_pretrained("bert-base-uncased")
max_length = 512
num_virtual_tokens = 20

def tokenize_function(examples):
    return tokenizer(
        examples["text"],
        padding="max_length",
        truncation=True,
        max_length=max_length - num_virtual_tokens
    )
```

This will also tokenize our incoming examples so you should note that the maximum length of any example will now be reduced by the number of virtual tokens. So, for example, with BERT, we have a maximum length of 512, but if we're going to have 20 virtual tokens, then the sequence maximum length will now be 492.

We'll now load a subset of the data and try with 5,000 examples, instead of 25,000. You can experiment with this number and trade off smaller amounts for faster training against larger amounts for better accuracy.

First, we'll create the indices that we want to take from the dataset for training, and we'll test them. Think of these as pointers to the records we're interested in. We're randomly sampling here:

```
# Use only 5000 examples for training
train_size = 5000
np.random.seed(42)
train_indices = np.random.choice(len(dataset["train"]), train_size,
                                                    replace=False)
test_indices = np.random.choice(len(dataset["test"]), train_size, replace=False)

tokenized_train = dataset["train"].map(tokenize_function, batched=True)
tokenized_test = dataset["test"].map(tokenize_function, batched=True)
```

Then, the last two lines define the mapping function, which simply takes the values from the dataset we're interested in and tokenizes them. We'll see that in the next step.

Creating the Data Loaders

Now that we have sets of tokenized training and test data, we want to turn them into data loaders. We'll do this by first selecting the raw examples from the underlying data that match the content in our indices:

```
# Create subset for training
tokenized_train = tokenized_train.select(train_indices)
tokenized_test = tokenized_test.select(test_indices)
```

Then, we'll set the format of the data that we're interested in. There may be many columns in a dataset, but you won't use them all for training. In this case, we'll want the `input_ids`, which are the tokenized versions of our input content; the `attention_mask`, which is a set of vectors that tells us which tokens in the `input_ids` we should be interested in (this has the effect of filtering out padding or other nonsemantic tokens); and the label:

```
tokenized_train.set_format(type="torch", columns=["input_ids",
                                        "attention_mask", "label"])
tokenized_test.set_format(type="torch", columns=["input_ids",
                                        "attention_mask", "label"])
```

Now, we can specify the DataLoader that takes these training and test sets. I have a large batch size here because I was testing on a 40Gb GRAM GPU in Colab. In your environment, you may need to adjust these:

```
train_dataloader = DataLoader(tokenized_train, batch_size=64, shuffle=True)
eval_dataloader = DataLoader(tokenized_test, batch_size=128)
```

Now that the data is processed and loaded into DataLoaders, we can go to the next step: defining the model.

Defining the Model

First, let's see how to instantiate the model, and then we can go back to the raw definition. Typically, in our code, once we've set up our DataLoaders, we'll want to create an instance of the model. We'll use code like this:

```
# Define the model
model = PromptTuningBERT(num_virtual_tokens=num_virtual_tokens,
                        max_length=max_length)

device = torch.device('cuda' if torch.cuda.is_available() else 'cpu')

model.to(device)
```

This keeps it nice and simple, and we'll encapsulate the underlying BERT in an override for a prompt-tuning version. Now, as nice as it would be for transformers to have one, they don't, so we need to create this class for ourselves.

As we would with any PyTorch class that defines a model, we'll create it with an __init__ method to set it up and a forward method that PyTorch's training loop will call during the forward pass. So, let's start with the __init__ method and the class definition:

```python
class PromptTuningBERT(nn.Module):
    def __init__(self, model_name="bert-base-uncased",
                       num_virtual_tokens=50,
                       max_length=512):
        super().__init__()
        self.bert = AutoModelForSequenceClassification.from_pretrained(
                       model_name,
                       num_labels=2)
        self.bert.requires_grad_(False)

        self.n_tokens = num_virtual_tokens
        self.max_length = max_length - num_virtual_tokens

        vocab_size = self.bert.config.vocab_size
        token_ids = torch.randint(0, vocab_size, (num_virtual_tokens,))
        word_embeddings = self.bert.bert.embeddings.word_embeddings
        prompt_embeddings = word_embeddings(token_ids).unsqueeze(0)
        self.prompt_embeddings = nn.Parameter(prompt_embeddings)
```

There's a lot going on here, so let's break it down little by little. First of all, I set the defaults for the num_virtual_tokens to 50 and the max_length default to 512. If you don't specify your own defaults when you instantiate the class, you'll get these values. In this case, the calling code sets them to 20 and 512, respectively, but you're free to experiment.

Next, the code sets up the transformers AutoModelForSequenceClassification class to get BERT:

```python
self.bert = AutoModelForSequenceClassification.from_pretrained(
                model_name,
                num_labels=2)
```

As with fine-tuning for IMDb, we're interested in training the model to recognize two labels, so they're set up here. However, one difference from fine-tuning is that we're not going to change any of the weights within the BERT model itself, so we set that we don't want gradients and freeze it like this:

```python
self.bert.requires_grad_(False)
```

The secret sauce in generating the soft prompts that we're going to use comes at the end of the init. We'll create a vector to contain our number of virtual tokens, and I just initialized it with random tokens from the vocabulary. There are smarter things that we might do here to make training more efficient over time, but for the sake of simplicity, let's go with this:

```python
token_ids = torch.randint(0, vocab_size, (num_virtual_tokens,))
```

The pretrained BERT model in transformers comes with embeddings, so we can use them to turn our list of random tokens into embeddings:

```
word_embeddings = self.bert.bert.embeddings.word_embeddings
prompt_embeddings = word_embeddings(token_ids).unsqueeze(0)
```

Importantly, we should now specify that the `prompt_embeddings` are parameters of the neural network. This will be important later, when we define the optimizer. We recently specified that all of the BERT parameters were frozen, but *these* parameters are not part of that and thus are not frozen, so they will be tweaked by the optimizer during training:

```
self.prompt_embeddings = nn.Parameter(prompt_embeddings)
```

We have now initialized a subclassed version of the tunable BERT, specified that we don't want to amend its gradients, and created a set of soft prompts that we will append to the examples as we're training—and we'll tweak only those soft prompts to soft-tune the two output neurons.

Now, let's look at the `forward` function that will be called during the forward pass at training time. Given that we've set up everything, this is pretty straightforward:

```
def forward(self, input_ids, attention_mask, labels=None):
    batch_size = input_ids.shape[0]
    input_ids = input_ids[:, :self.max_length]
    attention_mask = attention_mask[:, :self.max_length]

    embeddings = self.bert.bert.embeddings.word_embeddings(input_ids)
    prompt_embeddings = self.prompt_embeddings.expand(batch_size, -1, -1)
    inputs_embeds = torch.cat([prompt_embeddings, embeddings], dim=1)

    prompt_attention_mask = torch.ones(batch_size, self.n_tokens,
                                       device=attention_mask.device)
    attention_mask = torch.cat([prompt_attention_mask, attention_mask], dim=1)

    return self.bert(
        inputs_embeds=inputs_embeds,
        attention_mask=attention_mask,
        labels=labels,
        return_dict=True
    )
```

Let's look at it step-by-step. During the forward pass in training, this function will be passed batches of data. Therefore, we need to understand what the size of this batch is and then extract the `input_ids` (the tokens for the values read from the dataset) and the attention mask for that particular ID:

```
batch_size = input_ids.shape[0]
input_ids = input_ids[:, :self.max_length]
attention_mask = attention_mask[:, :self.max_length]
```

We'll also need to convert the `input_ids` into embeddings:

```
embeddings = self.bert.bert.embeddings.word_embeddings(input_ids)
```

Our soft prompts are also tokenized sentences. Originally, they were initialized to random words, and we'll see over time that they'll adjust appropriately. But for this step, these tokens need to be converted to embeddings:

```
prompt_embeddings = self.prompt_embeddings.expand(batch_size, -1, -1)
```

The expand method just adds the batch size to the prompt embeddings. When we defined the class, we didn't know how large each batch coming in would be (and the code is written to let you tweak that based on the size of your available memory), so using expand(batch_size, -1, -1) turns the vector of prompt embeddings, which was of shape [1, num_prompt_tokens, embedding_dimensions], into [batch_size, num_prompt_tokens, embedding_dimensions].

Our soft prompt tuning involved prepending the soft embeddings to the embeddings for the actual input data, so we do that with this:

```
inputs_embeds = torch.cat([prompt_embeddings, embeddings], dim=1)
```

BERT uses an `attention_mask` to filter out the tokens we don't want to worry about at training or inference time, which are usually the padding tokens. But we want BERT to pay attention to all of the soft prompt tokens, so we'll set the attention mask for them to be all 1s and then append that to the incoming attention mask(s) for the training data. Here's the code:

```
prompt_attention_mask = torch.ones(batch_size, self.n_tokens,
                                   device=attention_mask.device)

attention_mask = torch.cat([prompt_attention_mask, attention_mask],
                           dim=1)
```

Now that we've done all our tuning, we need to pass the data to the model to have it optimize and calculate the loss:

```
    return self.bert(
        inputs_embeds=inputs_embeds,
        attention_mask=attention_mask,
        labels=labels,
        return_dict=True
    )
```

We will see how this data is used in the training loop, next.

Training the Model

The key to this training is that we're going to do a full, normal training loop but in a special circumstance. In this case, we previously froze *everything* in the BERT model, *except* for the soft prompts that we defined as model parameters. Therefore, say we define the optimizer like this:

```
optimizer = AdamW(model.parameters(), lr=1e-2)
```

In this case, we're using standard code, telling it to tweak the model's parameters. But the only ones that are available to tune are the soft prompts, so this should be quick!

Note that the value for the learning rate is quite large. This helps the system learn quickly, but in a real system, you'd likely want the value to be smaller—or at least adjustable, starting large and then shrinking in later epochs.

So now, let's get into training. First, we'll set up the training loop:

```
num_epochs = 3

# Perform the training
for epoch in range(num_epochs):
    model.train()
    total_train_loss = 0
```

Managing data batches

For each batch, we'll get the columns (`input_ids` and `attention_masks`) as well as the labels and pass them to the model:

```
for batch in tqdm(train_dataloader,
                  desc=f'Training Epoch {epoch + 1}'):
    batch = {k: v.to(device) for k, v in batch.items()}
    labels = batch.pop('label')
    outputs = model(**batch, labels=labels)
```

This looks a little different from ones earlier in this book, but it's pretty much doing the same thing. The `tqdm` code just gives us a status bar because we're training. We read the data batch by batch, but we want the data to be on the same device as the model. So, for example, if the model is running on a GPU, we want it to access data in the GPU's memory. Therefore, this line will iterate through each column, reading the key and passing the value to the device:

```
batch = {k: v.to(device) for k, v in batch.items()}
```

It redefines the batch that was read in to ensure that the data is on the same device as the model. But we don't want the labels to be in the batch because the model expects them to be fed in separately, so we remove them from the batch with the `pop()` method:

```
labels = batch.pop('label')
```

Now, we can use the shorthand of **batch to pass the set of input values (in this case, the input_ids and the attention_mask) to the forward method of the model and unpack the dictionary along with the labels, like this:

```
outputs = model(**batch, labels=labels)
```

Handling the loss

The forward pass sends the data to the model and gets the loss back. We use this to update our overall loss, and we can then backward pass:

```
loss = outputs.loss
total_train_loss += loss.item()

loss.backward()
```

With the gradients flowing back, the optimizer can now do its job.

Optimizing for loss

Remembering that the model.parameters() will only manage the *trainable unfrozen* parameters, we can now call the optimizer. I added something called *gradient clipping* here to make the training a little more efficient, but the rest is just calling the optimizer's next step and then zeroing out the gradients so we can use them next time:

```
clip_grad_norm_(model.parameters(), max_grad_norm)  # Add here
optimizer.step()
optimizer.zero_grad()
```

> The idea behind *gradient clipping* is that sometimes, during back-propagation, the gradients can be too large and the optimizer might take very large steps. This can lead to a problem called *exploding gradients*, in which the value changes hide the nuances of what might be learned. But clipping scales the gradients down if their values grow too large, and in a situation like this one, they may not even be necessary.

Evaluation During Training

We also have a set of test data, so we can evaluate how the model performs during the training cycle. In each epoch, once the forward and backward passes are done and the model parameters are reset, we can switch the model into evaluation mode and then start passing all of the test data through it to get inference. We'll also compare the results of the inference against the actual labels to calculate accuracy:

```
model.eval()
val_accuracy = []
total_val_loss = 0
```

Then, we'll have similar code—but this time, it will be to read the eval batches, turn them into outputs with labels, and get predictions and loss values from the model:

```python
with torch.no_grad():
    for batch in tqdm(eval_dataloader, desc='Validating'):
        batch = {k: v.to(device) for k, v in batch.items()}
        labels = batch.pop('label')

        outputs = model(**batch, labels=labels)
        total_val_loss += outputs.loss.item()

        predictions = torch.argmax(outputs.logits, dim=-1)
        val_accuracy.extend((predictions == labels).cpu().numpy())
```

Once we've calculated these values, then at the end of each epoch, we can report on them and on training loss.

Reporting Training Metrics

During training in each epoch, we calculated the training loss, so we can now get the average across all records. We can do the same thing with the validation loss and (of course) with the validation accuracy and then report on them:

```python
avg_train_loss = total_train_loss / len(train_dataloader)
avg_val_loss = total_val_loss / len(eval_dataloader)
val_accuracy = np.mean(val_accuracy)

print(f"\nEpoch {epoch + 1}:")
print(f"Average training loss: {avg_train_loss:.4f}")
print(f"Average validation loss: {avg_val_loss:.4f}")
print(f"Validation accuracy: {val_accuracy:.4f}")
```

Running this training for three epochs gives us this:

```
Training Epoch 1: 100%|██████████| 79/79 [01:01<00:00, 1.28it/s]
Validating: 100%|██████████| 40/40 [00:27<00:00, 1.44it/s]

Epoch 1:
Average training loss: 0.6559
Average validation loss: 0.6037
Validation accuracy: 0.8036
Training Epoch 2: 100%|██████████| 79/79 [01:01<00:00, 1.28it/s]
Validating: 100%|██████████| 40/40 [00:27<00:00, 1.44it/s]

Epoch 2:
Average training loss: 0.6112
Average validation loss: 0.5854
Validation accuracy: 0.8386
Training Epoch 3: 100%|██████████| 79/79 [01:01<00:00, 1.28it/s]
Validating: 100%|██████████| 40/40 [00:27<00:00, 1.44it/s]

Epoch 3:
```

```
Average training loss: 0.5799
Average validation loss: 0.5270
Validation accuracy: 0.8736
```

This was done on an A100 in Colab with 40 Gb of GRAM, and as you can see, each epoch only took about 1 minute to train and 30 seconds to evaluate.

By the end, the average training loss had dropped from about 0.65 to 0.58. The accuracy was 0.8736. So, it's likely overfitting because we only trained for three epochs.

Saving the Prompt Embeddings

What's really nice about this approach is that you can simply save out the prompt embeddings when you're done. You can also load them back in for inference later, as you'll see in the next section:

```
torch.save(model.prompt_embeddings, "imdb_prompt_embeddings.pt")
```

What I find really cool about this is that this file is relatively small (61 K), and it doesn't require you to amend the underlying model in any way. Thus, in an application, you could potentially have a number of these prompt-tuning files and hot-swap and replace them as needed so that you can have multiple models that you can orchestrate, which is the basis for an agentic solution.

Performing Inference with the Model

To perform inference with a prompt-tuned model, you'll simply define the model with the soft prompts and then, instead of training them, load the pretrained soft prompts back from disk. We'll explore that in this section. If you don't want to train your own model, then in the download (*https://github.com/lmoroney/PyTorch-Book-FIles*), I've provided soft prompts from a version of the model that was trained for 30 epochs instead of 3.

For tidier encapsulation, I created a class that is similar to the one we used for training but that is just for inference. I call it `PromptTunedBERTInference`, and here's its initializer:

```python
class PromptTunedBERTInference:
    def __init__(self, model_name="bert-base-uncased",
                       prompt_path="imdb_prompt_embeddings.pt"):
        self.tokenizer = AutoTokenizer.from_pretrained(model_name)
        self.model = \
                AutoModelForSequenceClassification.from_pretrained(
                                    model_name, num_labels=2)
        self.model.eval()
        self.prompt_embeddings = torch.load(prompt_path)
        self.device = \
            torch.device('cuda' if torch.cuda.is_available() else 'cpu')
        self.model.to(self.device)
```

It's very similar to the initializer for the trainable one, except for a couple of important points. The first is that because we're *only* using it for inference, I've set it into eval mode:

```
self.model.eval()
```

The second is that we don't need to train the embeddings and do all the associated plumbing—instead, we just load them from the specified path:

```
self.prompt_embeddings = torch.load(prompt_path)
```

And that's it! As you can see, it's quite lightweight and pretty straightforward. It won't have a `forward` function because we're not training it, but let's add a `predict` function that encapsulates doing inference with it.

The predict function

The job of the `predict` function is to take in the string(s) that we want to perform inference with, tokenize it (them), and then pass it to the model with the soft tokens prepended. Let's take a look at the code, piece by piece.

First, let's define it and have it accept text that it will then tokenize:

```
def predict(self, text):
    inputs = self.tokenizer(text, padding=True,
                            truncation=True,
                            max_length=512-self.prompt_embeddings.shape[1],
                            return_tensors="pt")
    inputs = {k: v.to(self.device) for k, v in inputs.items()}
```

The text will be tokenized up to the maximum length, less the size of the soft prompt, and then each of the items in the input will be loaded into a dictionary. Note that the tokenizer will return multiple columns for the text—usually, the tokens and the attention mask—so we'll follow this approach to turn them into a set of key-value pairs that are easy for us to work with later.

Now that we have our inputs, it's time to pass them to the model. We'll start by putting `torch` into `no_grad()` mode because we're not interested in training gradients. We'll then get the embeddings for each of our tokens:

```
with torch.no_grad():
    embeddings = self.model.bert.embeddings.word_embeddings(
                                inputs['input_ids'])

    batch_size = embeddings.shape[0]

    prompt_embeds = self.prompt_embeddings.expand(
                        batch_size, -1, -1).to(self.device)

    inputs_embeds = torch.cat([prompt_embeds, embeddings], dim=1)
```

We have an attention mask for the input that's generated by the tokenizer, but we don't have one for the soft prompt. So, let's create one and then append it to the input attention mask:

```
attention_mask = inputs['attention_mask']
prompt_attention = torch.ones(batch_size, self.prompt_embeddings.shape[1],
                              device=self.device)
attention_mask = torch.cat([prompt_attention, attention_mask], dim=1)
```

Now that we have everything in place, we can pass our data to the model to get our inferences back:

```
outputs = self.model(inputs_embeds=inputs_embeds,
                     attention_mask=attention_mask)
```

The outputs will be the logits from the two neurons, one representing positive sentiment and the other negative. We can then Softmax these to get the prediction:

```
probs = torch.nn.functional.softmax(outputs.logits, dim=-1)
return {"prediction": outputs.logits.argmax(-1).item(),
        "confidence": probs.max(-1).values.item()}
```

Usage example

Using this class for a prediction is then pretty straightforward. We create an instance of the class and pass a string to it to get results. The results will contain a prediction and a confidence value, which we can then output:

```
# Usage example
if __name__ == "__main__":
    model = PromptTunedBERTInference()
    result = model.predict("This movie was great!")
    print(f"Prediction: {'Positive'
                          if result['prediction'] == 1 else 'Negative'}")
    print(f"Confidence: {result['confidence']:.2f}")
```

One note you might see with prompt tuning is low confidence values that can lead to mis-predictions, especially with binary classifiers like this one. It's good to explore your inference to make sure that it's working well, and there are also techniques that you could explore to ensure that the logits are giving the values you want. These include setting the temperature of the Softmax, using more prompt tokens to give the model more capacity, and initializing the prompt tokens with sentiment-related words (instead of random tokens, like we did here).

Summary

In this chapter, we explored different methods for customizing LLMs with our own data. We looked at two main approaches: traditional fine-tuning and prompt tuning.

Using the IMDb dataset, you saw how to fine-tune BERT for sentiment analysis and walked through all the steps—from data preparation, to model configuration, to training and evaluation. The model achieved an impressive 95% accuracy in sentiment classification in just a few epochs.

However, fine-tuning may not be appropriate in all cases, and to that end, you explored a lightweight alternative called prompt tuning. Instead of modifying model weights, the idea here was to prepend trainable soft prompts to inputs, which are optimized during training. This approach provides significant advantages in that it can be much faster and it doesn't change the underlying model. In this case, the tuned prompts could be saved (and they were only a few Kb) and then reloaded to program the model to perform the desired task. You then went through a full implementation, showing you how to create, train, and save these soft prompts, plus load them back to perform inference.

In the next chapter, we'll explore how you can serve LLMs, including customized ones. I'll explain how to do this in your own data center by using Ollama, which is a powerful tool for handling the serving and management of LLMs. You'll learn how to take models and turn them into services, and we'll also explore how to set up Ollama and use it over HTTP to talk with models in your data center.

## CHAPTER 17

Serving LLMs with Ollama

We've explored how to use transformers to download a model and put together an easy pipeline that lets you use it for inference or fine-tuning. However, I'd be remiss if I didn't show you the open source Ollama project, which ties it all together by giving you an environment that gives you a full wrapper around an LLM that you can either chat with in your terminal or use as a server that you can HTTP POST to and read the output from.

Technologies like Ollama will be the vanguard of the next generation of LLMs, which will let you have dedicated servers inside your data center or dedicated processes on your computer. That will make them completely private to you.

At its core, Ollama is an open source project that simplifies the process of downloading, running, and managing LLMs on your computer. It also handles nonfunctional difficult requirements, such as memory management and model optimization, and it provides standardized interfaces for interaction, such as the ability to HTTP POST to your models.

Ollama is also a key strategic tool you should consider because it bridges the gap between cloud-based third-party services like GPT, Claude, and Gemini and locally deployed services. It goes beyond giving you a local development environment to giving you one that you could, for example, use within your own data center to serve multiple internal users.

By running models locally, you can ensure the complete privacy of your data, eliminate network latency, and work offline. This is especially crucial in scenarios involving sensitive data or applications that require consistent, low-latency responses.

Ollama also supports a growing library of popular open source models, including Llama, Mistral, and Gemma, and it also supports various specialized models that are optimized for specific tasks. Each model can be pulled and run with simple commands, in a way that's similar to how Docker containers work. The platform handles model quantization automatically, optimizing models to run efficiently on consumer hardware while maintaining good performance.

In this chapter, we'll explore Ollama in three ways: installing it and getting started, looking at how you can instantiate specific models and use them, and exploring the RESTful APIs that let you build LLM applications that preserve privacy.

Getting Started with Ollama

The Ollama project is hosted at ollama.com. It's pretty straightforward to get up and running, and the home screen gives download options for macOS, Linux, and Windows. Note also that the Windows version needs Windows Subsystem for Linux (WSL). For this chapter, I'm using the macOS version.

When you navigate to the website, you'll see a friendly welcome to download (see Figure 17-1).

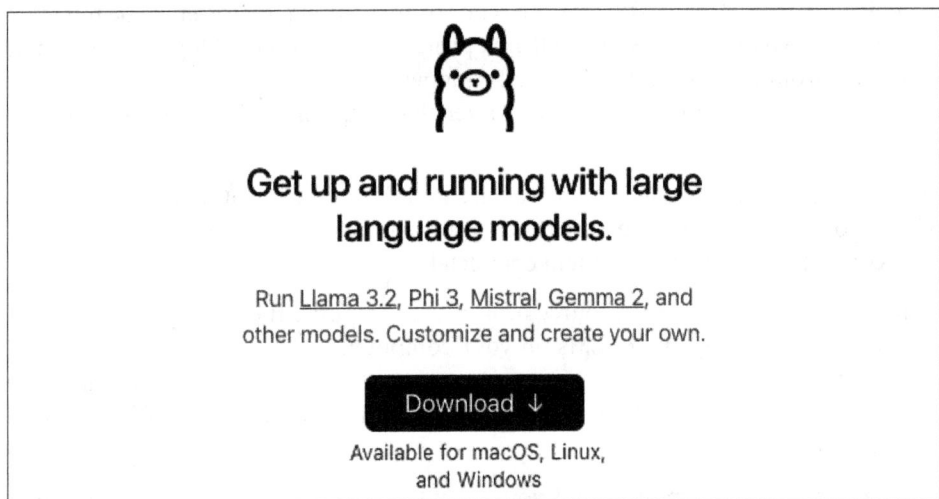

Get up and running with large language models.

Run Llama 3.2, Phi 3, Mistral, Gemma 2, and other models. Customize and create your own.

Download ↓

Available for macOS, Linux, and Windows

Figure 17-1. Getting started with Ollama

Once you've downloaded and installed Ollama, you can launch it, and you'll see it in the system bar at the top of the screen. Your main interface with Ollama will be the command line.

Then, with the `ollama run` command, you can download and use models. So, for example, if you want to use Gemma, then from Google, you can do the following:

```
>ollama run gemma2:2b
```

You'll want to be sure to note the parameters used, which you can find in the model's documentation page on Ollama (*https://oreil.ly/VMLKO*). While Ollama can and will quantize models that are optimized to run locally, it can't perform miracles, and only models that will fit in your system resources—most importantly, memory—will work. In this case, I ran the gemma2:2b (2-billion parameter) version, which requires about 8 GB of GPU RAM. On macOS, the shared RAM with the M-Series chips works well, while running on an M1 Mac with 16 GB, the Gemma 2B is fast and smooth with Ollama.

You can see me chatting with Gemma in Figure 17-2. These responses took less than one second to receive on my two-year-old laptop!

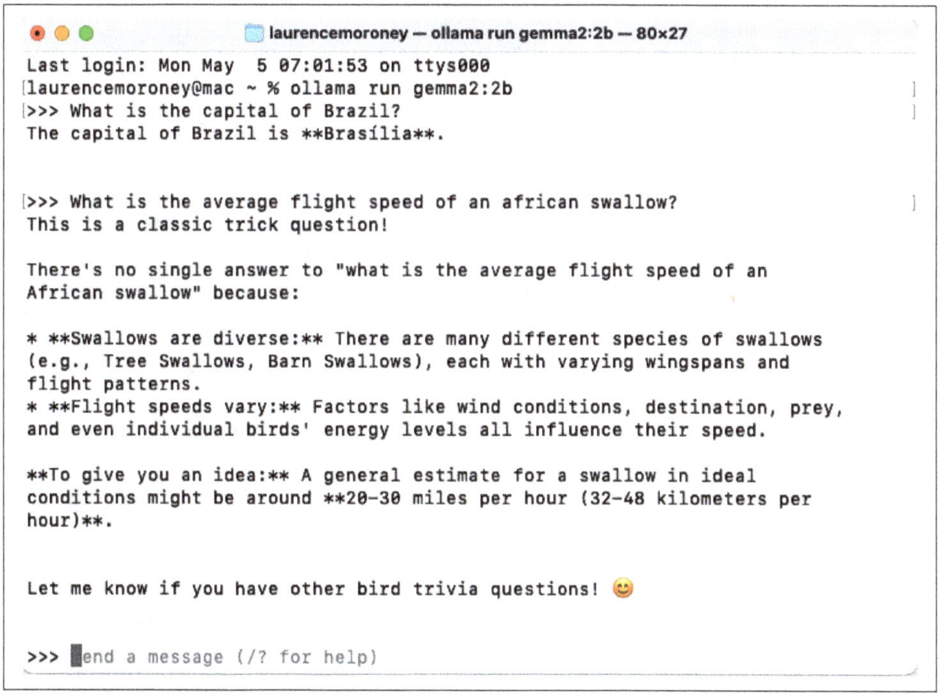

```
Last login: Mon May  5 07:01:53 on ttys000
[laurencemoroney@mac ~ % ollama run gemma2:2b
>>> What is the capital of Brazil?
The capital of Brazil is **Brasília**.

>>> What is the average flight speed of an african swallow?
This is a classic trick question!

There's no single answer to "what is the average flight speed of an
African swallow" because:

* **Swallows are diverse:** There are many different species of swallows
(e.g., Tree Swallows, Barn Swallows), each with varying wingspans and
flight patterns.
* **Flight speeds vary:** Factors like wind conditions, destination, prey,
and even individual birds' energy levels all influence their speed.

**To give you an idea:** A general estimate for a swallow in ideal
conditions might be around **20-30 miles per hour (32-48 kilometers per
hour)**.


Let me know if you have other bird trivia questions! 😊

>>> █end a message (/? for help)
```

Figure 17-2. Using Ollama in a terminal

It's great to have a localized chat like this, and you can experiment with different models, including multimodal ones like Llama 3.2.

So, for example, you could issue the following command:

```
ollama run llama3.2-vision
```

Then, within the terminal, you could do multimodal processing. For example, if your terminal supported it, and you dragged and dropped an image into the terminal, you could ask the model what it could see in the image. The multimodal power of Llama would parse the image for you, and Ollama would handle all the technical difficulties.

In my case, all I had to do was give a prompt and then drag and drop the image onto it. So, I opened an Ollama chat window with the preceding command and then entered this prompt:

```
Please give me a detailed analysis of what's in this image.
Call out any major or minor features and tell me everything you know about it.
Are there any interesting and fun facts?
Maybe even estimate when this picture was taken.
```

And then I just dragged and dropped the image into the chat window, and Ollama did the rest.

You can see in Figure 17-3 how detailed the results were. This is a photo I took of Osaka castle one morning while on a run in 2018. While Llama couldn't guess the date, it was able to predict the season based on the foliage in the image. It got everything else correct and gave very detailed output!

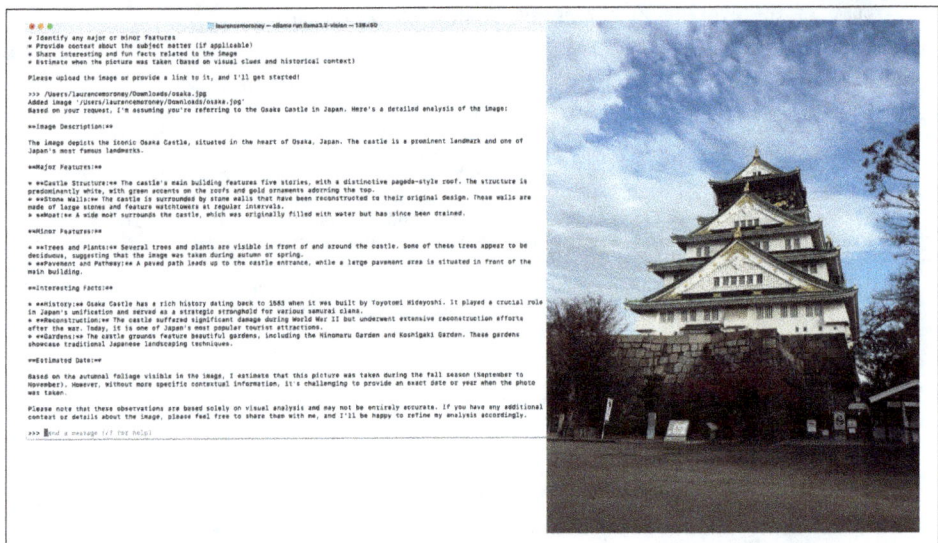

Figure 17-3. Using Ollama for a multimodal model

While it's really cool to have a local LLM that you can chat with in a privacy-preserving way, I think the real power in Ollama is in using it as a server that can then be the foundation of an application. We'll explore that next.

Running Ollama as a Server

To put Ollama into server mode, you simply issue the following command:

```
ollama serve
```

This will run the Ollama server on port 11434 by default, so you can hit it and ask it to do inference with a `curl` command to test it.

In a separate terminal window, you issue a `curl` command. Here's an example:

```
curl http://localhost:11434/api/generate -d '{
  "model": "gemma2:2b",
  "prompt": "Why is the sky blue?",
  "stream": false
}'
```

Note the `stream` parameter. If you set it to true, you'll get an active HTTP connection that will send the answer word by word. That will give you a faster time to the first word, which is very suitable for chat applications. And because the answer will appear little by little and usually faster than a person can read, it will make for a better user experience.

On the other hand, if you set the `stream` parameter to `false`, as I have done here, it will take longer to send something back, but when it does, you'll get everything at once. The time to the last token will probably be about the same as in streaming, but given that there will be no output for a little while, it will feel slower.

The preceding `curl` to Gemma gave me this response:

```
{"model":"gemma2:2b","created_at":"2024-12-09T18:10:05.711484Z",
  "response":"The sky appears blue because ... phenomena! \n",
  "done":true,
  "done_reason":"stop",
  "context":[106,1645,108,4385,603,573,...,235248,108],
  "total_duration":7994972625,
  "load_duration":820325334,
  "prompt_eval_count":15,
  "prompt_eval_duration":2599000000,
  "eval_count":282,
  "eval_duration":4573000000}%
```

I trimmed the response text and the context for brevity. Ultimately, you'd use the response text in an application, but I wanted to also show you how you can build more robust applications showing you everything else that the model provided.

The `done` parameter demonstrates that the prompt returned successfully. When the value is streaming, this parameter will be set to `false` until it has finished sending the text. That way, you can keep your UI updating the text word by word until the message is complete.

The `done_reason` parameter is useful in checking for errors, particularly when streaming. It will usually contain `stop` for normal completion, but in other circumstances, it might say `length`, which indicates that you've hit a token limit; `canceled` if the user cancels the request (by interrupting streaming, for example); or `error`.

The count values are also useful if you want to manage or report on token usage. The `prompt_eval_count` parameter tells you how many tokens were used in your prompt, and in this case, it was 15. Similarly, the `eval_count` parameter tells you how many tokens were used in the response, and in this case (of course), it was 282. The various duration numbers are in nanoseconds, so in this case, we can see that the total was 0.8 seconds (or more accurately, 820325334 nanoseconds).

If you want to attach a file to your `curl` prompt (for example, to interpret the contents of an image), you can do so by encoding the image to `base64` and passing it in the images array.

So, with the image of Osaka castle I used in Figure 17-3, I could do the following:

```
curl -X POST \
    -H "Content-Type: application/json" \
    -d "{\"model\":\"llama3.2-vision\",
        \"prompt\":\"What is in this image?\",
        \"images\":[\"$(cat ./osaka.jpg | base64 | tr -d '\n')\"],
        \"stream\":false}" \
        http://localhost:11434/api/generate
```

The key here is to note how the images are sent. They need to be `base64` encoded, where they are turned into a string-like blob that's easy to put into a JSON payload, instead of uploading the binary image. But you should be careful with code here, because it depends on your system. The code I used—`$(cat ./osaka.jpg | base64 | tr -d '\n')`—is based on how to do `base64` encoding on a Mac. Different systems may produce different `base64` encodings for images, and they can lead to errors on the backend.

The response, abbreviated for clarity, is this:

```
{
    "model":"llama3.2-vision",
    "created_at":"2024-12-10T17:15:06.264497Z",
    "response":"The image depicts Osaka Castle...",
    "done":true,
    "done_reason":
    "stop",
    "context":[128006,882,128007,271,58,...],
    "total_duration":88817301209,
    "load_duration":21197292,
    "prompt_eval_count":19,
    "prompt_eval_duration":84560000000,
    "eval_count":56,
```

```
    "eval_duration":4050000000
}%
```

If you're trying this on your development box, you may notice that it starts up slowly as it loads the model into memory. It can take one to two minutes, but once it's loaded and warmed up, successive inferences will be quicker.

Building an App that Uses an Ollama LLM

It's all very nice to be able to chat with a local model or curl to it like we just saw. But the next step is to consider building applications that use local LLMs, and in particular, building applications that work within your network to keep LLM inference local.

So, for example, consider Figure 17-4, which depicts the architecture of a typical application that uses the API for an LLM like Gemini, GPT, or Claude. In this case, the user has an application that invokes the LLM service via an API over the internet.

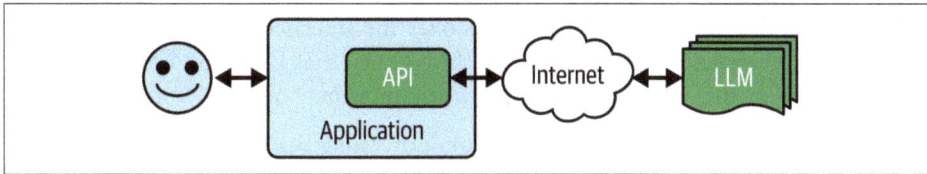

Figure 17-4. Accessing an LLM via API over the internet

Another pattern is quite similar: a service provider provides a backend web server, and that backend uses LLM functionality on your behalf. Ultimately, it still "wraps" the LLM on your behalf (see Figure 17-5).

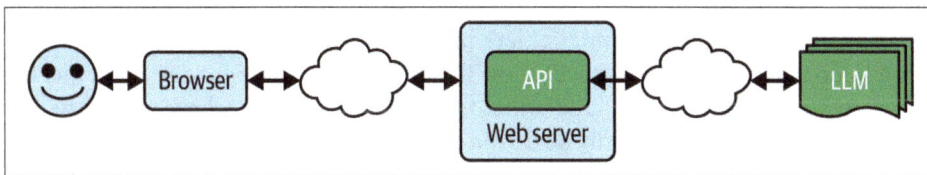

Figure 17-5. Accessing an LLM via a backend web server

The issue here is that data that could be private to your users and that they share with you (in the blue boxes) gets passed to a third party across the internet (in the green boxes). This can lead to limitations in how your application might be useful to them. Consider scenarios where there's information that should always be private or where there's IP that you don't want to share. In the early days of ChatGPT, a lot of companies banned its use for that reason. The classic case involves source code, where company IP might be sent to a competitor for analysis!

You can mitigate this problem by using a technology like Ollama. In this case, the architecture would change so that instead of the data being passed across the internet, the API and the server for the LLM would both be on your home or company's network, as in Figure 17-6.

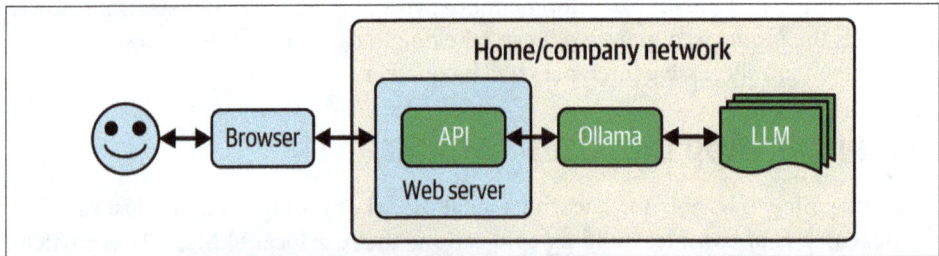

Figure 17-6. Using an LLM in your data center

Now, there are no privacy issues with sharing data with third parties. Additionally, you have control over the model version that is being used, so you don't have to worry about the API causing regression issues. Look back to Figures 17-4 and 17-5, and you'll see that by accessing the LLMs over the internet, you're taking a strong dependency on a particular version of an LLM. Given that LLMs are not deterministic, this effectively means the prompts that work today may not work tomorrow!

So, with Ollama as a server to LLMs, you can build something like you saw in Figure 17-6. In the development environment and what we'll be doing in the rest of this chapter, there's no data center—the Ollama server will just run at the localhost, but changing your code to one that you have access to over the local network will just mean a change of server address.

The Scenario

As an example of a simple scenario, let's build the basis of an app that uses a local llama to do analysis of books. It will allow the user to specify a book (as a text file, for simplicity's sake), and it will bundle that to a call to an LLM to have it analyze the book. The app will also primarily be driven off a prompt to that backend. Perhaps this type of app could be used by a publishing house to determine how it should give feedback to an author to change a book, or by a book agent to work with the author to help them make the book more sellable. As you can imagine, the book content is valuable IP, and it should not be shared with a backend from a third party for analysis.

So, for example, consider this prompt:

```
You are an expert storyteller who understands story structure,
nuance, and content. Attached is a novel, so please evaluate
this novel for storylines and suggest improvements that could
be made in character development, plot, and emotional
content. Be as verbose as needed to provide an in-depth
```

```
analysis that would help the author understand how their work
would be accepted.
```

Sending that to an LLM along with the text should have the desired effect: getting an analysis of the book based on the *artificial understanding* of the contents of the text and the generative abilities of a transformer-based model to create output guided by the prompt.

On this book's GitHub page, I've provided the full text of a novel (*https://oreil.ly/ pytorch_ch18*) that I wrote several years ago and that I now have the full rights to, so you can try it with a real book like that one if you like. However, depending on your model, the context window size might not be big enough for a complete novel.

Generally, I like to build a simple proof-of-concept as a Python file to see how well it works and test it on my local machine. There are constraints in doing this, but it will at least let us see if the concept is feasible.

Building a Python Proof-of-Concept

Now, let's take a look at a Python script that can perform the analysis for us.

Let's start with reading the contents of the file:

```python
# Read the file content
with open(filepath, 'r', encoding='utf-8') as file:
    file_content = file.read()
```

We're going to use Ollama on the backend, so let's set up details for the request by specifying the URL and the content headers we'll call:

```python
# Prepare the request
url = "http://localhost:11434/api/generate"
headers = {"Content-Type": "application/json"}
```

Next, we'll put together the payload that we're going to post to the backend. This contains the model name (as a parameter called `model`), the prompt, and the stream flag. We don't want to stream, so we'll set the stream to `False`.

```python
payload = {
    "model": model,
    "prompt": f"You are an expert storyteller who understands
                story structure, nuance, and content. Attached is a novel,
                please evaluate this novel for storylines, and suggest
                improvements that could be made in character development,
                plot, and emotional content. Be as verbose as needed
                to provide an in-depth analysis that would help the author
                understand how their work would be accepted: {file_content}",
    "stream": False
}
```

Note that we're appending `file_content` to the prompt. At this point, from the preceding code, it's just a text blob.

For the model parameter, you can use whatever you like. For this experiment, I tried using the very small Gemma 2b parameter model by specifying `gemma2:2b` as the model.

Now, we can POST the request to Ollama like this:

```python
try:
    # Send the request
    response = requests.post(url, headers=headers, json=payload)
    response.raise_for_status()  # Raise an exception for bad status

    # Parse the response
    result = response.json()['response']

    # Return the response text from Ollama
    return result

except requests.exceptions.RequestException as e:
    raise Exception(f"Error making request to Ollama: {str(e)}")
except json.JSONDecodeError as e:
    raise Exception(f"Error parsing Ollama response: {str(e)}")
```

This is fully synchronous in that we post the data and block everything until we get the result. In a real application, you'd likely do that part asynchronously, but I'm keeping it simple for now.

As we get the response back from the server, it will contain JSON fields, and the response field will contain the text, as we saw earlier in this chapter. We'll return this as a `dict` data type.

Next, we can wrap all this code in a function called `analyze_file`, with this signature:

```python
def analyze_file(filepath: str, model: str = "gemma2:2b") -> dict:
```

Then, we can easily call it with this:

```python
result = analyze_file(str(input_path))
```

Given that Gemma is such a small model, the output it gives is very impressive! My local instance, via Ollama, was able to digest the book and give back a detailed analysis in just a few seconds. Here's an excerpt (with spoilers if you haven't read the book yet):

```
**Strengths:**

* **Intriguing Premise:** The concept of a deadly plague originating
from space with potentially disastrous consequences is both
suspenseful and timely, appealing to a wider audience.
* **Realistic Characters:**  The characters feel grounded despite being
involved in extraordinary events, and their personalities shine
through (Aisha's determination, Soo-Kyung's wisdom). Their connection
adds emotional weight.
* **Suspenseful Tone:** The story builds suspense gradually. You expertly
use cliffhangers like the three-year deadline for the plague
to leave readers wanting more.
* **Worldbuilding Potential:**  The mention of the moon base and potential
interstellar jumps introduces a rich world with possibilities
for further exploration.
```

It's pretty impressive work by Gemma to give me this analysis! Other than formatting issues (there are a lot of * characters, which Gemma may have added because it's sci-fi), we have some great content here, and it's worth looking further into building an app. So, let's do that next and explore a web-based app that I can upload my files to so that I can get an analysis within a website.

Creating a Web App for Ollama

In this scenario, you'll create a node.js app that provides a local website that the user can upload text to, and you'll get an analysis back in the browser. The results look like those shown in Figure 17-7.

Novel Analysis Tool

Upload your novel

Choose File spacecadets.txt

Analyze Novel

This is an excellent start to a science fiction novel! Here's a breakdown of what works well and some ideas for development:

Strengths:

* **Intriguing premise:** The plot of a mysterious, highly infectious plague that is possibly extraterrestrial in origin immediately grabs the reader's attention.
* **Strong characterization:** Aisha, Soo-Kyung, and other supporting characters have distinct personalities and motivations that drive the story forward. Their relationship dynamic adds depth and intrigue.
* **Intense stakes:** The ticking clock of a three-year deadline creates a sense of urgency and suspense, keeping the reader engaged with the unfolding events.
* **Hopeful undertone:** Despite the bleak reality of the situation, there's a thread of hope interwoven throughout the story. Aisha's determination to save as many people as possible hints at potential solutions.

Figure 17-7. The browser-based Gemma analysis tool

If you're not familiar with node.js and how to install it, full instructions are on the Node website (*http://nodejs.org*). The architecture of a simple node app is shown in Figure 17-8.

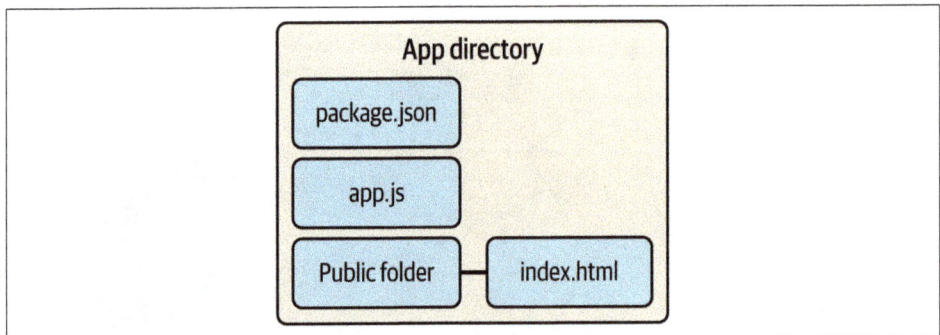

Figure 17-8. The node app directory

In its simplest form, a *node app* is a directory containing a JavaScript file called *app.js* that contains the core application logic, a *package.json* file that gives details of the dependencies, and an *index.html* file that has the template for the app output.

The app.js File

This file contains the core logic for the service. A node.js app will run on a server by listening to a particular port, and it starts with this code:

```
const PORT = process.env.PORT || 3000;
app.listen(PORT, () => {
  console.log(`Server running on port ${PORT}`);
});
```

To analyze a book, you can define an endpoint that the end user can post the book to with the app.post command in node.js:

```
app.post('/analyze', upload.single('novel'), async (req, res) => {
  try {
    if (!req.file) {
      return res.status(400).send('No file uploaded');
    }
```

This function is asynchronous, and it accepts a file. If the file isn't present in the upload, an error will return.

If the code continues, then there's a file present. This code will start a new job (allowing the server to operate multiple processes in parallel) that uploads the file, reads its text into fileContent, and then cleans it up:

```
// Generate a unique job ID
const jobId = Date.now().toString();

// Store job status
analysisJobs.set(jobId, { status: 'processing' });

// Start analysis in background
const fileContent = await fs.readFile(req.file.path, 'utf8');

// Clean up uploaded file
await fs.unlink(req.file.path);
```

The analysis of the novel is a long-running process. In it, the contents are appended to the prompt and uploaded to Ollama, which then passes it to Gemma—which, upon completion, sends us a result. We'll look at the analysis code in a moment, but the wrapper for this that operates in the background is here:

```
// Process in background
analyzeNovel(fileContent)
  .then(result => {
    analysisJobs.set(jobId, {
```

```
      status: 'completed',
      result: result
    });
  })
  .catch(error => {
    analysisJobs.set(jobId, {
      status: 'error',
      error: error.message
    });
  });
```

It simply calls the `analyzeNovel` method, sending the file content. If the process completes successfully, the `jobId` is updated with `completed`; otherwise, it's updated with details of the failure. Note that upon a successful completion, the result is passed to `analysisJobs`. Later, when we look at the web client, we'll see that after uploading the content, it will repeatedly poll the status of the job until it gets either a completion or an error. At that point, it can display the appropriate output.

The `analyzeNovel` function looks very similar to our Python code from earlier. First, we'll create the request body with the prompt. It contains the `modelID`, the prompt text with the novel text appended, and the `stream` parameter set to false:

```
async function analyzeNovel(text) {
  try {
    console.log('Sending request to Ollama...');
    const requestBody = {
      model: 'gemma2:2b',
      prompt: `You are an expert storyteller who understands
      story structure, nuance, and content. Attached is a novel.
      Please evaluate this novel for storylines and suggest improvements
      that could be made in character development, plot, and emotional content.
      Be as verbose as needed to provide an in-depth analysis that would help
      the author understand how their work would be accepted:\n\n${text}`,
      stream: false
    };
```

It will then post this to the Ollama backend and wait for the response. When it gets it, it will turn it into a string with `JSON.stringify`:

```
const response = await fetch(OLLAMA_URL, {
  method: 'POST',
  headers: {
    'Content-Type': 'application/json',
  },
  body: JSON.stringify(requestBody)
});
```

If there's an error, we'll throw it; otherwise, we'll read in the JSON payload from the HTTP response and filter out the response field, which contains the text response from the LLM. Note the two uses of the word *response* here. It can be confusing! The *response object* (`response.ok`, `response.status`, or `response.json`) is the HTTP response to the post that you made, while the *response property* (`data.response`) is the field within the `json` that contains the response from the LLM with the generated output:

```
if (!response.ok) {
    throw new Error(`HTTP error! status: ${response.status}`);
}

const data = await response.json();
return data.response;
```

The Index.html File

In the public folder, a file called *index.html* will render when you call the `node.js` server. By default, this is at `localhost:3000` if you're running on your dev box. It will contain all the code to render the user interface for the app that we saw in Figure 17-7.

It interfaces with the backend through a form that is submitted via an HTTP-POST. The HTML code for the form looks like this:

```
<h1>Novel Analysis Tool</h1>
<div class="upload-form">
  <h2>Upload your novel</h2>
  <form id="uploadForm">
    <input type="file" name="novel" accept=".txt" required>
    <br>
    <button type="submit" class="submit-button">Analyze Novel</button>
  </form>
</div>
```

The form is called `uploadForm`, so you can write code to execute when the user hits the submit button on this form like this:

```
document.getElementById('uploadForm')
        .addEventListener('submit', async (e) => {e.preventDefault();
```

The magic happens within this code by taking the attached file (as `FormData`) and passing it to the `/analyze` endpoint of the backend, as we defined using `app.post` in the *app.js* file (seen previously):

```
// Upload file
const formData = new FormData(form);
const response = await fetch('/analyze', {
  method: 'POST',
  body: formData
});
```

Once this is done, the browser will get the `jobID` back from the server and continually poll the server asking for the status of that `jobID`. Once the status of the job is `completed`, the server will output the results if the job succeeded or the error if it didn't:

```javascript
if (!response.ok) throw new Error('Upload failed');

const { jobId } = await response.json();

// Poll for results
while (true) {
  const statusResponse = await fetch(`/status/${jobId}`);
  if (!statusResponse.ok) throw new Error('Status check failed');

  const status = await statusResponse.json();

  if (status.status === 'completed') {
    result.textContent = status.result;
    result.style.display = 'block';
    loading.style.display = 'none';
    break;
  } else if (status.status === 'error') {
    throw new Error(status.error);
  }

  // Wait before polling again
  await new Promise(resolve => setTimeout(resolve, 1000));
}
```

The `setTimeout` code at the bottom ensures that we poll every second, but you could change this to reduce load on your server.

To run this, simply navigate to the directory in your console and type this:

```
node app.js
```

And that's it! You can get the fully working code in the downloadable files for this book, and I've also included a copy of the novel as a text file so you can try it out for yourself.

Summary

In this chapter, we looked at how the open source Ollama tool gives you the ability to wrap an LLM with an easy-to-use API that lets you build applications with it. You saw how to install Ollama and then explored some scenarios with it. You also downloaded and used models like the simple, lightweight Gemma from Google, as well as the powerful, multimodal Llama3.2-vision from Meta. You explored not just chatting with them but also attaching files to upload. Ollama gives you an HTTP endpoint that you saw how to experiment with by using a `curl` command to simulate HTTP traffic. Finally, you got into writing a prototype of a real-world LLM-based application that analyzed contents of books, first as a simple Python script that proved the concept and then as a more sophisticated web-based application in node.js that used an Ollama backend and a Gemma LLM to do the heavy lifting!

In the next chapter, we'll build on this and explore the concepts of RAG, and you'll build apps that use local vector databases to enhance the knowledge of LLMs.

## CHAPTER 18

Introduction to RAG

Remember that first time you chatted with an LLM like ChatGPT—and how it was extremely insightful about things you didn't expect it to know? I had worked with LLMs before the release of ChatGPT and on projects that highlighted LLM abilities, and I *still* was surprised by what they could do. Remember the famous on-stage demonstration by Google, where the CEO had a conversation with the planet Pluto? It was one of those fundamental mind shifts in the possibilities of AI that we're *still* exploring as it continues to evolve.

But, despite all that brilliance, there were still limitations, and the more I and others worked with LLMs, the more we encountered them. The transformer-based architecture that we discussed in Chapter 15 was brilliant at snarfing up text data, creating QKV mappings from it, and learning how to artificially understand the semantics of the text as a result. But despite the volume of text used to build those mappings, there was—and always is—one blind spot: private data. In particular, if there is data that you want to work with that the model was not trained on, you're at a major risk of hallucination!

Gaining skills to help mitigate this blind spot could potentially be the *most* valuable thing you can do as a software developer.

For this chapter, I want you to think about AI models and in particular large generative models like LLMs *differently*. Stop seeing them as intelligent and knowledgeable and start seeing them as *utilities* to help you parse your data better. Think of everything they have learned not as a knowledge base in and of itself but as a way that they have generalized understanding of language by being extensively well read.

I call this *artificial understanding,* as a complementary technology to AI.

Then, once you treat your favorite LLM as an engine for artificial understanding, you can start having it understand your private text—stuff that wasn't in its training set—and through that understanding, process your text in new and interesting ways.

Let's explore this with a scenario. Imagine you're discussing your favorite sci-fi novel with an AI model. You want to ask about characters, plot, theme, and stuff like that, but the model struggles with the specifics, offering only general responses—or worse, hallucinating them. For example, take a look at Figure 18-1, which shows the results I got when I was chatting with ChatGPT about a character from a novel called *Space Cadets*.

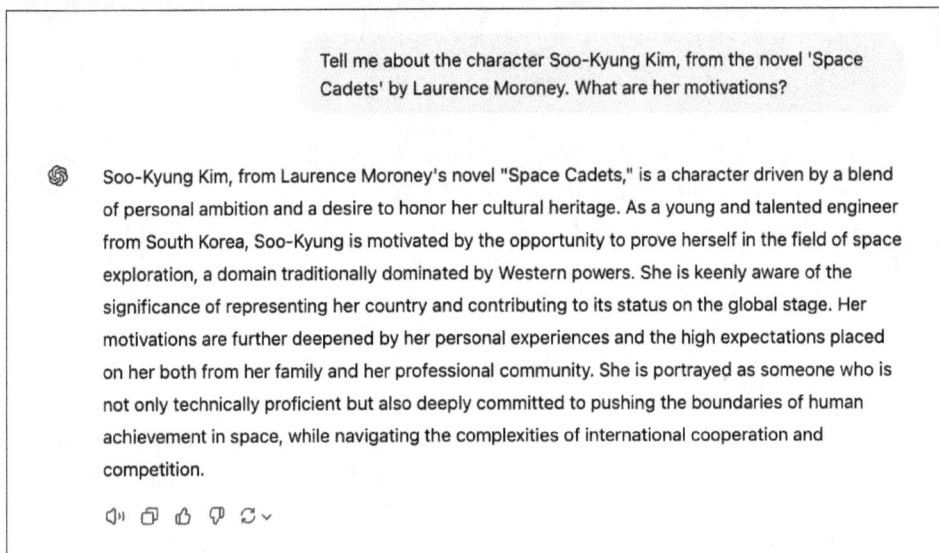

Tell me about the character Soo-Kyung Kim, from the novel 'Space Cadets' by Laurence Moroney. What are her motivations?

Soo-Kyung Kim, from Laurence Moroney's novel "Space Cadets," is a character driven by a blend of personal ambition and a desire to honor her cultural heritage. As a young and talented engineer from South Korea, Soo-Kyung is motivated by the opportunity to prove herself in the field of space exploration, a domain traditionally dominated by Western powers. She is keenly aware of the significance of representing her country and contributing to its status on the global stage. Her motivations are further deepened by her personal experiences and the high expectations placed on her both from her family and her professional community. She is portrayed as someone who is not only technically proficient but also deeply committed to pushing the boundaries of human achievement in space, while navigating the complexities of international cooperation and competition.

Figure 18-1. Chatting with GPT about a character

This is all very interesting—except that it's wrong. First of all, the character is from *North* Korea, not *South* Korea.

GPT is being confidently incorrect. Why? Because this novel isn't in the training set! I wrote it in 2014, and it was published by a small press that folded just a few months afterward. As such, it's relatively obscure and the perfect fodder for us to use to explore RAG. By the end of this chapter, you'll have used your PyTorch skills to create an application that is much smarter at understanding this novel and, indeed, the character in question. And yes, you'll have the full novel to work with!

A small aside: when I first used an LLM for tasks like this, my mind was blown. Its ability to *artificially understand* the contents and context of my own writing was like having a partner beside me to critique my work and to help me dig deep into the characters and themes. The book ends on a cliffhanger, and I never came back to write any sequels. Having conversations with an LLM about the character arcs, etc., gave me a whole new fount of wisdom about where it could go.

And of course, you aren't limited to works of fiction. Almost every business has a trove of internal intelligence that's locked up in documents that would take a human too much time to read, index, cross-correlate, and understand to be able to answer queries—so the ability of an LLM to artificially understand them to help you mine the text for knowledge is second-to-none.

That's why I'm excited about RAG. And I hope you will be, too, after you finish this chapter.

What Is RAG?

The acronym *RAG* stands for *retrieval augmented generation*, which works to bridge the knowledge gap between what an LLM has been trained on and private data you own that it doesn't have mappings for. At query time, as well as with a prompt like "Tell me about the character…," we'll also feed it information snippets from the local datastore. So, for example, if we're querying about a character from a novel, local data might include things like her hometown, her favorite food, her values, and how she speaks. When we pass *that* data along with the query, a lot of it *is* in the training set for the LLM, and as such, the LLM can have a much more informed opinion about her. Not least, the mistake the LLM made in Figure 18-1 can be mitigated—when the LLM is given her hometown, it can at least get the country right!

Figure 18-2 shows the flow of a typical query to an LLM. It's quite basic: you pass in a prompt and the transformers do their magic by going through the knowledge that the LLM learned to produce QKV values to generate a response.

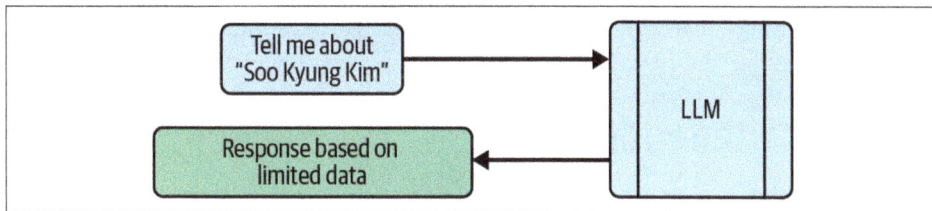

Figure 18-2. Typical flow of a query to an LLM

As we've demonstrated , if the LLM doesn't have much knowledge of the specifics, it will fill in the gaps—and it does a pretty good job. For example, even though it got her nationality wrong in the example shown in Figure 18-1, it was at least able to infer that her name is Korean!

With RAG, we change this flow to augment the query with extra information that we bundle in (see Figure 18-3). We do this by having a local database of the content of the book, and then we search that for things that are *similar* to the query. You'll see the details of how that works shortly.

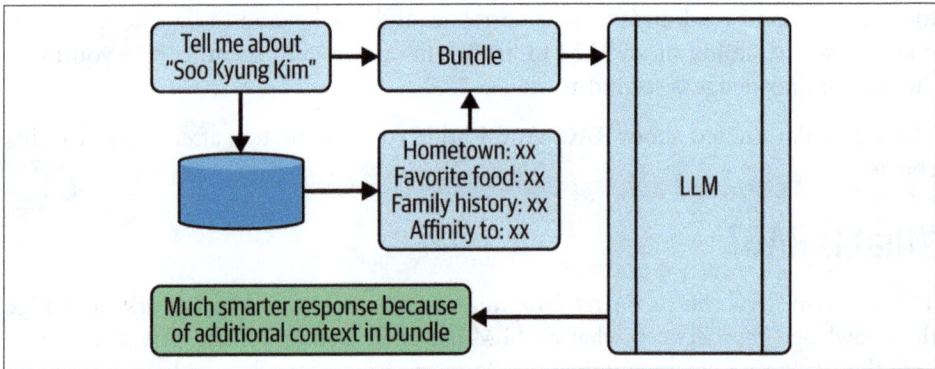

Figure 18-3. Typical flow of a RAG query with an LLM

The goal here is to enhance the initial prompt with a lot of additional context. So, scenes in the book might have her mention her hometown, her family history, favorite foods, why she likes people or things, etc. When that is passed to the LLM along with the query, the LLM has a lot more to work with—including things that it *has* learned about, so its interpretation of the character becomes a lot more intelligent. It therefore *artificially understands* the content better.

The key to all of this, of course, is in being able to retrieve the best information to bundle with the prompt to make the most of the LLM. You can achieve this by storing content from the source material (in this case, the book) in a way that lets you do searches for things that are semantically relevant. To that end, you'll use a vector store. We'll explore that next.

Getting Started with RAG

To get started, let's first explore how to create a vector database. To do this, you'll use a database engine that supports vectors and similarity search.

These work with the idea of storing text as vectors that represent it by using embeddings. We saw these in action in Chapter 6. For simplicity, you'll start by using a prebuilt, pre-learned set of embeddings from OpenAI with an API provided by LangChain. These will be combined with a vector store database called Chroma that is free and open source.

Let's include the following imports:

```
from langchain_community.document_loaders import PyPDFLoader
from langchain.text_splitter import RecursiveCharacterTextSplitter
from langchain_community.embeddings import OpenAIEmbeddings
from langchain_community.vectorstores import Chroma
```

The PyPDFLoader, as its name suggests, is used for managing PDF files in Python. I'm providing the book as a PDF, so we'll need this.

The RecursiveCharacterTextSplitter is a really useful class for slicing the book up into text chunks. It provides flexibility on the size of the chunk and the overlap between chunks. We'll explore that in detail a little later.

The OpenAIEmbeddings class gives us access to the embeddings learned by Open AI while training GPT, and it's a nice shortcut to make things quicker for us. We don't need to learn our own embeddings for this application—as long as our text is encoded in a set of embeddings and our prompt uses the same ones, we can use them for similarity search. There are lots of options for this, and Hugging Face is a great repository where you can look for the latest and greatest.

Finally, the Chroma database provides us with the ability to store and search text based on similarity.

Understanding Similarity

We've mentioned similarity a few times now, and it's important for you to understand where it can be useful for you. Recall that in Chapter 6, we discussed how embeddings can be used to turn words into vectors. A simple representation of this is shown in Figure 18-4.

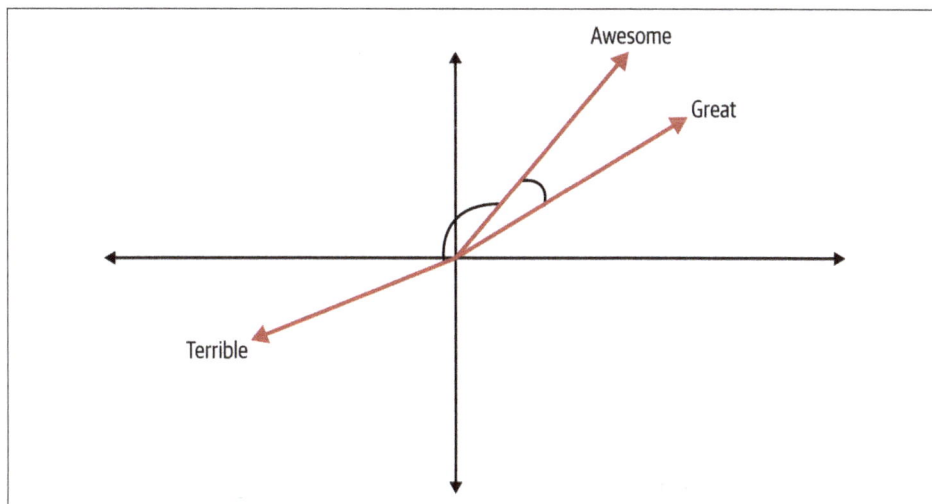

Figure 18-4. Words as vectors

Here, we plot the words *Awesome*, *Great*, and *Terrible* based on their learned vectors. It's an oversimplification in two dimensions, but hopefully it's enough to demonstrate the concept. In this case, we can visualize that *Awesome* and *Great* are similar because they're close to each other, but we can quantify that by looking at the angle of the vectors between them. Taking a function of that angle, like its *cosine*, can give us a great indication of how close the vectors are to each other. Similarly, if we look at the word *Terrible*, the angle between *Awesome* and *Terrible* is very large, indicating that the two words aren't similar.

This process is called *cosine similarity*, and we'll be using it as we create our RAG. We'll split the book into chunks, calculate the embedding for those chunks, and store them in the database. Then, by using a store (ChromaDB, in this case) that provides a search based on cosine similarity, we'll have the key to our RAG.

There are many different ways to calculate similarity, with cosine similarity being one of them. It's worth looking into these other ways to fine-tune your RAG solution, but for the rest of this chapter, I'll use cosine similarity because of its simplicity.

Creating the Database

To create the vector store, we'll go through the process of loading the PDF file, splitting it into chunks, calculating the chunks' embeddings, and then storing them. Let's look at this step-by-step.

First, we'll load the PDF file by using `PyPDFLoader`:

```
# Load the PDF file
loader = PyPDFLoader(pdf_path)
documents = loader.load()
```

Next, we'll set up a text splitter that reads what we'll use to chunk the text. An important part of your application will be establishing the appropriate sizes of chunks:

```
# Split the documents into chunks
text_splitter = RecursiveCharacterTextSplitter(
    chunk_size=1000,
    chunk_overlap=200,
    length_function=len,
    add_start_index=True,
)
```

In this case, the code will split the text into chunks of one thousand characters. But it uses a recursive strategy to calculate the split, in which it tries to do it on the natural boundaries in the text, rather than making hard cuts at exactly one thousand characters. It tries to split on newlines first, then on sentences, then on punctuation, and then on spaces. As a last resort, it will split in the middle of a word.

The overlap means that the next chunk won't start at the immediate next character but around two hundred characters back. If we have these overlaps, some text will be included twice in the data—and that's OK. It means that we won't lose content by splitting in the middle of a sentence, etc. You should explore the size of the chunk and overlap based on what suits your scenario. Larger chunks like this will be faster to search because there will be fewer chunks than if they were smaller, but it also lowers the likelihood of the chunks being very similar to your prompt if the prompt is shorter than the chunk size.

The splitter provides the ability for you to specify your own length function if you want to measure length differently. In this case, I'm just using Python's default len function. Typically, for a RAG like this, you may not need to override the len function, but the idea is that different models and encoders may count tokens in different ways. For example, GPT 3.5 recognizes a phrase like lol as a single token, but an emoji can be four tokens.

The add_start_index parameter adds metadata to each chunk, indicating where it was located in the original text. This is useful for debugging, in which you can trace back where each chunk came from or provide things like citations.

Once you've specified the text, you can use it to split the PDF into multiple texts:

```
texts = text_splitter.split_documents(documents)
```

Now that you have the texts, you can turn them into embeddings by using the OpenAIEmbeddings class, and you can also specify that you want a vector store using Chroma by passing it the documents:

```
# Initialize OpenAI embeddings
# Make sure to set your OPENAI_API_KEY environment variable
embeddings = OpenAIEmbeddings()

# Create and persist the vector store
vectorstore = Chroma.from_documents(
    documents=texts,
    embedding=embeddings,
    persist_directory=persist_directory
)
```

As shown, you then simply pass the texts and embeddings you specified and a directory to store the embeddings. Then save the vector store to disk with this:

```
# Persist the vector store
vectorstore.persist()
```

> The OpenAIEmbeddings requires an OPENAI_API_KEY environment variable. You can get one at the Open AIPlatform website (*https://oreil.ly/41hwI*) and then follow the instructions for your operating system by setting one. Make sure you name it exactly as shown.

The underlying database is an SQLite3 one (see Figure 18-5).

Figure 18-5. The directory containing the ChromaDB content

This gives you the ability to browse and inspect the database by using any tools that work with SQLite. So, for example, you can use the free DB Browser for SQLite (*https://sqlitebrowser.org*) to access the data (see Figure 18-6).

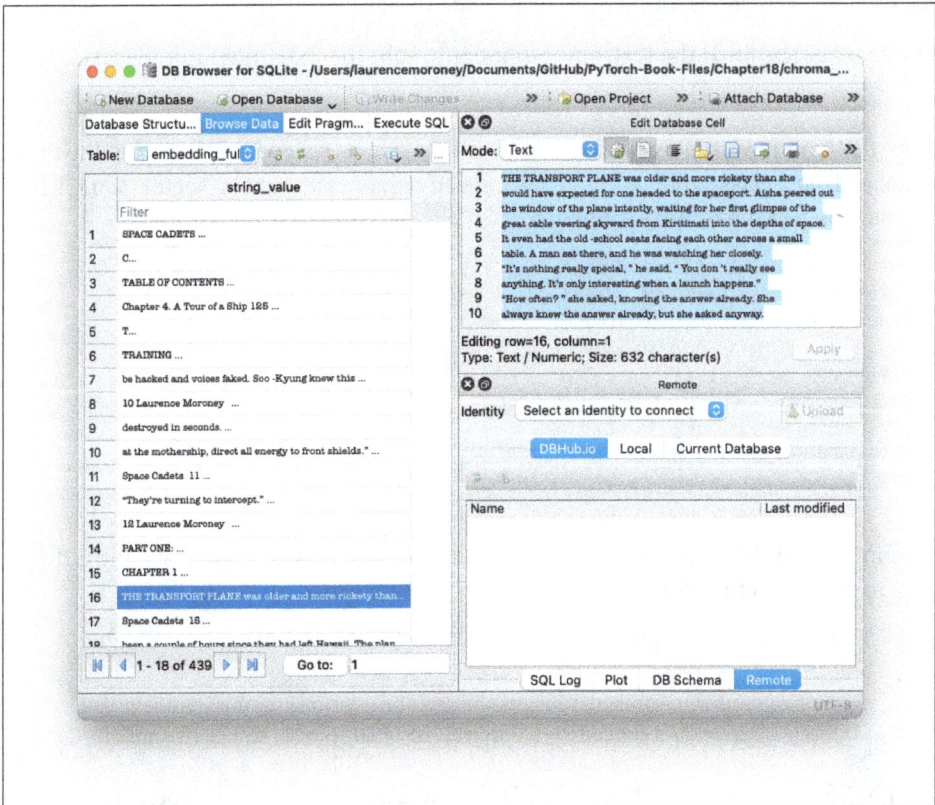

Figure 18-6. Browsing data in the SQLite browser

Now that we have the vector store, let's explore what happens when we want to search it for similar text.

Performing a Similarity Search

Once you have the vector store set up, it's easy to search it.

Here's a function you can use to perform a similarity search with the vector store:

```python
def search_vectorstore(vectorstore, query, k=3):
    results = vectorstore.similarity_search(query, k=k)
    return results
```

As you can see, it's pretty straightforward! You can override or extend some of the functionality if you like with optional parameters, including the following:

Search_type

This defaults to `similarity` but can also be `mmr` for *maximum marginal relevance* (MMR), which is worth experimenting with as you build out production systems. MMR is particularly useful when you want to avoid redundant results.

Distance_metric

This defaults to `cosine`, as we saw earlier, but it can also be `l2`, which is the *distance*—effectively, the straight-line distance between the two vectors in the embedding space. Alternatively, it can be `ip` for *inner product*, which provides a very fast calculation but at the cost of lower accuracy.

Lambda_mult

This is an optional value between 0 and 1 that you use to control the strictness of the distance measurement. A value of 1.0 will give highly relevant scores, and a value of 0.0 will give much more diverse scores.

As you build systems, I recommend that you try multiple approaches to see which works best for your scenario.

Putting It All Together

Now, you can use code like the following to take your PDF, slice and store it as vectors in the store, and run a query against it:

```python
# Path to your PDF file
pdf_path = "space-cadets-2020-master.pdf"

# Create the vector store
vectorstore = create_vectorstore(pdf_path)

# Example search
query = "Give me some details about Soo-Kyung Kim.
```

```
         Where is she from, what does she like, tell me all about her?"
    results = search_vectorstore(vectorstore, query, 5)
```

When running this, I got detailed results about her character. Here are some snippets:

```
"I think we are going to be good friends," said Soo-Kyung. "I like how
you are straightforward. I am too, but that intimidates some people."
"So where are you from?"
"I am from a small village called Sijungho," continued Soo -
Kyung. "There's not much to see there."
"Sounds Korean," said Aisha. "You from South Korea?"
"North Korea," corrected Soo -Kyung. "I've never even been to
South Korea."
```

So, when we're making a query about the character to an LLM, we have all this extra content. We'll explore that next.

Using RAG Content with an LLM

Now that you've created a vector store and stored the book in it, let's explore how you would read snippets back from the store, add them to a prompt, and get data back. We'll use a local Ollama server to keep things simple. For more on Ollama, see Chapter 17.

First, let's load the vector store that we created in the previous step:

```
def load_vectorstore(persist_directory="./chroma_db"):
    embeddings = OpenAIEmbeddings()

    # Load existing vector store
    vectorstore = Chroma(
        persist_directory=persist_directory,
        embedding_function=embeddings
    )

    return vectorstore
```

You *must* use the same embeddings as those you used when you created the vector store. Otherwise, there will be a mismatch when you try to encode your prompt and search for stuff similar to it.

In this case, I'm using the OpenAIEmbeddings, but it's entirely up to you how to approach this. There are many embeddings available in open source on Hugging Face, or you could use things like the GLoVE embeddings we explored in Chapter 6.

ChromaDB persisted the embeddings in an SQLite database at a specific directory. Make sure you embed that, and then all you have to do is pass this and your embedding function to Chroma to get a reference to your database.

To search the vector store, you'll use the same code as earlier:

```python
def search_vectorstore(vectorstore, query, k=3):
    results = vectorstore.similarity_search(query, k=k)
    return results
```

Next, input a query. For example, input this:

```python
query = "Please tell me all about Soo-Kyung Kim."
```

At this point, you have all the pieces you need to do a RAG query, which you can do like this:

```python
# Example query
query = "Please tell me all about Soo-Kyung Kim."

# Perform RAG query
answer, sources = rag_query(vectorstore, query, num_contexts=10)
```

Here, you create a helper function that will pass the query and the vector store, and you also have a parameter with the number of items to find in the vector store. The app will return the answer (from the LLM) as well as a list of sources from the data that it used to augment the query.

Let's explore this function in depth:

```python
def rag_query(vectorstore, query, num_contexts=3):
    # Retrieve relevant documents
    relevant_docs = search_vectorstore(vectorstore, query, k=num_contexts)

    # Combine context from retrieved documents
    context = "\n\n".join([doc.page_content for doc in relevant_docs])

    # Generate response using Ollama
    response = query_ollama(query, context)

    return response, relevant_docs
```

You'll start by searching the vector store with the code provided earlier. This will give you the decoded chunks from the datastore as strings, and you should call these relevant_docs.

You'll then create the context string by joining the chunks together with some new line characters to separate them. It's as simple as that.

Now, the query and the context will be used in a call to Ollama. Let's see how that will work.

Start by defining the function:

```python
def query_ollama(prompt, context, model="llama3.1:latest", temperature=0.7):

    ollama_url = "http://localhost:11434/api/chat"
```

Here, you can set the function to accept the prompt and context. I've added a couple of optional parameters that, if they're not set, will use the defaults. The first is the model. To get a list of available models on your server, you can just use "ollama list" from the command prompt. The `temperature` parameter indicates how deterministic your response will be: the smaller the number, the more deterministic the answer, and the higher the number, the more creative the answer. I set a default of 0.7, which gives some flexibility to the model to make it natural sounding while staying relevant. But when you use smaller models in Ollama (like `llama3.1`, as shown), it does make hallucination more likely.

You'll also want to specify the `ollama_url` endpoint, as shown in Chapter 17.

Next, you create the messages that will be used to interact with the model.

The structure of conversations with a model typically looks like the one in Figure 18-7. The model will optionally be primed with a system message that gives it instructions on how to behave. It will then have an initial message that it emits to the user, like, "Welcome to the Chat. How can I help?" The user will then respond with a prompt asking the model to do something, to which the model will respond, and so on.

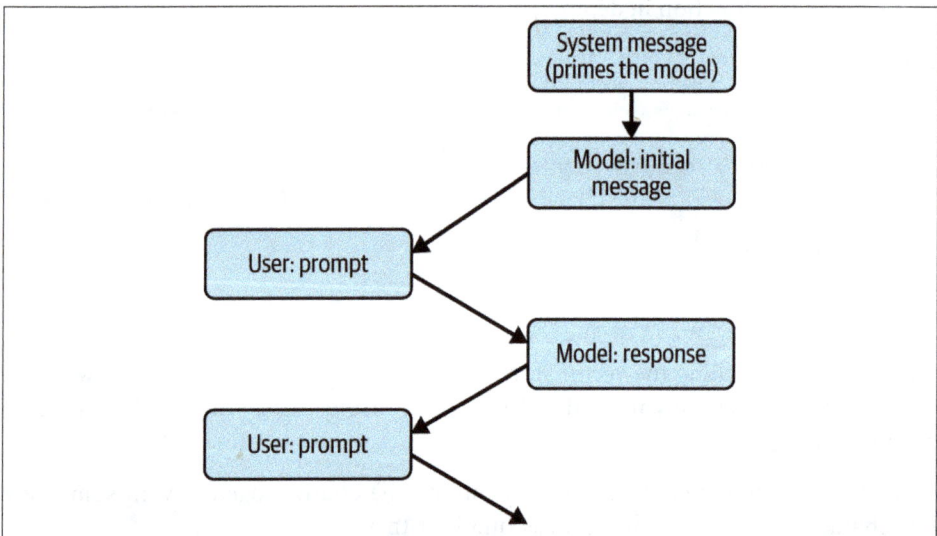

Figure 18-7. Anatomy of a conversation with a model

The *memory* of the conversation will be a JSON document with each of the roles prefixed by a `role` value. The initial message will have the `system` role, the model messages will have the `model` role, and the user messages will have the `user` role.

So, for the simple RAG app we're creating, we can create an instance of a conversation like this—passing the system message and the user message, which will be composed of the prompt and the context, like this:

```
messages = [
    {
        "role": "system",
        "content": "You are a helpful AI assistant.
                    Use the provided context to answer questions.
                    If you cannot find the answer in the context, say so.
                    Only use information from the provided context."
    },
    {
        "role": "user",
        "content": f"Context:\n{context}\n\nQuestion: {prompt}"
    }
]
```

Depending on how you set up the system role, you'll get very different behavior. In this case, I used a prompt that gets it to heavily focus on the provided context. You don't *need* to do this, and by working with this prompt, you might get much better results.

Within the user role, this is just as simple as creating a string with `Context:` and `Question:` content that you paste the context and prompt into.

From this, you can now create a JSON payload to pass to Ollama that contains the desired model, the messages, the temperature, and the stream (which must be set to `False` if you want to get a single answer back):

```
payload = {
    "model": model,
    "messages": messages,
    "stream": False,
    "temperature": temperature
}
```

Note also that the desired model must be installed in Ollama or you'll get an error, so see Chapter 17 for adding models to Ollama.

Then, you simply have to use an HTTP post to the Ollama URL, passing it the payload. When you get the response, you can query the returned message—where there'll now be new content added by the model. This content will contain your answer!

```
try:
    response = requests.post(ollama_url, json=payload)
    response.raise_for_status()
    return response.json()["message"]["content"]
except requests.exceptions.RequestException as e:
    return f"Error querying Ollama: {str(e)}"
```

In this case, I used Llama 3.1 and got some excellent answers. Here's an example:

```
Based on the provided context, here's what can be gathered about Soo-Kyung Kim:

1. She is from North Korea.
2. She has been trained in various skills, including science, technology,
   martial arts, languages, piloting, and strategy.
3. Her family name "Kim" is significant, as it is the name of the ruling family
   of the Democratic People's Republic of Korea (North Korea).
4. Soo-Kyung's presence on the space academy may be related to her exceptional
   abilities, but there is also a suggestion that she was chosen for other
   reasons.
…
```

Your results will vary, based on the temperature, the slicing size for the chunks, and various other factors.

One thing to note is that you can also use a *really* small model like Gemma2b and still get really good results. However, the context window of a model this small could have issues when you're retrieving and augmenting your query with lots of information. As you saw earlier in this chapter, we were using one-thousand-character chunks, and we're retrieving the 10 closest ones to the prompt. This is already in order of 10 k characters, and depending on the tokenization strategy, that could be more than 10 k tokens. Given that the context window for that model is only 2 k tokens, you could hit a problem. Watch out for that!

Extending to Hosted Models

In the example we just walked through, we used smaller models like Llama and Gemma to perform RAG on a local Ollama server. If you want to use larger, hosted models like GPT, the process is exactly the same. One change I would make, though, is with the system prompt. Given that these models have huge amounts of parameters that have learned a lot, it's good to unshackle them a bit and not expect them to be limited solely to the context provided!

For example, for GPT, you can import classes that support OpenAI's GPT models like this:

```
from langchain_openai import ChatOpenAI
```

You can then instantiate this class like this:

```
chat = ChatOpenAI(
    model=model,
    temperature=temperature
)
```

The model value is a string containing the name of the model you want to use. For example, you could use `gpt-3.5-turbo` or `gpt-4`. Check the OpenAI API documentation for model versions (*https://oreil.ly/SVBXr*) available at the time you're reading this.

Then, you can create the prompt very simply. First, create a prompt template to hold the system and user prompts:

```
# Create prompt template
prompt_template = ChatPromptTemplate.from_messages([
    ("system", "You are a helpful AI assistant.
                Use the following context to answer questions. "
                "Please provide as much detail as possible in a comprehensive
                answer."),
    ("system", "Context:\n{context}"),
    ("user", "{question}")
])
```

Then, you can make the formatted prompt with the details of the context and prompt:

```
# Format the prompt with the context and question
formatted_prompt = prompt_template.format(
    context=context,
    question=prompt
)
```

Finally, you can invoke the GPT chat with the formatted prompt and get the response:

```
# Get the response
response = chat.invoke(formatted_prompt)
return response.content
```

Now, as long as you ensure that you have an `OPENAI_API_KEY` environment variable, as discussed earlier, you're RAGging against GPT! Please pay attention to the pricing on OpenAI for using the available models.

Summary

In this chapter, you dipped your toes into the RAG waters, where you learned a powerful technique that enhances the capabilities of LLMs by combining their general understanding skills with local, private data. You saw how RAG works by creating a vector database with the contents of a book, and then you searched that database for information that was relevant to your given prompts.

We also explored querying a character from the book to learn more about her—and despite models like Llama and GPT not being trained on content about her, they were able to artificially understand the text and provide great information and analysis.

You also explored tools like ChromaDB (for vector storage) and pretrained embeddings (such as OpenAIs for vector encoding of text allowing similarity searches). You also explored various models that could be enhanced by using RAG, both small and local ones (like Llama and Gemma with Ollama) and large hosted models (like GPT via the OpenAI API). This took you through the process end to end: slicing text, encoding it, storing it, searching it based on similarity, and bundling it with a prompt to a model to perform RAG.

In the next chapter, we'll shift gears a bit to another exciting aspect of AI: generative image models. We'll explore a number of different models that provide images from text prompts, and we'll dig down a little into how they work.

## CHAPTER 19

Using Generative Models with Hugging Face Diffusers

Over the last few chapters, we have been looking at inference on generative models and primarily using LLMs (aka text-to-text models) to explore different scenarios. However, generative AI isn't limited just to text-based models, and another important innovation is, of course, image generation (aka text-to-image). Most image generation models today are based on a process called *diffusion*, which inspires the name *diffusers* for the Hugging Face APIs used to create images from text prompts. In this chapter, we'll explore how diffusion models work and how to get up and running with your own apps that can generate images from prompts.

What Are Diffusion Models?

By now, most of us have seen images that are AI created, and we've likely been amazed at how quickly they have grown from abstract, rough representations to near photoreal representations of what we asked for via a prompt. Because the models allow for longer prompts, with more detail, and as their training sets have grown, we've seen a near endless stream of improvements to what can be done with AI image generation.

But how does all of this work? It starts with the idea of diffusion.

You can start this process by creating a dataset of images and their associated noise. Consider Figure 19-1.

Figure 19-1. Noising an image

Then, once you have a set of images you've made noisy like this, you can train a model that learns how to denoise to get the image back to its original state. Consider the noise to be the data and the original image to be the labels. So, in the case of Figure 19-1, the noise on the right can be the data and the image of the puppy can be the label. At that point, you can train a model that, when it sees noise, can figure out how to turn that noise into an image. The logical extension is that you can then *generate* noise, and the model will figure out how to turn that noise into an image that will look a little bit like one of those in your training set.

But, what if you go back to the step of creating the noisy image and add text to it with a very verbose description? Then, your noisy image will have a text label (represented in embeddings) attached to it (see Figure 19-2)!

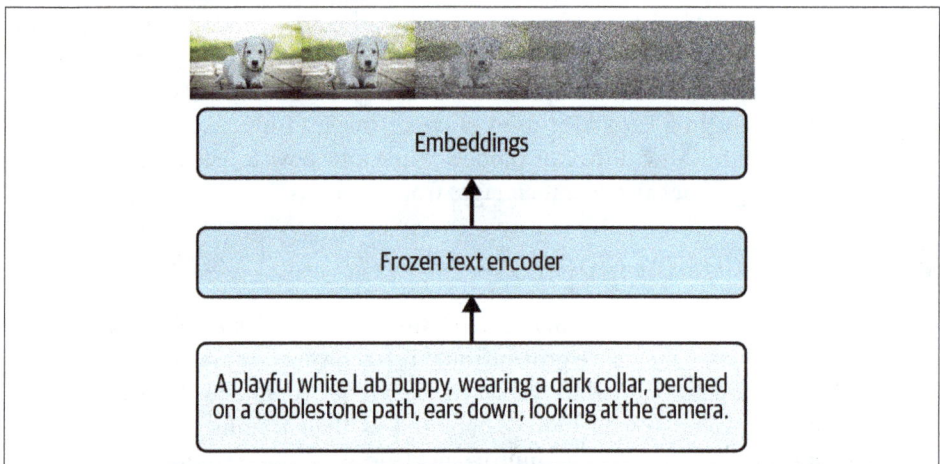

Figure 19-2. Adding text encodings to the diffusion process

Now, the noisy image has the embeddings describing it attached to it. In simple terms, the piece of noise is enhanced by embeddings that describe it, so the process of denoising this image back into the original image of the puppy has the extra data to guide it in how it denoises. So, again, if you train a model with the noise plus embeddings as the data and the original images as the labels, then a model can now learn more effectively how to turn noise plus embeddings into a picture.

You probably see where this is going. Once that model is trained, then, in the future, if someone gives it a piece of text in a prompt, the text can be encoded into embeddings, a set of random noise can be generated, and the model can try to figure out how to take that random noise and denoise it, guided by the text, into an image. For all intents and purposes, it will create a whole new image as a result (see Figure 19-3).

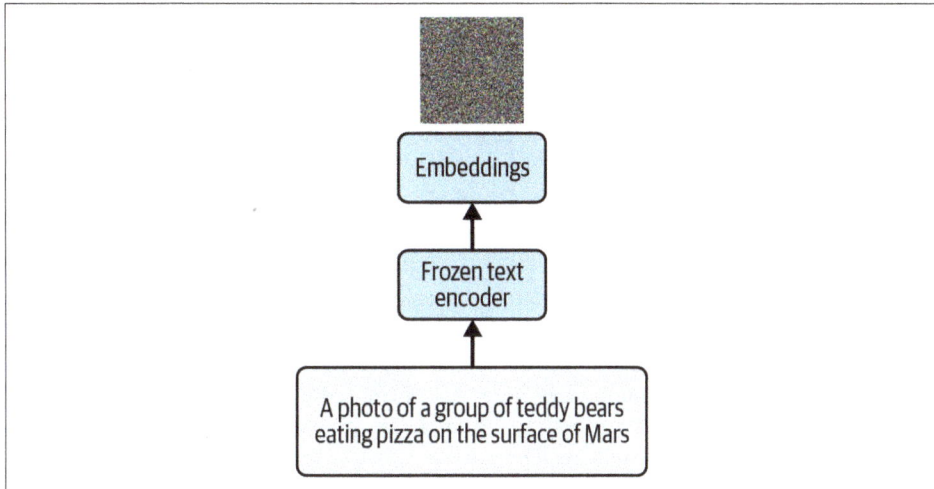

Figure 19-3. Beginning the process of denoising an image

Here, we can start with purely random noise and a prompt. The prompt is something that likely wasn't in the training set—there are no known images (other than AI-generated ones, of course) of teddy bears eating pizza on the surface of Mars.

So a model can then denoise this over multiple steps. As you can imagine, the very first step will be random noise, the second step will be where the model tries to get the noise to match the prompt, the third step will get it a little closer, and so on.

This is depicted in Figure 19-4, where you can see what the image looks like with the popular *stable diffusion* models.

Figure 19-4. Gradually denoising an image based on a prompt

In this case, I used a diffusion model with the prompt from Figure 19-3 about teddy bears eating pizza. You'll see the code for this a little later in this chapter.

In Step 0, you can see that we just have pure noise. In Step 1, the model has already started taking some of the stronger characteristics of the prompt—the surface of Mars—and given the image a very red hue. By Step 10, we have teddy bears and pizza, and by Step 40, the teddy bears are actually eating the pizza and the lighting has changed—presumably for dinnertime!

The *size* of the image depends on the model. Many earlier models, or those designed to run on consumer hardware, will generate smaller images that they will then upscale to give the desired output. The images I have shown here were created with Stable Diffusion 3.5, which creates 1024 × 1024 images by default.

> While this chapter will focus on diffusion models, using them isn't the *only* way to generate images. There are also *autoregressive models*, which learn the mappings between the tokens for the text in the description of the image and the tokens that represent the visual contents of the image. With lots of examples of these mappings, you can train a model on them. Then, you can give the model a piece of text, and it will be able to predict the tokens for that text and reassemble them into an image.

Using Hugging Face Diffusers

Just as Hugging Face offers a transformers library (as we explained in Chapter 15), it also offers a diffusers library to make it easier for you to use diffusion models. Diffusers abstract the complexities of using various models into an easy-to-use API.

To get started with diffusers, you simply install them like this:

```
pip install -U diffusers
```

The diffusers library manages the pipelining of model inference in the same way we experienced in earlier chapters with transformers. There are many steps involved in getting a model to render an image based on a prompt: encoding the prompt, making embeddings, passing the embeddings to the model along with any hyperparameters it needs, grabbing the output tensors, and turning them into an image. But diffusers encapsulate this for you into a pipeline, and there are a number of open source pipelines for many different models.

So, for example, in the image of teddy bears on Mars, I used Stable Diffusion 3.5 Medium, which you can find on the Hugging Face website (*https://oreil.ly/liUY-*).

This model has limited access, so at the top of the Hugging Face page, you'll see a form that you need to fill out to get permission. You'll also need to configure your Hugging Face secret key in Colab (if you're using Colab), which we demonstrated how to do back in Chapter 14.

If you aren't using Colab, your code will need to be signed in to Hugging Face using their API. You can do this with the following code:

```
from huggingface_hub import login

login(token="<YOUR TOKEN HERE>")
```

Once you're signed in (or if you're using a model that doesn't require signing in), the process of generating an image is as follows:

1. Create a Generator object, which allows you to specify the seed.

2. Create an instance of the appropriate pipeline for the model you require.

3. Send that pipeline to the appropriate accelerator.

4. Generate the image with the pipeline, giving it the appropriate parameters.

Let's look at this step-by-step.

First, you specify the generator using `torch.Generator`, where you will specify the accelerator for the generator and set the seed. You use the seed value to create the initial noise with a level of determinism. If you want to be able to *replicate* the images that are generated, despite the noise being random, you do so by guiding the noise with the seed. In other words, when the noise is generated with a seed value, the *same* noise will be generated subsequent times with the same seed. So effectively, the noise will be pseudo-random, as there will be a deterministic seed at play. On the other hand, if you don't specify a seed, you'll get a random value for it. Here's an example:

```
# Set your seed value
seed = 123456  # You can use any integer value you want

# Create a generator with the seed
generator = torch.Generator("cuda").manual_seed(seed)
```

Next, you'll specify the pipeline and instantiate it with a model:

```
pipe = StableDiffusion3Pipeline.from_pretrained(
            "stabilityai/stable-diffusion-3.5-medium",
            torch_dtype=torch.bfloat16)
pipe = pipe.to("cuda")
```

Here, we're using Stable Diffusion 3.5, which uses the `StableDiffusion3Pipeline` class. The diffusers API is open source, with new pipelines being added all the time. You can inspect them on GitHub (*https://oreil.ly/uYGnJ*).

You can also browse the different models on the Hugging Face website (*https://oreil.ly/R4dKT*). Often, their landing pages will include source code about which pipeline to use and the address of the model.

Once you have the pipeline, you can use it to create an image by specifying the prompt and some other model-dependent parameters that you'll find in the model document. So, for example, for stable diffusion, you'll specify the number of inference steps and the generator that you specified earlier. You should also specify *where* you want the pipe to execute—(in this case, it's cuda, as you can see in the previous code, which uses the GPU accelerator in Colab):

```
image = pipe(
    "A photo of a group of teddy bears eating pizza on the surface of mars",
    num_inference_steps=40,
    generator=generator   # Add the generator here
).images
image[0].save("teddies.png")
```

I've found that the best way to experiment with this is to explore the pipeline's source code and see the parameters that it supports. For example, with the Stable Diffusion 3 pipeline, there's a *negative prompt* that dictates things that you do *not* want to see in the image. Often, you can use this to make images better. For example, you may have heard that image generators, particularly early ones, were very bad at drawing hands. You could use the negative prompt to have the image generator avoid this problem by saying "deformed hands" or something similar in that prompt.

You can also specify things you don't want to see in the image that are more trivial! For example, every instance of the image I drew had the teddy bears eating *pepperoni* pizza. I could remove the pepperoni from this image with this code:

```
image = pipe(
    "A photo of a group of teddy bears eating pizza on the surface of mars",
    negative_prompt="pepperoni",
    num_inference_steps=40,
    generator=generator   # Add the generator here
).images
```

The resulting image is shown in Figure 19-5.

The teddy on the left doesn't look thrilled about it, but the others seem more content!

In this case, we used text-to-image to create these images—but diffusion models have become a little more advanced with add-ons for *image-to-image*. With such add-ons, instead of starting with random noise, we can begin with an existing image and then perform *inpainting*, in which we can have the model fill in new details in an existing image. We'll explore this next.

Figure 19-5. Teddies that don't like pepperoni

Image-to-Image with Diffusers

When inspecting the source code for the pipeline, you may have discovered other classes in there, such as this one: `StableDiffusion3Img2ImgPipeline`.

As its name suggests, this class allows you to start with one image to create another. You can initialize it in a way that's very similar to initializing the text-to-image pipeline:

```python
from diffusers import StableDiffusion3Img2ImgPipeline

# Set your seed value
seed = 123456

# Create a generator with the seed
generator = torch.Generator("cuda").manual_seed(seed)

# Load the model
pipe = StableDiffusion3Img2ImgPipeline.from_pretrained(
    "stabilityai/stable-diffusion-3.5-medium",
    torch_dtype=torch.bfloat16)

pipe = pipe.to("cuda")
```

Then, you'll specify an image to use as the source image:

```python
from PIL import Image
# Load and preprocess the initial image
init_image = Image.open("puppy1.jpg").convert("RGB")
```

I'm starting with an image of a puppy (see Figure 19-6).

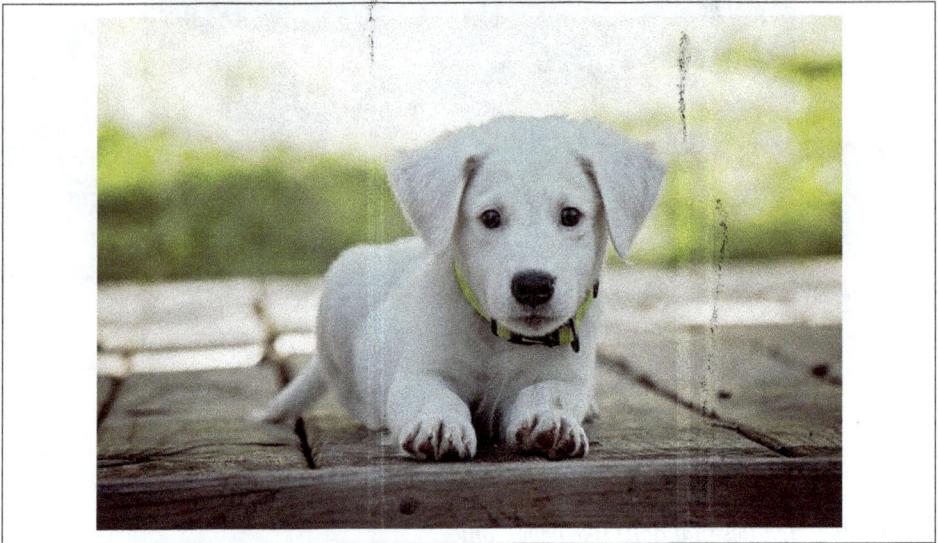

Figure 19-6. Source image of a puppy

We'll use this as the initialization image in an image-to-image pipeline with the following code. Note that the prompt is specifying a highly detailed photograph of a baby *dragon*:

```python
# Generate the image
image = pipe(
    prompt="A highly detailed photograph of a baby dragon",
    image=init_image,
    strength=0.7,
    num_inference_steps=100,
    generator=generator
).images
```

The strength parameter specifies how closely the generated image should follow the input image. At 0.0, the model won't do anything and the output will be the input image. At 1.0, it will effectively *ignore* the input image and will just act as a text-to-image model.

Under the hood, it does this with the following process.

Given that the code specified a strength of 0.7, the model will add noise to the image until the image has had 70% of its pixels replaced by noise (and thus only 30% of the image is the original values).

The model will then run 70 denoising steps (70% of the 100 specified), which will give an image like Figure 19-7.

Typically, if you use strength 0.2 to 0.4, you'll get style transfer and other minor modifications. At 0.5 to 0.7, you'll have basic composition maintained, but major element changes, like puppy to dragon, will be seen. Above 0.8, you'll see almost complete regeneration, but some slight influence from the original may be retained.

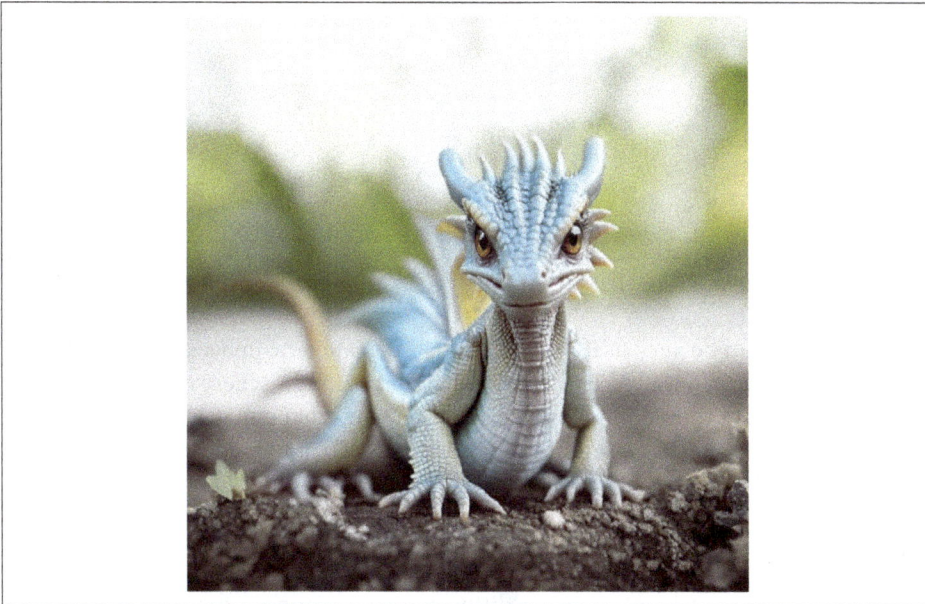

Figure 19-7. Using image-to-image to turn a puppy into a dragon

You can see that the basic pose has been maintained, but the computer has imagined a dragon to replace the puppy as required. There's also new foreground and background, as we didn't specify anything about them, but they're pretty close to the originals.

As an example of a different strength level, Figure 19-8 shows the strength at 0.4. We can also see that the basic shape of the puppy has been maintained, but it has become more dragon-like, with scaly skin and the beginnings of claws!

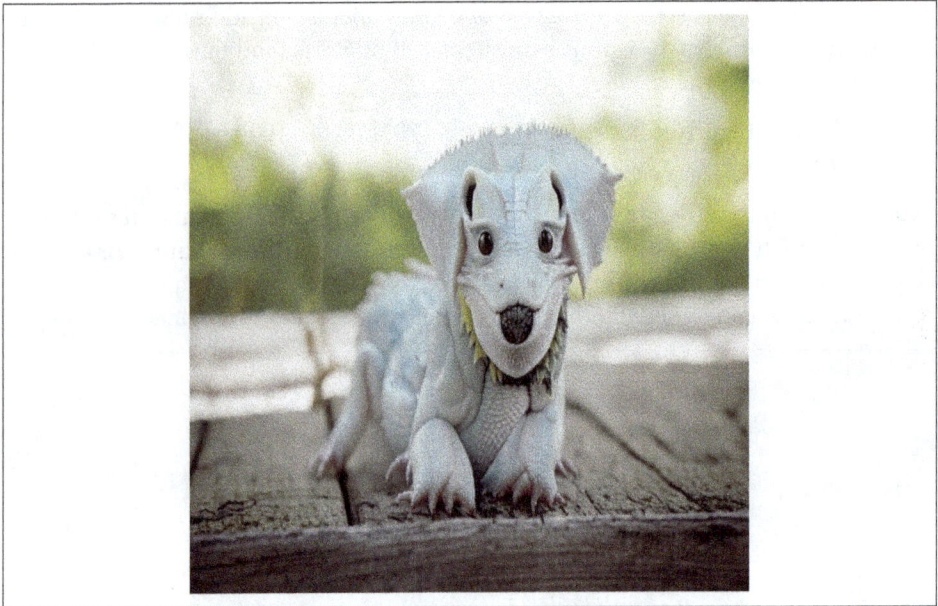

Figure 19-8. Strength level of 0.4 for the puppy to dragon image-to-image

This technique can be very useful in helping you create new images by starting from existing ones. I've seen it used in scenarios like filmmaking—where one can start with existing video that's filmed in a basic, cheap locale but then enhanced with image-to-image frame by frame to get a different outcome. It's a much cheaper way of doing postproduction by adding special effects!

Inpainting with Diffusers

Another scenario that involves using diffusers that is supported by some models—including stable diffusion models—is the idea of *inpainting*, in which you can take an image and replace parts of it with AI-generated content. So, for example, consider the puppy from Figure 19-6 and how you can change the image so the little pooch is on the moon, as in Figure 19-9.

Figure 19-9. Using inpainting to put our puppy on the moon

You can do this by using a pattern that's similar to the previous one. First, you'll set up the pipeline for inpainting:

```python
from diffusers import StableDiffusion3InpaintPipeline

# Load the inpainting pipeline
pipe = StableDiffusion3InpaintPipeline.from_pretrained(
    "stabilityai/stable-diffusion-3.5-medium",
    torch_dtype=torch.bfloat16
)
pipe = pipe.to("cuda")
```

The parameters to initialize it are the same as earlier. Next, you'll need the generator:

```python
# Set seed for reproducibility
generator = torch.Generator("cuda").manual_seed(42)
```

Then, you'll specify the source image, which in this case is the original image of the puppy:

```python
# Load the original image and mask
original_image = Image.open("puppy.jpg").convert("RGB")
```

The complicated step is the next one, in which you specify the *mask* for the image:

```python
mask_image = Image.open("puppymask.png").convert("L")
```

A *mask* is simply an image that corresponds to the original one, in which pieces to be *replaced* are in white and pieces to be *preserved* are in black. Figure 19-10 shows the mask image used for the puppy. I like to think of this as similar to the green-screen process used in making movies. The white part of the image is the screen, and the black is the stuff that's in front of it! The model will then replace the white with whatever you prompt it for.

Figure 19-10. Mask for the image

There are many ways to create masks. For this one, I used the Acorn 8 tool for the Mac. This tool gives you the ability to remove the background and paint it all in white, and then, for what's left, it lets you select the pixels with a magic wand and paint them all in black. Every tool does this differently, so be sure to check the appropriate documentation.

Once you have the image and the mask, you can easily use the pipeline to have the model inpaint the areas that correspond to the white part of the mask. Given that the puppy is already present, I didn't mention it in the prompt, and I just used "on the surface of the moon" to get the image in Figure 19-9:

```
# Generate the inpainted image
image = pipe(
    prompt="on the surface of the moon",
    image=original_image,
    mask_image=mask_image,
    num_inference_steps=50,
    generator=generator,
    strength=0.99  # How much to inpaint the masked area
).images[0]
```

The diffusers API, as you can see, gives you a very consistent approach to managing image creation, be it directly from a text prompt, starting from a source image, or inpainting a particular area.

Summary

In this chapter, you explored how to use generative models for image creation by using the Hugging Face diffusers library. You started by looking at the fundamental underlying concepts, seeing how the idea of denoising to create new content works.

You also looked into practical code-based implementation of image generation by using the diffusers API, and you focused on three main approaches:

1. You explored text-to-image by converting text prompts directly into images using the Stable Diffusion 3.5 model. You also looked at how you can control this process with parameters like the seed value and the number of inference steps.

2. You explored image-to-image by starting with an existing image and transforming it by using a prompt. In particular, you saw how the `strength` hyperparameter controls the overall transformation

3. You explored inpainting by preserving parts of the original image by using a mask, which allows for targeted modifications while preserving some elements.

You also explored hands-on, concrete code examples of each of these approaches, which showed you how to do the pipeline setup, generator initialization, and basic parameter tuning. You also saw how *negative* prompts can help you get images closer to what you really want.

In the next chapter, you'll look at LoRA (low-ranking adaptation), which lets you fine-tune diffusion models to achieve more controlled and customized images. LoRA is a powerful technique that allows for efficient model adaptation by only fine-tuning a small number of parameters, thus helping you guide the model toward specific styles, subjects, or artistic directions. You'll explore how to implement LoRA with the diffusers library, and you'll customize these models to create *specialized* image generators for your needs.

## CHAPTER 20

Tuning Generative Image Models with LoRA and Diffusers

In Chapter 19, you explored the idea of diffusers and how models trained with diffusion techniques can generate images based on prompts. Like text-based models (as we explored in Chapter 16), text-to-image models can be fine-tuned for specific tasks. The architecture of diffusion models and how to fine-tune them is enough for a full book in its own right, so in this chapter, you'll just explore these concepts at a high level. There are several techniques for doing this, including *DreamBooth, textual inversion,* and the more recent *low-ranking adaptation* (LoRA), which you'll go through step-by step in this chapter. This last technique allows you to customize models for a specific subject or style with very little data.

As with transformers, the diffusers Hugging Face library is designed to make using diffusers, as well as fine-tuning them, as easy as possible. To that end, it includes pre-built scripts that you can use.

We'll go through a full sample of creating a dataset of a fictitious digital influencer called Misato, using LoRA and diffusers to fine-tune a text-to-image model called Stable Diffusion 2 for her. Then, we'll perform text-to-image inference to demonstrate how to create new images of Misato (see Figure 20-1).

Figure 20-1. LoRA-tuned Stable Diffusion 2 images

Training a LoRA with Diffusers

To train a LoRA with diffusers, you'll need to perform the following steps. First, you'll need to get the source code for diffusers so you can have access to its premade training scripts. Then, you'll get or create a dataset that you can use to fine-tune Stable Diffusion. After that, you'll run the training scripts to get a fine-tune for the model, publish the fine-tune to Hugging Face, and run inference against the base model with the LoRA layers applied. Once you're done, you should be able to create images like those shown in Figure 20-1. Let's walk through each of these steps.

Getting Diffusers

To get started with LoRA, I have found the best thing to do is to first clone the source code for diffusers to get the training scripts.

You can do this quite simply by git-cloning it, changing into the directory, and running `pip install` at the current location:

```
git clone https://github.com/huggingface/diffusers
cd diffusers
pip install .
```

If you're using Colab or another hosted notebook, you'll use syntax like this:

```
!git clone https://github.com/huggingface/diffusers
%cd diffusers
!pip install .
```

This will give you a local version of diffusers that you can use. The text-to-image LoRA fine-tuning scripts are in the */diffusers/examples/text_to_image* directory, and you'll need to install their dependencies like this:

```
%cd /content/diffusers/examples/text_to_image # or whatever your dir is
!pip install -r requirements.txt
```

These dependencies include the specific versions of tools like accelerate, transformers, and torchvision. It's good to git-clone from source so that you get the latest versions of the *requirements.txt* to make your life easier!

Finally, you'll also need the xformers library, which is designed to make transformers more efficient and thus speed up the process for you. You can get it like this:

```
!pip install xformers
```

Now, you have a diffusers environment that you can use for fine-tuning. In the next step, you'll get the data.

Getting Data for Fine-Tuning a LoRA

The two main ways in which you'll fine-tune a LoRA are for *style* and for *subject*. In the former case, you can get a number of images of the specific style that you want and train the model so that it will output in that style. I would urge caution when doing this because many artists earn their livelihood from their style of creation, and you should respect that. Similarly, you should consider the impact of training models based on commercial styles. Unfortunately, many of the tutorials I have seen online ignore this, and such practices bring down the overall impact of AI and drive the narrative of generative AI away from being *creative* and toward *stealing IP*. So, please be careful with that.

Similarly, when it comes to the subject, I see many tutorials that use examples of doing a Google Image search for a celebrity so you can create a LoRA of them. Again, I would urge you *not* to do this. Please only create a LoRA for someone whose likeness you have permission to use.

So that you can have something you *can* use, I created a dataset for a digital influencer. I call her Misato, after my favorite character in a popular anime. All of the images were rendered by me using the popular Daz 3D rendering software.

You can find this dataset on the Hugging Facewebsite (*https://oreil.ly/Y1qeY*).

If you want to create a dataset like this, I would recommend that you use images of the same figure from multiple angles that also focus on specific segments. For example, you can use these:

- 3–4 portrait headshots (passport-style photos)
- 3–4 three-quarters headshots from each side
- 3–4 profile pictures, showing the side of the face
- 3–4 full-length body shots

For each of these images, you also need a prompt that describes the image. You'll use this in training to give context to the image and how it should be represented.

So, for example, consider Figure 20-2, which is a portrait shot that I generated for Misato.

Figure 20-2. Portrait shot of Misato from the dataset

This image is paired with the following prompt: "Photo of (lora-misato-token), high-quality portrait, clear facial features, neutral expression, front view, natural lighting."

Note the use of (lora-misato-token), where we indicate the subject of the image. Later, when we create prompts to generate new images, we can use the same token—for example, "(lora-misato-token) in food ad, billboard sign, 90s, anime, Japanese pop, Japanese words, front view, plain background." This prompt will give us what you can see in Figure 20-3. We have an entirely new composition, with Misato as the model in a fast-food campaign!

Once you have a set of images, you'll need to create a *metadata.jsonl* file that contains the images associated with their prompts in a standard format that you can use when fine-tuning. It's JSON with a link to the filename and the prompt for that image. The one for Misato is on the Hugging Face website (*https://oreil.ly/MfmGh*).

Figure 20-3. Inference from a LoRA token

A snippet of the *metadata.jsonl* file is here:

```
{ "file_name": "rightprofile-smile.png",
           "prompt": "photo of (lora-misato-token),
           right side profile, high quality, detailed features,
           smiling, professional photo "}

{ "file_name": "rightprofile-neutral.png",
           "prompt": "photo of (lora-misato-token),
           right side profile, high quality, detailed features,
           professional photo "}
```

That's pretty much all you need. For training with diffusers, I've found it much easier if you publish your dataset on Hugging Face. To do this, when logged in, visit the Hugging Face website (*https://oreil.ly/Ez3Gp*). There, you'll be able to specify the name of the new dataset and whether or not it's public. Once you've done this, you'll be able to upload the files through the web interface (see Figure 20-4).

Once you've done this, your dataset will be available at *https://huggingface.co/datasets/ <yourname>/<datasetname>*. So, for example, my username (see Figure 20-4) is "lmoroney," and the dataset name is "misato," so you can see this dataset at *https:// huggingface.co/datasets/lmoroney/misato*.

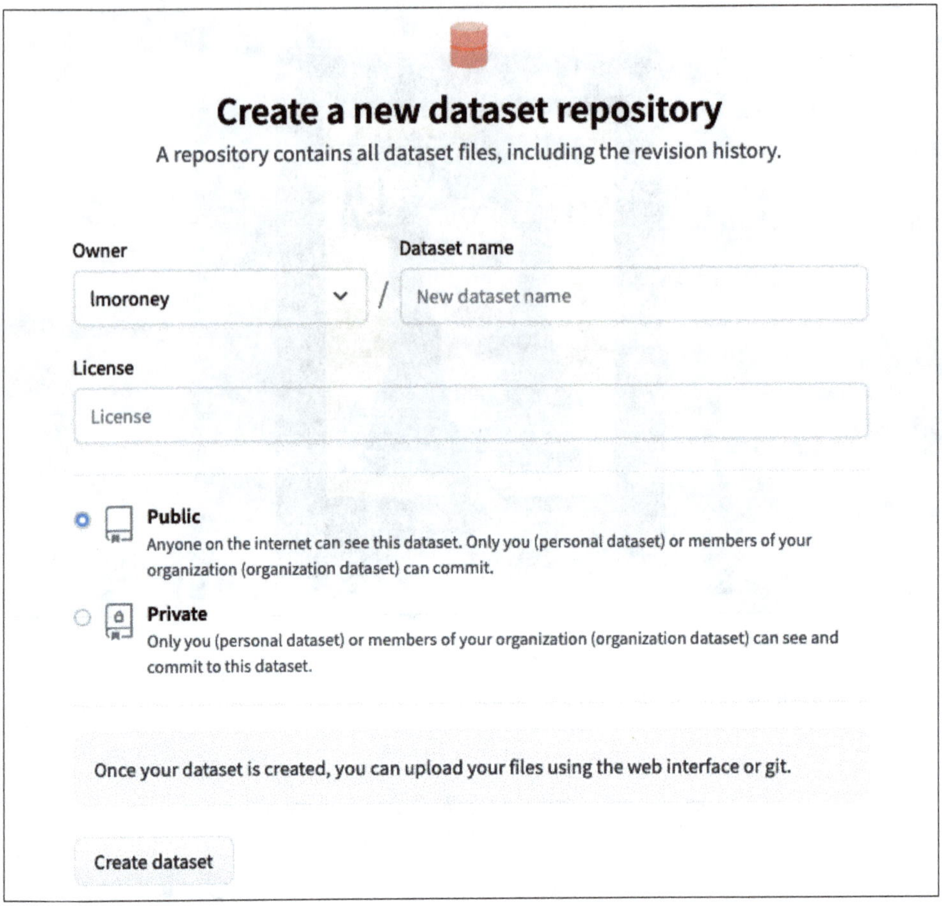

Figure 20-4. Creating a new dataset on Hugging Face

Fine-Tuning a Model with Diffusers

As mentioned earlier, when you clone the diffusers repo, you get access to a number of example pre-written scripts that give you a head start in various tasks. One of these is training text-to-image LoRAs. But before running the script, it's a good idea to use `accelerate`, which abstracts underlying accelerator hardware, including distribution across multiple chips. With `accelerate`, you can define a configuration. Find the details on the Hugging Face website (*https://oreil.ly/TnaII*).

For the purposes of simplicity, when you're using Colab, here's how you can set up a basic `accelerate` profile:

```
from accelerate.utils import write_basic_config

write_basic_config()
```

Then, once you have that, you can use `accelerate launch` to run the training script. Here's an example:

```
!accelerate launch train_text_to_image_lora.py \
  --pretrained_model_name_or_path="stabilityai/stable-diffusion-2" \
  --dataset_name="lmoroney/misato" \
  --caption_column="prompt" \
  --resolution=512 \
  --random_flip \
  --train_batch_size=1 \
  --num_train_epochs=1000 \
  --checkpointing_steps=5000 \
  --learning_rate=1e-04 \
  --lr_scheduler="constant" \
  --lr_warmup_steps=0 \
  --seed=42 \
  --output_dir="/content/lm-misato-lora"
```

Note that running this is very computationally intensive. With the preceding set of hyperparameters (I'll explain each one in a moment), using an A100 in Google Colab took me about 2 hours (or 17 compute units) to train. Compute units cost money (at the time of publication, about 10 cents each), so be sure to understand how this all works and that it does cost money!

The script takes the following hyperparameters:

Pretrained_model_name_or_path
> This can be a local folder (for example, */content/model/*) or the location on *huggingface.co*—so for example, *http://huggingface.co/stabilityai/stable-diffusion-2* is the location of the model called Stable Diffusion 2. You can also specify this without the *huggingface.co* part of the URL.

Dataset-name
> Similarly, this can be a local directory containing the dataset or the address of it on *huggingface.co*. As you can see, I'm using the Misato dataset here.

Caption_column
> This is the column in the *jsonl* file that contains the caption for the images. You can specify the caption here.

Resolution
> This is the resolution that we'll train the images for. In this case, it's 512×512.

Random_Flip
> This is image augmentation (as in Chapter 3). As the Misato dataset already has multiple angles covered, this probably isn't needed.

Train_batch_size

This is the number of images per batch. It's good to start with 1 and then tweak it as you see fit. When I was using the A100 GPU in Colab, I noticed that training was only using about 7 GB of the 40 GB, so this could be safely turned up to speed up training.

Num_training_epochs

This is how many epochs to train for.

Checkpointing_steps

This is how often you should save a checkpoint.

Learning_rate

This is the LR hyperparameter.

LR_scheduler

If you want to use an adjustable learning rate, you can specify the scheduler here. The nice thing with an adjustable LR is that the best LR later in the training cycle isn't always the same as the best one from earlier in the cycle, so you can adjust it on the fly.

LR_Warmup_steps

This is the number of steps you'll take to set the initial LR.

Seed

This is a random seed.

Output_dir

This is where you save the checkpoints as training happens.

Then, when training, you'll see a status that looks something like this:

```
Resolving data files: 100% 22/22 [00:00<00:00, 74.14it/s]
12/30/2024 19:23:48 - INFO - __main__ - ***** Running training *****
12/30/2024 19:23:48 - INFO - __main__ -   Num examples = 21
12/30/2024 19:23:48 - INFO - __main__ -   Num Epochs = 1000
12/30/2024 19:23:48 - INFO - __main__ -   Instantaneous batch size per device...
12/30/2024 19:23:48 - INFO - __main__ -   Total train batch size (w. parallel...
12/30/2024 19:23:48 - INFO - __main__ -   Gradient Accumulation steps = 1
12/30/2024 19:23:48 - INFO - __main__ -   Total optimization steps = 1000
Steps:  10% 103/1000 [05:03<44:00,  2.94s/it, lr=0.0001, step_loss=0.227]
```

Once the model is trained, in its directory folder, you'll see a structure like the one depicted in Figure 20-5.

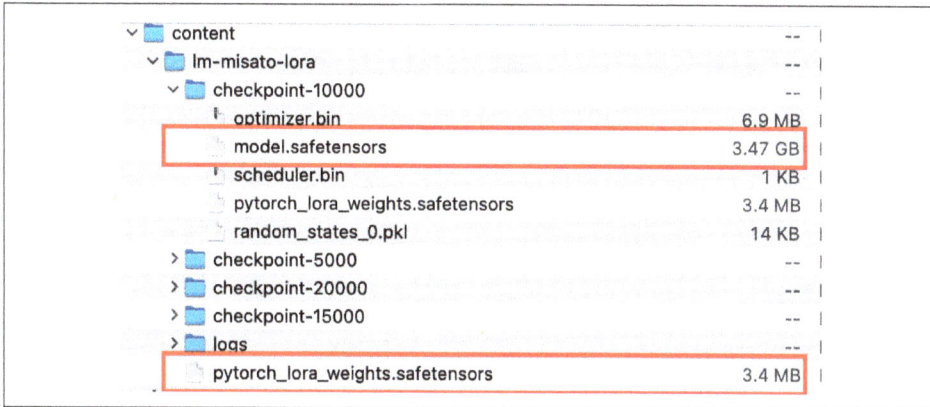

Figure 20-5. A trained directory

The original `model.safetensors` model is highlighted, and you can see that it is 3.47 GB in size! The fine-tuned LoRA, on the other hand, is much smaller at just 3.4 MB.

You can use this in the next step, where you upload the model to the Hugging Face repository to make it very easy for inference to use it.

Publishing Your Model

The fine-tuned directory that you've saved while training contains a lot more information than you need, including clones of the base model. As a result, if you try to publish and upload the model, you'll end up taking a lot longer because you'll have to upload lots of unneeded gigabytes!

Therefore, you should edit your directory structure to remove the *model.safetensors* files from the checkpoint directories and keep the rest.

Then, when you're signed into Hugging Face, you can visit *huggingface.co/new* to see the "Create New Model Repository" page (see Figure 20-6).

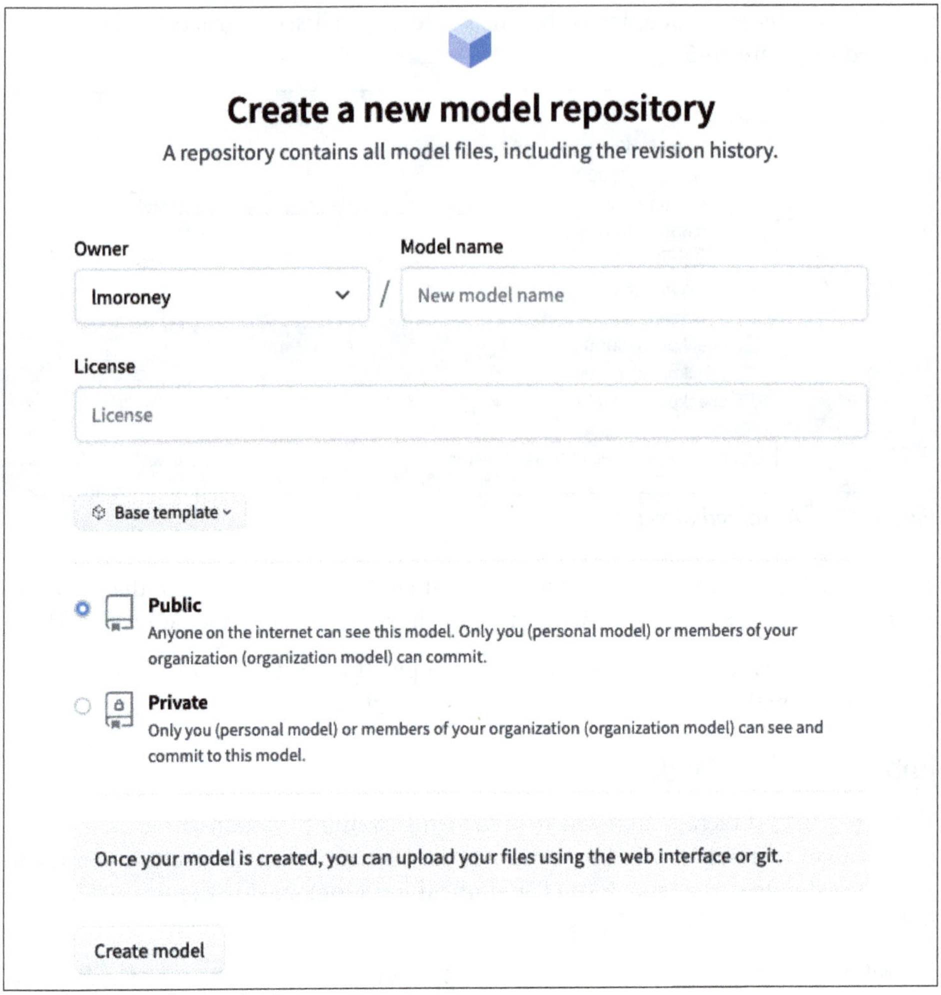

Figure 20-6. Creating a new repository

Follow the steps, and be sure to select a license. Then, when you're done, you can upload the files via the web interface in the next step. When you're done with that, you should see something like the screen depicted in Figure 20-7, where I named the model "finetuned-misato-sd2," given that the data was "misato" and the model I tuned was Stable Diffusion 2.

You can see this for yourself on the Hugging Face website (*https://oreil.ly/zmlal*).

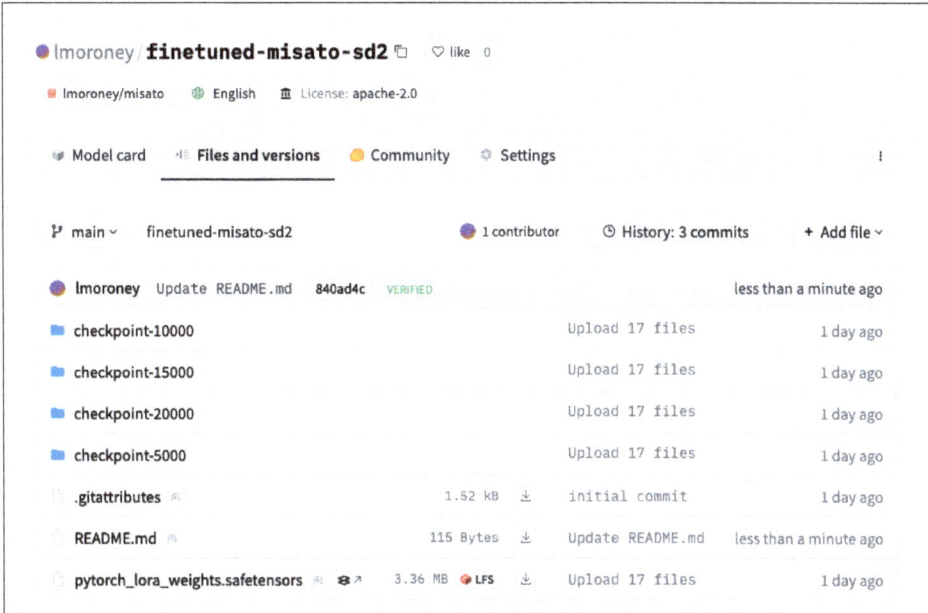

Figure 20-7. The fine-tuned Misato LoRA for Stable Diffusion 2

Now that the dataset and the model are both published on Hugging Face, using diffusers to do an inference with it is super simple. We'll see that in the next step.

Generating an Image with the Custom LoRA

To create an image using the custom LoRA, we'll go through a process that's similar to the one in Chapter 19. You'll use diffusers to create a pipeline, but you'll also add a scheduler. In stable diffusion, the role of the scheduler determines how the image evolves from random noise to the final image. Not all schedulers work with LoRA, and you'll have to ensure that the scheduler you use works with the base model you're working with.

There are lots of schedulers you can use, and you can find them on the Hugging Face website (*https://oreil.ly/SUlZl*).

In this case, you can experiment with using the `EulerAncestralDiscreteScheduler`:

```
import torch
from diffusers import (
    StableDiffusionPipeline,
    EulerAncestralDiscreteScheduler,
)
```

Then, specify our `model_id` and pick the appropriate version of the scheduler for it:

```
model_id = "stabilityai/stable-diffusion-2"

# Choose your device
device = "cuda" if torch.cuda.is_available() else "cpu"

# 1. Pick your scheduler
scheduler = EulerAncestralDiscreteScheduler.from_pretrained(
    model_id,
    subfolder="scheduler"
)
```

Once you've done that, you can create the pipeline from the `StableDiffusionPipe line` class and load it to the accelerator device:

```
# 2. Load the pipeline with the chosen scheduler
pipe = StableDiffusionPipeline.from_pretrained(
    model_id,
    scheduler=scheduler,
    torch_dtype=torch.float16
).to(device)
```

The next step is to assign the new LoRA weights, which are the retrained layers that determine the new behavior of the model:

```
# 3. (Optional) Load LoRA weights
pipe.load_lora_weights("lmoroney/finetuned-misato-sd2")
```

Stable diffusion supports both a prompt *and* a negative prompt, where the first prompt defines what you want in the image and the second prompt defines what you *do not* want. Here's an example:

```
# 4. Define prompts and parameters
prompt = "(lora-misato-token) in food ad, billboard sign, 90s, anime,
          japanese pop, japanese words, front view, plain background"

negative_prompt = (
    "(deformed, distorted, disfigured:1.3), poorly drawn, bad anatomy,
    wrong anatomy, "
    "extra limb, missing limb, floating limbs, (mutated hands and
    fingers:1.4), "
    "disconnected limbs, mutation, mutated, ugly, disgusting, blurry,
    amputation"
)
```

The negative prompt is very useful in helping you avoid some of the issues with AI-generated visuals, such as deformed hands and faces.

Next up is to define the hyperparameters, such as the number of inference steps, the size of the image, and the seed. There's also a parameter called *guidance scale*, which controls how imaginative your model is. A guidance scale value of less than 5 gives the model more creative freedom, but the model may not follow your prompt closely. A guidance scale value that's higher than 7 will make the model adhere more strongly to your prompt, but it can also lead to strange artifacts. The guidance scale value in the middle—6—is a nice balance between freedom and adherence. There's no hard and fast rule, so feel free to experiment:

```
num_inference_steps = 50
guidance_scale = 6.0
width = 512
height = 512
seed = 1234567
```

Next, you just generate the image as usual:

```
# 5. Create a generator for reproducible results
generator = torch.Generator(device=device).manual_seed(seed)

# 6. Run the pipeline
image = pipe(
    prompt,
    negative_prompt=negative_prompt,
    width=width,
    height=height,
    num_inference_steps=num_inference_steps,
    guidance_scale=guidance_scale,
    generator=generator,
).images[0]

# 7. Save the result
image.save("lora-with-negative.png")
```

As an experiment, you can try using a different scheduler with the same hyperparameters to yield similar results (see Figure 20-8):

```
# For DPMSolver, use:
from diffusers import DPMSolverMultistepScheduler

scheduler = DPMSolverMultistepScheduler.from_pretrained(model_id,
            subfolder="scheduler", algorithm_type="dpmsolver++")
```

Figure 20-8. The same prompt and hyperparameters with different schedulers

Note that the text in the image is entirely made up, but given that the prompt is about advertisements, the tone is similar. In the picture on the left, the characters represent "loneliness" and "no," while in the image on the right, they suggest "husband split?"

What's most interesting is the consistency in the character! For example, consider Figure 20-9, in which Misato was painted in the styles of Monet and Picasso. We can see that the features learned by LoRA were consistent enough to (mostly) survive the restyling process.

Figure 20-9. Character consistency across styles

This example used Stable Diffusion 2, which is an older model but one that's easy to tune with LoRA. As you use more advanced models and tune them, you can get much better results, but the time and costs of tuning will be much higher. I'd recommend starting with a simpler model like this one and working on your craft. From there, you can build up to the more advanced models.

Additionally, Misato's synthetic nature has triggered different features in the LoRA retraining, leading to the new images that have been created from her having a low-res, highly synthetic look. While the images have been close to photoreal to the human eye, they clearly haven't been to the model, which learned a LoRA that was very CGI in nature and lower resolution than the ones in the training set!

Summary

In this chapter, you had a walk-through of how to fine tune a text-to-image model like stable diffusion by using LoRA and the diffusers library. This technique allows you to customize models for a specific subject or style with a small custom file. In this case, you saw how to tune Stable Diffusion 2 for a synthetic character. In this chapter, you also went through all the steps—from cloning diffusers to creating a training environment for them that included a fully custom dataset. You learned how to use the training scripts to create a new LoRA based on the synthetic character and how to publish that to Hugging Face. Finally, you saw how to apply the LoRA to the model at inference time to create novel images using the LoRA for the Misato character!

Index

About the Author

Laurence Moroney is an award-winning researcher and best-selling author. A veteran in the programming and machine learning industry, Laurence has authored over 20 books, including top-selling programming books, science-fiction novels, and even movie-related comic books. He teaches popular AI specializations on Coursera with Andrew Ng of DeepLearning.AI. A recognized speaker all over the world, Laurence is passionate about demystifying AI and ML for software developers.

Colophon

The animal on the cover of *AI and ML for Coders in PyTorch* is a hobby bird (*Falco subbuteo*). These falcons—which can be found throughout the United Kingdom, Europe, Africa, and Asia—are small, swift, opportunistic hunters that are known for their agility and predatory prowess.

Although hobby birds are often mistaken for large swifts, their distinctive appearance sets them apart. Adults typically measure between 28 and 36 centimeters in length, with a wingspan of 68 to 84 centimeters; they weigh 131 to 340 grams. Their plumage features a slate-gray back and head, a white throat, and heavily streaked underparts that transition to reddish-brown "trousers" on their thighs. Their long, pointed wings and short tail aid in their agility, making them skilled at high-speed chases.

Hobby birds are diurnal hunters, primarily active during dawn and dusk. Their diet mostly consists of large insects, such as dragonflies, moths, and beetles, but they will occasionally eat other birds, such as swallows and small waders. Hobbies are known for performing spectacular dives to pursue their prey; they can even catch insects mid-air with their talons and transfer them to their beak to eat while still in flight.

Hobbies are very adaptable; they live in different habitats, including open country that has small patches of forest, farmland areas with surrounding trees, and even urban areas with wooded parks. They are migratory birds, often spending summers in the UK to breed before migrating to Africa for the winter. While their conservation status is currently labeled as "Least Concern" by the International Union for Conservation of Nature (IUCN), they do face some threats, such as illegal hunting.

Many of the animals on O'Reilly covers are endangered; all of them are important to the world.

The cover illustration is by Monica Kamsvaag, based on an antique line engraving from *Animate Creation*. The series design is by Edie Freedman, Ellie Volckhausen, and Karen Montgomery. The cover fonts are Gilroy Semibold and Guardian Sans. The text font is Adobe Minion Pro; the heading font is Adobe Myriad Condensed; and the code font is Dalton Maag's Ubuntu Mono.

O'REILLY®

Learn from experts.
Become one yourself.

60,000+ titles | Live events with experts | Role-based courses
Interactive learning | Certification preparation

**Try the O'Reilly learning platform
free for 10 days.**